The Audience Commodity in a Digital Age

Steve Jones
General Editor

Vol. 94

The Digital Formations series is part of the Peter Lang Media and Communication list.
Every volume is peer reviewed and meets
the highest quality standards for content and production.

PETER LANG
New York • Washington, D.C./Baltimore • Bern
Frankfurt • Berlin • Brussels • Vienna • Oxford

The Audience Commodity in a Digital Age

Revisiting a Critical Theory of Commercial Media

Edited by Lee McGuigan & Vincent Manzerolle

PETER LANG
New York • Washington, D.C./Baltimore • Bern
Frankfurt • Berlin • Brussels • Vienna • Oxford

Library of Congress Cataloging-in-Publication Data

The audience commodity in a digital age: revisiting a critical theory
of commercial media / edited by Lee McGuigan, Vincent Manzerolle.
pages cm. — (Digital formations; volume 94)
Includes bibliographical references and index.
1. Digital media—Social aspects. 2. Mass media—Audiences.
I. McGuigan, Lee, editor of compilation.
II. Manzerolle, Vincent, editor of compilation.
HM742.A93 302.23—dc23 2013018176
ISBN 978-1-4331-2360-3 (hardcover)
ISBN 978-1-4331-2359-7 (paperback)
ISBN 978-1-4539-1157-0 (e-book)

Bibliographic information published by **Die Deutsche Nationalbibliothek**.
Die Deutsche Nationalbibliothek lists this publication in the "Deutsche
Nationalbibliografie"; detailed bibliographic data is available
on the Internet at http://dnb.d-nb.de/.

The paper in this book meets the guidelines for permanence and durability
of the Committee on Production Guidelines for Book Longevity
of the Council of Library Resources.

Printed in the United States of America

Acknowledgments

We would be remiss to proceed beyond even one sentence without thanking Eileen Meehan and Janet Wasko, without whom this volume would not exist. We offer our sincerest thanks for their mentorship and generosity.

Thanks to everyone at Peter Lang, especially Steve Jones, editor of the Digital Formations series, for making this experience uniquely pleasant and painless, and Mary Savigar, acquisitions editor, for taking a chance on two greenhorns and shepherding us through the process. Jackie Pavlovic and the production teams also deserve thanks.

We thank the outstanding contributors for flattering us with their wisdom and making our jobs easy. These scholars have shaped our intellectual development, and we are truly honoured to call them colleagues.

Lee reserves special comment for Edward Comor and Robert Babe. If scholarship is a boat, Professor Babe taught me to build it and Professor Comor taught me to row (and then to swim when it capsizes). Most importantly, Lee thanks his family for undying (and undeserved) support. Parents John and Diane, brother Scott, Aunt Maureen, Janet and Mike, Margie and Jeff, Sheila and Ken, and Grandma Mona—you mean everything to me. Love, also, to passed grandparents: Jim McGuigan, and B.K. and Irene Lee. I work solely in the hope of making you proud.

Vincent thanks his loving and supportive partner Stacy Manzerolle. He also owes thanks to parents Donald and Gianna, sister Veronica, Uncle Ray and Aunt Jo, and to grandparents Clorinda Mamolo and Loretta Manzerolle. He dedicates his efforts here to Valmont Manzerolle and Ian Bruno who both sadly passed away during the course of this project. Despite their generational differences, they both lived their lives kindly, honourably, and with a profound sense of adventure.

Together, we thank the Faculty of Information and Media Studies at the University of Western Ontario, especially Sandra Smeltzer, Nick Dyer-Witheford, Daniel Robinson, Jonathan Burston, Keir Keightley and James Compton.

For allowing us to reprint materials vital to this collection, we thank Arthur and Marilouise Kroker, editors and publishers of *CTheory* and the *Canadian Journal of Political and Social Theory*; Taylor & Francis, publishers of *Critical Studies in Media Communication*; Peter Lang, publishers of *The Spectacle of Accumulation: Essays in Media, Culture & Politics* (by Sut Jhally).

The following materials are reprinted with permission:

Dallas Smythe, "Communications: Blindspot of Western Marxism" was first published in the *Canadian Journal of Political and Social Theory*, Vol 1. No. 3, 1977

Graham Murdock, "Blindspots About Western Marxism: A Reply to Dallas Smythe" was first published in the *Canadian Journal of Political and Social Theory*, Vol. 2 No. 2, 1978.

Both the essay and the response are published with permission of the editors Arthur and Marilouise Kroker.

Eileen R. Meehan, "Ratings and the Institutional Approach: A Third Answer to the Commodity Question" was first published in *Critical Studies in Mass Communication*, Vol. 3 No. 4, 1986. Published with permission of Taylor & Francis and the author.

Sut Jhally and Bill Livant, "Watching as Working: The Valorization of Audience Consciousness" was previously published in *The Spectacle of Accumulation: Essays in Culture, Media, and Politics*, New York: Peter Lang, 2006. Published with permission of Peter Lang and the author.

Portions of the reprinted material have been altered to conform to stylistic guidelines; some portions have been left unaltered to preserve the integrity of the original compositions.

Foreword

Vincent Mosco
Emeritus Professor of Sociology
Queen's University

Dallas Smythe had a profound impact on communications studies and helped to establish the foundation for a political economy of communication. So it is a great pleasure to see Lee McGuigan and Vincent Manzerolle produce this excellent volume that brings together key research in the history of debates over Dallas's work and new scholarship that provides a reassessment of his legacy, particularly for the burgeoning field of digital labor studies. The time is certainly ripe to assess the importance of Dallas's writing on the intersection of audiences and labor because scholars are increasingly documenting how digital technologies, and especially social media, blur the lines between what it means to consume and to labor in the media world. McGuigan's introduction provides a fine map of Dallas's work and a guide to how scholars have used it. I will limit this preface to a few personal reflections on my relationship with Dallas and follow this with a brief discussion of what I believe are his vital contributions to research, particularly in the field of political economy.

I first learned about Dallas's work in 1978 when as a professor at Georgetown University I happened upon "Communications: Blindspot of Western Marxism" in the *Canadian Journal of Political and Social Theory*. Since I was doing research on the relevance of new forms of "Western" Marxism, including state theory, world systems theory, labor theory, and theories of alienation, for the expanding worlds of media and information technology, I had to track down this article. I searched widely but it was nowhere to be found in Washington, D.C., not even at the venerable Library of Congress. So I contacted Dallas who responded promptly with the article and included a note with a caustic and quite understandable reference to the utter irrelevance of Canada in American intellectual life, evidenced in the failure of one of the nation's major research libraries to carry an important Canadian journal. I devoured the article and the ensuing debate, and made use of it in my research. It was wonderful to know that a leading Canadian scholar shared my interest in documenting the importance of Marxism for communication studies, at a time when most Marxists, of every tendency, ignored the significance of the media. Or when they did give it attention, it was only to see the media as the ideological reflection of an economic base.

A year later, I had the pleasure of meeting Dallas who was the featured speaker at a workshop on critical theory and communication studies that took place at the University of Illinois. The event also included several people who would go on to become key figures in the emergence of a political economy of communication. Dallas impressed me at the Illinois meeting and I left feeling that I had found my intellectual community. His work on the audience commodity inspired the chapter "Home Sweet Marketplace" in my book *Pushbutton Fantasies* and when an opportunity arose to actually work with Dallas, I seized it. Dallas had joined Janet Wasko, Dan Schiller, and John Lent at Temple University and it was wonderful to spend however brief a time with these and other remarkable critical media scholars. At Temple, Dallas was a great teacher in everything he did. For example, we were all deeply impressed, and not a little bit concerned, by his decision to live in a residence on the Temple campus because it was one of the most dangerous in the United States, located in a neighborhood that suffered from decades of government neglect. We were therefore not surprised when Dallas was injured in a mugging but, as always, impressed that he lectured on the political economy of the occurrence. Dallas and I had countless discussions as he brought *Dependency Road* to completion and he provided important counsel as we formed the Union for Democratic Communication and published our own work in political economy.

When in 1983 I chaired the annual Telecommunications Policy Research Conference that brought together communication policy professionals and academics in the Washington, D.C. area, of course I asked him to give a major talk. Few people could match his ability to connect serious theoretical reflection with practical policy advice. Never fearful, he also enjoyed attacking Canadian policy makers for what he saw as obsequious pleading to their American counterparts for some understanding of the Canadian need to protect its national culture.

Dallas and I remained in contact when we both left for Canada, me to Queen's University and Dallas back to Simon Fraser. We had a memorable reunion at an ICA meeting in Honolulu in 1985. Living up to his reputation as someone intensely keen to argue the world, whatever the time of day or night, he asked if we could meet at dawn every morning to walk the beach and debate the state of political economy. A year later I hosted his visit to Queen's where, never shy of controversy, he gave a tour de force seminar on the topic "The State is a Myth." By that time Dallas had become active in the peace and anti-nuclear movements and in one of the most memorable dinners I had the honor to host, my partner and I welcomed Dallas to our home to meet Canadian Major General Leonard Johnson who was courageously leading Generals for Peace and Disarmament.

We honored Dallas at the 1991 International Communication Association conference by presenting him with the *festschrift* papers written in his honor by what had become, with his help, a genuine community of critical political economists. At 79 and in declining health, Dallas nevertheless took the time to send contributors a critique of their contributions. The resulting published book demonstrated not only our wish to honor him but his desire to honor us with his detailed recommendations for chapter improvements.

I had one more significant encounter with Dallas before his passing when about a year before his death he agreed to meet me for a series of interviews about his life, his work, and the fields of political economy and communication studies that I used in the first edition of *The Political Economy of Communication*. Viewing political economy as a governing paradigm for all the social sciences, he started with a definition of the field, "the study of control and survival in social life," and we proceeded to discuss and debate for about twelve hours. These sessions proved invaluable to the book and to my future work in the political economy of communication.

Dallas would find me remiss if I made no mention of our disagreements about such issues as the merits of Western Marxism, the meaning of monopoly capital, and the limits of institutional political economy. But all of these were overshadowed by three profound qualities that have lived on in my thinking well after his passing in 1993. For the first person to teach a course in the United States on the political economy of communication there was a great deal of work to be done. So we could let it pass if Dallas had focused only on the communication power structure that was mobilizing to control the media and telecommunications industries in the post-World War II period. His work on U.S. hegemony in world media, and particularly on its relationship to his native Canada is well documented. But Dallas was consistently committed to a *dialectical* analysis of communication and power and therefore did not stop at domination and exploitation. One of the primary examples is the relationship between consumption and labor, a key theme in this collection. There was nothing especially novel in demonstrating that the audience was the primary commodity in the media business. Given his experience as an economist for the Federal Communications Commission (FCC), Dallas would be familiar with the regular pitches that broadcasters made to advertisers about the ability of this or that program to capture a particular slice of the demographic pie. The trade press was filled with ads promising to deliver men ages 18–49 to your brand's car, beer, or cigarette. His work was conceptually significant because it situated the audience in the dialectical triangle between advertisers and media companies and because it documented the ways audience activity could be characterized as labor. It is evidence of Dallas's conceptual vision that his position attracted considerable debate and criticism. This is to be expected from someone who wanted to do more than just apply an off-the shelf theory to an empirical problem. Dallas was committed to changing both the shelf and its contents. In doing so, he constantly searched for the contradictions inherent in the exercise of power that might bring about social transformation, whether this was the peasantry in China developing its own technological solutions or peace movements in North America facing down the military-industrial complex.

A second characteristic that is essential for understanding Dallas's work is his commitment to *internationalism*, the belief that you cannot understand society or change it without comprehending and working on the complex global forces that directly or indirectly shape the world. As a Canadian, Dallas understood clearly that his own society was incomprehensible without an intimate knowledge of the elephant next door. Drawing on his commitment to the dialectic, Dallas was careful to document both U.S. domination and the struggles of Canadians to build their own cultural, media, and telecommunications networks. Concepts, like nationalism, are important, but quite shaky substitutes for the constant fluidity of domination and resistance that resulted in the tenuous border between the two societies. Important as was the Canada-U.S. relationship, Dallas also recognized that new bases of power were emerging, especially in Asia. In fact, long before most media scholars paid any attention to China and well before it became fashionable to visit, Dallas travelled to China and wrote extensively about Mao's revolution, the role of technology and mass media, and the prospects for a challenge to U.S. hegemony. It is hard to say how Dallas would respond to the new China, probably by reiterating the admonition to take the long view, but one of his many remarkable attributes was a fearlessness and tenacity to take on an issue that practically no one in the West dared to address. He demonstrated similar qualities when the prospects for democratic socialism in Chile led him to that country to do what he could to provide advice on media reform to the government of Salvatore Allende. Given that the Federal Bureau of Investigation (FBI) had already developed a file on him, with information provided by faculty at the University of Illinois, Dallas had to be concerned about the implications of visiting a nation whose government the United States was actively working to topple. But nothing got between Dallas and his principles, especially not the American security establishment.

Dallas's work in Chile displayed another characteristic of his research and his life. From the time he worked as an economist with the FCC, Dallas demonstrated the importance of *praxis* or the unity of theoretical reflection and practical action. Unlike many academics who build a wall around their professional lives so as not to risk "tainting" it with political activity, Dallas welcomed activism, saw it as intimately connected to his scholarly work and supported it in his students and colleagues. There were times when this was demonstrated in his policy analysis and advice, as when he served as the first economist hired by the FCC. Then there were occasions when he demonstrated this in his academic activism, such as his work to help establish the political economy of communication at the International Association for Media and Communication Research. Dallas also demonstrated praxis with his involvement in the peace and anti-nuclear movements, as well as in his work for organizations genuinely committed to media democracy.

Dallas Smythe was an extraordinary scholar, policy advisor, and activist. I am honored to have been his friend. My only regret is that he is not with us to join the debate that this excellent book will undoubtedly inspire.

Part Four. Toward a Materialist Theory of Commercial Media in a Digital Age

CHAPTER ONE

After Broadcast, What? An Introduction to the Legacy of Dallas Smythe

Lee McGuigan[1]
Annenberg School for Communication
University of Pennsylvania

Dallas Smythe (1907–1992) is foundational to the critical tradition in communications and media studies: He was the first Chief Economist at the Federal Communications Commission (1943–1948). In 1948 at the University of Illinois, he introduced the first university course on the political economy of communications (Schiller 1981; Melody 1993). He has been called the "first great scholar" of television (Miller 2010). His content analyses of TV programming in major American markets broke ground for a holistic paradigm of mass media research and provided empirical support for a policy to mandate educational broadcasting (Smythe 1954). As a teacher and colleague, he influenced a generation of critical scholars, including Thomas Guback, William Melody, Gertrude Robinson, Robert Babe, Robin Mansell, Vincent Mosco, Janet Wasko, Yuezhi Zhao, Nick Dyer-Witheford, and Manjunath Pendakur. Just by introducing James Carey to the work of Harold Innis, Smythe altered the study of communication and culture in North America (Carey and Grossberg 2006; Buxton 2012). His impact is amplified if we reckon the disciples and dissenters of such notable colleagues as Herbert Schiller and George Gerbner, in the United States, and Graham Murdock and Kaarle Nordenstreng, in Europe (see McChesney 2007; Mosco 2009). In the pages and chapters that follow, we introduce scholarship from across these generations.

But Smythe's legacy is not without debate, even among sympathetic critics. Robert Babe (2009, 18–23) traces the lineage of the political economy of communication and culture to Harold Innis and Theodor Adorno.[2] Dan

Schiller (1999b) identifies Robert A. Brady, an economist and mentor to Smythe, as a progenitor of a political economy approach to communication. Smythe's discussion of the Consciousness Industry—a term borrowed from Hans Magnus Enzensberger (Smythe 1981, 5)—appears derivative of Adorno and Horkheimer's "culture industry" concept (Babe 2009, 50, n. 64). Smythe exaggerates his materialist break from ideology, remaining focused on the mediation of *consciousness* (Jhally 1990, 73; Babe 2000, 132–134). Furthermore, his recognition that the audience is the commodity product of commercial media, the "discovery" for which he is most famous, is of questionable originality. In *Cable Television and Telecommunications in Canada*, for which Smythe wrote a foreword, Babe (1975, 5) reproduces a statement from Harry Boyle, a former vice-chairman of what is now the Canadian Radio-Television and Telecommunications Commission: "[S]o much broadcasting is merely a frantic effort to secure numerically for one set of broadcasters a certain total of audience." Smythe himself cites a 1970 judgement of the Canadian Special Senate Committee on Mass Media: "What the media *are* selling, in capitalist society, is an audience, and the means to reach that audience with advertising messages" (1981, 8). Non-Marxist economists (Melody 1973), historians (Barnouw 1978), and media critics (Brown 1971) reached similar conclusions almost concurrently.[3]

These ideas have even longer histories in industry practice. By the late-1800s, business enterprise understood that market-making through communications media was necessary to curb prospective crises of over-production. Before the arrival of radio broadcasting, trade journals touted the capacity of a commercial press to command the attention of valuable audiences (Marchand 1985) and to manufacture consumer consciousness (Ewen 1976). Advertising agency N.W. Ayer & Son collected information about audience markets for regional newspapers as early as 1879 (Beniger 1986, 378), and by 1900 the Lord & Thomas agency was using sales figures to infer returns on advertising investments (384). Radio and television inherited and developed within similar commercial institutions (Williams 1974).

Perhaps it is obvious to scholars, professionals, and policymakers that media organizations sell audiences to advertisers. Many researchers today acknowledge the commodity existence of media audiences as a matter of course. Why, then, is it important to understand that the audience is *the principal commodity product of advertiser-supported media*? What is the purchase of Smythe's thesis—one formulated to explain mass broadcasting—when digital information-communication technologies (ICTs) facilitate ubiquitous connectivity to media and markets?

Smythe concedes, in the often-ignored first sentence of his most controversial essay, "The argument presented here [...] is an attempt to start a debate, not to conclude one" (1977, 1). This introduction provides evidence of his success. I am without either the proximity for biography or the wisdom for critique;[4] my fortune is perspective. Dallas Smythe helped pioneer a provocative scholarship that continues to examine communications, capital, consciousness, and power relations in social processes. Intended to be less a coronation than a resource for students, this chapter introduces some salient themes that punctuate Smythe's oeuvre and then refers to exemplary recent work on the political economy of media, culture and technology. It is a regrettably incomplete survey of literature attending to a range of questions that concerned Smythe. I make no claims that he invented these objects and methods of inquiry, or that he solely inspired the scholars and scholarship discussed. Yet, twenty years after his death, it is clear that Smythe's ideas remain formative to critical research probing blind spots along the road out of dependency and domination.[5]

On the Audience Commodity: Media, Consciousness, and Alienation

To appraise Smythe's thesis, we must appreciate its scope. A big thinker, Smythe confronted fundamental issues about where power resides in the mediated processes of (re)producing culture and society. In *Dependency Road* he proceeds from a maxim of twentieth-century Spanish philosopher Ortega y Gasset: "Living is nothing more or less than doing one thing instead of another" (quoted in Smythe 1981, 1). This notion pervades Smythe's understanding that political economy and ideology relate to survival, control and resource allocation—how humans make the world. He might equally have posed the question that guided Harold Innis in analyzing the bias of communication: "Why do we attend to the things to which we attend?" (Innis 1964, xvii).[6] As with Innis, Smythe brought the biases of an economist to the study of communication and culture; yet he was compelled to understand how people, institutions, and commodities fit within social processes. Smythe saw that markets betrayed promises of efficient provisioning, and economists neglected mediating influences of cultural production. In emphasising the importance of communications to political power and consciousness, and in portraying commodities and technologies as "teaching machines," Smythe is to be considered, like Innis, a preeminent theorist of media.

Embedded in a material history, media are both world-*making* and world-*made*. So, too, are institutions. Institutions are habituated ways of thinking and acting, structured within formal organizations or informal cultural patterns (Berger and Luckmann 1966). As mediators of social interaction, institutions are biased or coercive toward certain processes and outcomes. They tune consciousness—what we attend to or ignore—toward certain emphases, and they afford certain interests access to levers of power (Turow 1997b). In austere times, public schools teach students to excel at performing standardized tests. When campaign expenditures tally in the billions of dollars, career-politicians inherit values from financiers (Nichols and McChesney 2013). A society that worships "The Economy" nurtures a staple armaments industry and pardons financial corruption. A system of communication that sells audiences to advertisers operates differently than one that serves the information needs of public citizens using commonly shared resources. Certain values are bolstered or attenuated in each circumstance. As institutional economist John K. Galbraith writes, "The engines of mass communication, in their highest state of development, assail the eyes and ears of the community on behalf of more beer but not of more schools" (1969, 230). Smythe's insights into the pressures and limits of institutions, and media in general, have particular acuity for analyzing the commodification of communications and human life.

The audience commodity concept is dialectical: advertiser-supported media produce audiences as both *economic* and *social-cultural* products (Smythe 1981, 13). Economically, audiences are statistical abstractions representing aggregate individuals attending to the means of communication, shaped by a systemic priority on evaluating buying-power. Socially and culturally, audiences are real people living as producers and consumers in capitalism. In both respects, commercial broadcasting is implicated in the creation of *markets*—markets *of* consumers (for sale to advertisers) and markets *for* consumers (to buy advertisers' products). The audience commodity thesis is a cornerstone for understanding the economic functions of commercial media and an entrée into the pressing questions of economic sociology: what is to be produced, for whom, and how? (222–232).

Having the manufacture of commodity audiences as an organizing principle for a system of communication presupposes and reproduces institutional arrangements that empower industrial capitalists, advertisers, market researchers, and audience surveillance firms and impose constraints over the range of values and ideas conveyed, lionized or dismissed (Meehan 2005). The broadcast policies that accompany such arrangements produce and respond to forms of citizenship generally anchored in platitudes about consumer sovereignty and a "free marketplace" of ideas (Napoli 1999; Baker 2002), which tacitly sanction the ascendency of consumers and commerce over citizens and community. Smythe understood that communication systems belong to the bedrock of political power and that mass media are fundamental to why the world is as it is. Why do we do, attend to, and care about certain things and not others? Like Innis, Smythe sought answers at the nexus of media, markets, and culture.

What may be significant, then, is not the novelty of Smythe's discovery, but his elucidation of how commercial media and the manufacture of audiences (as commodity products, commodity producers, and commodity consumers) fit within the political economy of capitalism under conditions of corporate oligarchy, transnational commercialism, and consumer culture (Mattelart 1991, 195). A principal aim of this book is to acknowledge that Smythe's thesis is not confined to a debate about Marxist orthodoxy. Instead, processes of audience commodification—the economic and social production of consumers, and the commercial mediation of consciousness—should be understood as keystones of a broader, if not entirely precise, effort to theorize and historicize communications and culture in capitalism.

The linchpin of Smythe's argument is *alienation*, the denial of human essence or "free conscious activity" (Marx 1844/1961). For Smythe, free thought is restricted by the Consciousness Industry. We are not owners of our own social reproduction: we make our homes, our shared spaces, and our identities with branded products, both physical and symbolic, furnished by a covey of transnational corporations (TNCs). In cultural life, advertising slogans and film scripts infiltrate conversation, and peer groups congregate around particular brands, lifestyles, and celebrity personas. As astute historians argue, existence in consumer capitalism is mediated by "the discourse through and about objects" (Leiss, Kline, and Jhally 1990). In sustaining ourselves physically, public health and environmental crises owe in part to the power of the Consciousness Industry to market particular foods and household products. In one example germane to the history of broadcasting, the profitability of soft drinks and breakfast cereals made from corn by-products has perverted the economics of agriculture. Consortia of mass food brands hold sway over children's buying requests and legislative government in the United States. U.S. food and beverage marketers spend approximately $2 billion annually targeting children and teens (Federal Trade Commission 2012). From 2010 to 2011, Procter & Gamble set a record for annual marketing outlays, spending $9.3 billion to exhort people to make their homes with P&G products (Neff 2011). To paraphrase Galbraith (1967, 205), it is beyond drastic to ignore such vast investments in consciousness and demand management (Smythe 1981, 24).

This is not a simple matter of indoctrination by broadcasters. While Smythe (1954) was acutely aware of and troubled by the representation of reality as entertainment, he rejected qualitative textual analysis in favour of a more general appraisal of the commercial logic of media. We relate to certain forms of communication in certain contexts, and *time spent watching television is time spent not doing something else* (cf. Smythe 1977, 6–7). Despite his protests, Smythe was in partial agreement with Marshall McLuhan: it's not just *what* you watch on TV; it's *that* you watch TV.[7] As many critics have failed to appreciate, Smythe did not suggest that media manipulate audiences of cultural dupes; rather media are part of socialization in everyday lived experience whereby people reproduce themselves, in part, through the mediating institutions of consumer capitalism. Commercial media, and the messages they transmit, straddle culture and economy (Meehan 1986; McAllister 2005); they are significant not only as sales tools and productive industries, but also as social communication, setting a milieu for (often magical) expectations about human existence in a world of goods (Williams 1980; Jhally 1990).

Smythe (1981, 267) described television viewing as part of "a lifelong process" of living with commodities. This is a process of negotiation and contradiction. He emphasized the antagonisms between institutions of the family and institutions of the market (ibid, 270–271). It matters whether we cultivate our values from experiences with people or experiences with commodities, including commercial media. It matters how and what we eat. It matters how and what we desire. It matters how and what we communicate. When the lessons of humanity are taught in a commercial context saturated by a going concern we call "the military-industrial-entertainment complex," we are alienated from the reproduction of ourselves. As he repeats throughout *Dependency Road*, the work of the audience for the Consciousness Industry is toward the production of *people*—"people who live and work to perpetuate the capitalist system built on the commoditization of life" (1981, 9).

It is obvious that the reproduction of culture and society is imminently relevant. But, with communication central in these processes, we must fathom ongoing changes in media and technology. The concept of "audience labor" has sparked varied debate, so it is a fitting point of departure. Already subject to suspicion from cultural-pluralists who chide Smythe for neglecting the agency of media users, the "work" of atomized mass audiences seems increasingly dissimilar from experiences with mobile and Internet-enabled devices. In many ways, however, integrated communications media, and ubiquitous connectivity to markets and social networks, thrust ever more of human experience into commodity relations (Andrejevic 2004; Mosco 2009; Napoli 2010; McGuigan 2012a; Prodnik 2012). We arrive at a key question for this book: *In a digital age, what is the heuristic merit of the work of the audience?*

The Digital Labor Debate

Smythe's ideas have fueled ongoing efforts to understand labor in relation to digital media. Yet it is reasonable to wonder if a theory formulated to explain the economic functions of mass broadcasting has currency in the conditions of digitization and "informational" or "post-industrial" capitalism. These debates are especially salient today (Scholz 2013). Internet-enabled media, the bleeding edge of industry in an information society, have been the impetus for theorizing "free" and "immaterial" labor (Terranova 2000; Hardt and Negri 2000). Scholars consider digital labor in the context of exploitation (Andrejevic 2008, 2011; Fuchs 2010a; Rey 2012), revolutionary conflict and struggle (Dyer-Witheford 1999; Huws 2003; Mosco and McKercher 2007, 2008), and non-commodified social arrangements for production and communication (Arvidsson 2009). The Internet *in toto* is the focus of numerous volumes and reports employing approaches of political economy and critical social theory (e.g., Bermejo 2007; Fuchs 2008; Mansell 2012). Other central topics of study include the production of consumer data (Andrejevic 2002; Zwick and Knott 2009), online gaming (Grimes 2006; Zang and Fung 2013), the untethering of labor from the office through mobile devices (Manzerolle 2013), and the social and legal apparatuses that structure these relations, such as the institutions of private property and the wage labor contract (Comor 2011). One particularly notable effort to map the politics of labor in creative industries is being undertaken by Canadian researchers Nicole Cohen, Greig de Peuter and Enda Brophy.[8]

Social networking platforms and search engines have fermented inquiry. Scholars theorize immaterial and affective labor on Web 2.0 (Coté and Pybus 2007; Brown and Quan-Haase 2012). Several studies consider surveillance and the valorization of personal information on Facebook (Cohen 2008; Fuchs 2012e). Elaborating Richard Maxwell's (1991) critique of the audience commodity, Micky Lee (2011) uses the case of Google's advertising enterprise to refine and expunge themes from the "blindspot debate." Conversely, Kang and McAllister (2011) interpret evidence from the case of Google as an affirmation of the audience commodity thesis. Responding to literature toasting the freedom of users to participate in cultural production (e.g., Jenkins 2006; Tapscott and Williams 2008; see Spurgeon 2008), and more theoretically to futurists such as Alvin Toffler, Don Tapscott and Nicholas Negroponte, a robust scholarship discusses prosumers, prosumption, and user-generated content (e.g., van Dijck 2009; Fuchs 2011e; Ritzer and Jurgenson 2010; Kreiss, Finn, and Turner 2011; Duffy 2010; Zwick, Bonsu, and Darmody 2008).[9]

Marxist and Institutional Approaches

Smythe's thesis is inspired by Marx's discussion (1976, 90–94) of the dialectical, or mutually constituted, relationship between production and consumption (see page 42 in this volume). Materialism, for Smythe, denotes "the actual processes which link people together in social production and social consumption" (Smythe 1981, xvi). It is unsurprising, then, that he figures in debates about prosumption. Many writers have further interpreted Smythe's ideas in relation to Volume I of *Capital* (Marx 1973). Sut Jhally and Bill Livant (1986) deserve credit for their ambitious effort to articulate a labor theory of value for media through a more rigorous application of Marxist categories (see also Jhally 1982, 1990; Livant 1979, 1982). Some scholars, however, contest the very classification of audience activity as "labor." Among both Marxists and cultural-pluralists, doubts abound about the "exploitative" character of media use (e.g., Hesmondhalgh 2010). Brett Caraway (2011) suggests that a properly Marxist analysis of digital media should treat advertiser-support as a form of rent, rather than production of surplus value. Caraway indirectly echoes elements of Michael Lebowitz's (1986) critique of the "blindspot paradigm," in which Lebowitz argues that commercial media operate in the sphere of circulation (accelerating product turnover), not production. Ted Magder (1989) is persuaded by Lebowitz to agree that media organizations cannot own audiences' labor power.[10] Even Mosco (2009, 137) concedes that it is "contentious and doubtfully productive" to argue that audiences constitute labor. Analytical points about labor and value remain hotly debated, manifesting most recently in direct exchanges between Christian Fuchs (2010b, 2012f) and Adam Arvidsson and Eleanor Colleoni (2012).[11]

Others follow Eileen Meehan (1984) toward an "institutional approach." The institutional understanding posits that audiences, commodified as *ratings*, do not exist objectively in media use, but rather emerge from a tension between ways of defining valuable audiences and the formal procedures for manufacturing them as standardized commodities (see also Turow 1997b, Chapter Five). As Meehan (1993a) puts it, the commodity audience is not naturally occurring, like a tree; it is a construction, like a toothpick, shaped by business exigencies and an unequal political economy. Indeed, the very concept of the audience "was hatched largely out of the marketing departments of companies with a stake in selling products through the media" (Mosco and Kaye 2000, 32). The work of producing audiences as discrete packages of information is undertaken by a range of interests, including ratings agencies and market researchers (Maxwell 1991; Bolin 2009). Philip Napoli (2003a) offers the most thorough elaboration of

Meehan's research on the economics of audience information systems. Napoli (2011) and Fernando Bermejo (2009) identify the institutional political economy of audience production as intimately linked to both the development of information technologies and the form and content of media messages. Shawn Shimpach (2005) agrees that the audience emerges from institutionalized relations and the work of users. Following Martin Allor's (1988) analysis, Shimpach also counts cultural and creative activity as part of the work of watching. In helping to expose the ontological paradox in the commodity audience—that it is *the* tangible product exchanged between media organizations and advertisers, and yet it does not exist tangibly outside of these relationships—this institutional approach serves as a useful counterpoint to debates about Marxist theory. The conclusion that audiences cannot be owned is unsettled if we think of audiences as discrete packages of information that *exist only in market relations*.

Smythe's work demonstrates a vital synthesis of Marxist and institutional approaches to political economy. The Marxist influence is explicit in Smythe's discussion of labor and commodification; yet he equally inherits from the institutional traditions of Galbraith and Thorstein Veblen. Galbraith's influence is evident in Smythe's attention to formal/concrete institutions, including advertising, transnational corporations, state governments and public policy. Smythe's work is also sensitive to informal/abstract institutions—shared habits of thought and action—which Veblen animated adroitly in his portraits of pecuniary cultures. Wedding these heterodox tacks in economics, Smythe tacitly raises a compelling question: can Marxist and institutional approaches be complementary? An affirmative case is made in the collective works of Jhally, Meehan, Melody, Mosco, and Murdock.[12]

Data and Digital Surveillance: A Factory in the Marketplace

A field of research proximate to the institutional analysis of ratings focuses on the processes of producing saleable *data*. Engines of market research surged in the 1980s and 1990s, and investments in networked computing systems allowed businesses to accumulate vast databases of consumer information (Turow 1997a; Schiller 1999a). Indebted to the works of Oscar Gandy, Jr. and David Lyon, critical scholars examine various political-economic dimensions of surveillance and ICTs. Greg Elmer (2004) studies a personal information economy organized around surveillance and database technologies that capture feedback and construct consumer profiles. Mark Andrejevic (2004) argues similarly that, through surveillance, consumption becomes subject to scientific management techniques previously honed for production. Andrew McStay (2011) probes the commercial and regulatory relations enabling inspection of data collected by Internet service providers (see also Fuchs 2012c). Other exemplary studies add to the literature on electronic surveillance in the workplace (Clement 1992; Barney 2000: 156–163), consumer research databases (Chung and Grimes 2005; Zwick and Knott 2009), data-oriented mediation of identity (Turow 2006; Manzerolle and Smeltzer 2011), digital television technologies (Carlson 2006), mobile devices and location-aware applications (Humphreys 2011), targeted political advertising (Turow et al. 2012), and various forms of online shopping (Draper 2012). A spate of research questions arise as (mostly) online-based firms attempt to challenge existing oligopolies in audience manufacturing, and especially as telecommunications systems operators leverage their control of infrastructure—such as cable set-top boxes and Internet service provision—to gather data independently, by-passing existing institutions (McGuigan 2012b).

Perspectives on consumer surveillance provide evidence of interactive media acting as factory-like sites of data harvesting and boundless digital marketplaces (Tinic 2006; Andrejevic 2007, 2009b; McGuigan 2012a). Reliant on the information resources mined from users' "digital footprints"—records of clicks and transactions—interactive marketing strategies are increasingly oriented toward behavioural targeting and evaluations of actual outcomes (Napoli 2011). Beyond their commodity value, feedback data inform the administration of digital markets, serving as predictive indictors and sharpening marketing pitches to suit the known proclivities of individuals (Andrejevic 2013). Paramount to the conventional wisdom among advertisers and marketers is the coupling of tailored commercial solicitations with immediate purchasing opportunities. Smythe's thesis was a prescient forecast of the proliferation of *the means of consumption*; today, technologies and market institutions can facilitate commodity exchange instantly on almost any media platform (Barney 2000, 163–187).

These areas of research both complement and test the limits of Smythe's thesis. The work of the audience for advertisers and media organizations is more concrete now than in Smythe's day (Napoli 2010), and the production of commodity audiences further precedes the free lunch in the sphere of digital publishing, as automated advertising exchanges and programmatic buying practices divert subsidies from journalism and function almost exclusively to satisfy advertiser demand for particular demographics (Turow 2011; McChesney 2013). Thses cases illustrate that advertiser-supported media produce *consumers* as economic products—assemblages of data—and as social products—people existing in a ubiquitous digital marketplace. As business scholars are wont to boast, in the digital age the marketplace is "omnipresent" (Watson et al. 2002) and consumers are "always on" (Vollmer and Precourt 2008).

Monopoly and Dependency Beyond the Capitalist Core

Smythe viewed commercial media as bulwarks of "monopoly capitalism," a political economy wherein a handful of elite TNCs control global markets and set policy agendas for the capitalist system. The market for audience ratings tends toward monopoly because of efficiencies gained by adopting *one* authoritative measurement protocol (Meehan 1984; Napoli 2003a; Bermejo 2009). Sustained by patent protections, fixed capital, and the inertia of legacy business routines, the Nielsen Company has wielded power comparable to a government agency. The structural biases of computing networks—e.g., layered interconnectedness, reliance on compatible software standards and component parts—encourage consolidation in markets for digital telecommunications (Barney 2000, 114–120). Among the most spectacular examples, Microsoft and Intel, respectively, have dominated markets for computer operating systems and semiconductors. Technical dependency, product bundling and brand advertising reinforce their power to determine prices against the deflationary pressure of real improvements in computer processing (Schiller 1999a, 90–91). Despite much-ballyhooed "reversals" in corporate/consumer power relations, concentration in media ownership, in the form of the conglomerate, remains a pressing issue for democratic communications (Winseck 2008; Crawford 2013).[13] Columbia University law professor Tim Wu writes, "There is no understanding communications, or the American and global culture industry, without understanding the conglomerate." Wu continues, "the conglomerate is the dominant organizational form for information industries of the late twentieth and early twenty-first centuries; here and abroad it is inseparable from the production of the lion's share of culture" (2011, 219). In a recent spectacular example, Comcast's purchase of NBC Universal consolidated "an unprecedented combination of cable, Internet, studio and broadcast assets" (Johnson 2011; see Crawford 2013).[14]

In the decades following Smythe's work, the monopoly position of TNCs has been strengthened on a global scale through various "free market" interventions. Mergers and acquisitions nurtured by "market regulation" (Mosco 2009, 176) in the 1980s and 1990s catalyzed conglomeration of media and information technology assets (Schiller 1999a: 28–35, 63–69) and advertising/marketing properties (Turow and McAllister 2002). Telecommunications policies in Canada and the United States, enacted in 1993 and 1996, mandated in no uncertain terms that information infrastructure should mature according to market edicts, overseen by and for private enterprise (Barney 2000, 113–114; McChesney 1996, 2013; Babe 2011). Neoliberal imperatives to lubricate global trade increasingly colored policy debates over the thirty-year period bookended by UNESCO's forums on the New World Information and Communication Order, started in the mid-1970s, and the International Telecommunication Union's World Summit on the Information Society, in 2003 and 2005 (Pickard 2007).

Neoliberal influence is preponderate in the broader political economy (Schiller 1999a). The 1994 North American Free Trade Agreement institutionalized the Global South as the "workshop of the world." Sweeping de-industrialization of North America and erosion of organized labor were part of a bargain that overestimated the controllability of intellectual property and a "knowledge economy." In 1999, the repeal of legislation that cordoned commercial and investment banks sanctioned deleterious leveraging and speculation in securities markets. Further "modernization" of commodity futures in 2000 freed from oversight exotic and volatile instruments, such as credit default swaps, leaving bankers and analysts holding a monopoly of knowledge in these arcane exchanges. The influence of a global economic empire expanded throughout post-Soviet Europe, formalized in the powers of the International Monetary Fund, the World Bank, and the World Trade Organization. Digitized capital and computer networks marched in concert, as innovation and infrastructures were directed in no small part by corporate demand for the information technologies and services necessary to coordinate ever more spatially expansive and temporally precise supply chains. As wage relations and the price system penetrated new territories, commodity consumption has become the primary means of reproduction for more humans than ever in history (Schiller 2013).

A market-oriented system of higher education (Schiller 1999a, Chapter Four), as well as predatory consumer credit lending (Manzerolle 2010), have plunged unprecedented numbers of skilled and educated youths into debt and dependency. The prospective Comprehensive Economic and Trade Agreement between Canada and the European Union promises further liberal-modernization: "technology" and "knowledge" supplant manufacturing industries as economic staples; capital enjoys greater freedom to poach labor abroad; and Canada depends less on the U.S. and more on the E.U. (Shufelt 2013). Harbingers of a shift to post-Fordism, marked by flexibility and precarity in labor markets and the financialization of capital, are documented in David Harvey's comprehensive and prescient tome, *The Condition of Postmodernity* (1990), and further assessed in Nick Dyer-Witheford's *Cyber-*

Marx (1999). It should be noted that Dyer-Witheford gestures toward important terrain for class contest and global organization. Resistance to the World Trade Organization and G8/G20 summits, as well as the occupation of New York City's financial district, signals cleavages in neoliberal capitalism.

Smythe correctly observed that nations hitherto peripheral to the capitalist core would be engines of geo-political change in the twenty-first century. He showed a particular interest in China's communication systems and cultural policy. Scholars in the political economy tradition remain dutiful to international concerns. Numerous collections discuss the global political economy of communication (e.g., Comor 1996; Chakravartty and Zhao 2008). Yuezhi Zhao, Micky Lee, Guobin Yang, Dan Schiller, Dal Yong Jin, Gerald Sussman, and Paula Chakravarrty are notably erudite in analyzing power relations that structure communications, commerce, and politics in Asia. In relation to China's Cultural Revolution, Smythe argued that "the power of capitalism today rests on its success in developing capitalist *consumption* relations" (1973/1994, 239–240). Edward Comor (2008) examines consumption relations in India and China in historical detail and with a precise sociological treatment of *power* in globalization. Like Smythe, Comor expresses tempered optimism about the potential to resist a global hegemony in these places where capitalist consumption is not yet a fully naturalized institution (i.e., invisible and taken-for-granted). Marwan Kraidy (2005) examines and theorizes global expressions of cultural "hybridity." By integrating political economy and cultural studies, both Comor and Kraidy add critical rigour and nuance to debates about transnational flows of communication, culture, and power. Manjunath Pendakur (2003) has written extensively on India's film industry. Others have looked at media systems in Latin America (see Mosco and Schiller 2001) and the Arab world (Kraidy 2005), and class struggle and exploitation of the Global Worker (Dyer-Witheford 2010).

Exploitation in processes of globalization is not confined to human labor. Smythe also warned about pressing "ecological disasters" and the "rape of natural resources" (1981, 18). The assembly of consumer technologies and the mining of precious metals and component parts yield environmental hazard and human suffering, concentrated in the developing world (Maxwell and Miller 2012). Certain regions on the "periphery" of consumer capitalism, especially those rich in resources of minerals, unprotected labor, or potential landfill space, support affluent societies' dependency on cheap goods and receptacles for high-tech waste. A regime of R&D that focuses on bringing incremental innovations to market rather than revolutionizing production processes exacerbates the costs of high-tech consumerism. Proprietary science in industrial agriculture and the privatization of genetic material are trends that raise ethical questions about ownership of life as property. Global marketization of food science and technology also devastates local economies and ecosystems, with acute damage experienced in India and South and Central America. So-called "externalities" of sovereign debt held by the IMF—such as the near-annihilation of Brazil's rainforests—make student debt crises seem trivial. The Consciousness Industry's imperative of limitless economic growth is in violent tension with ecological balance and human survival in the *longue durée* (Jhally 2000; Babe 2006a). Recognition of such contradictions is confounded by the present-mindedness described grimly by Innis, and latent in Smythe's writing.[15]

Advertising, Branding and Commodification of Human Experience

Smythe attempted, in his words, "a Marxist theory of advertising and of branded commodities under monopoly capitalist conditions" (1977, 16; see page 43 in this volume). Commodification of culture, information, and natural and symbolic resources persist as focal points for many writers addressed herein. A notable contingent of scholars not engaged with Smythe's literature, and not Marxist in any formal sense, studies the sociology of markets and consumption, with particular emphases on consumer culture, advertising, and consciousness. Stuart Ewen, Zygmunt Bauman, Don Slater, and Juliet Schor, among others, understand consumption as a mediating sociological institution, and one married to communication, the cultural circulation of images, and power-laden and socially constructed habits of thought and action. Particularly consonant with Smythe's work, Bauman's book, *Consuming Life* (2007), foregrounds the commodification of human existence and the institutionalization of consumption as a linchpin of social organization. "The most prominent feature of the society of consumers," Bauman writes, "is the *transformation of consumers into commodities*" (2007, 12). A 2012 issue of the *Annals of the American Academy of Political and Social Science* (Shah et al. 2012) devoted to consumption and politics takes up many relevant themes, such as "capitalist realism" (cf. Smythe 1981, Chapter 9). Smythe's name is absent from this research, though not surprisingly, as his works, and political economy in general, have been persistently ignored by the mainstream of scholarship on media, culture, and economics (Mansell 2004; Artz 2008; Babe 2006b).

Influenced by Galbraith (1967), Smythe also endeavoured to theorize branding and advertising within processes of demand management, including both direct interventions into commodity markets (e.g., securing space on store shelves and inflating brand equity) as well as mediation of consumer consciousness (e.g., stimulating desires and shepherding tastes). Notable among recent critical scholarship on branding are the works of Adam Arvidsson (2006), Liz Moor (2007), and Sarah Banet-Weiser (2007). Alison Hearn's (2008) discussion on "self-branding" is especially germane. Similarly, journalistic books interrogating the political economy and spectacular culture of consumer capitalism, such as Naomi Klein's *No Logo* (2000) and Chris Hedges' *Empire of Illusion* (2009), have found popular reception.

William Melody (1973) was among the first critical scholars to study children's television from a political economy perspective, and this work, with an emphasis on the exploitation of audiences, resonates in Smythe's subsequent arguments. Marketers have long understood the importance of consolidating their share of children's consciousness and cultivating lifelong brand loyalty. Researchers continue to place children centrally in considering online advertising (Chung and Grimes 2005), commercialism and consumption (Asquith 2009; Hill 2011; Schor 2004; Wasko 2008), and cultural experiences with television (Livingstone 2009). Television remains the prevailing mass medium and the surest investment for marketers. Despite some discussion about the end of television (see Katz 2009), critical scholars continue to further our understanding of television as an industry (Meehan 2005; Jhally 2007; Lotz 2007; Mayer 2011), a milieu of culture and citizenship in postmodern times (Postman 1985; Miller 2007), and an important field of social sciences and humanities research (Wasko 2005a; Miller 2010). As television becomes an interactive marketplace with greater capacity to facilitate immediate purchasing of branded goods and services and to harvest information as the raw material of the commodity audience, Smythe's thesis may be more germane than ever.[16]

Changes in technologies are mutually constituted with the restructuring of media businesses. The social emergence of smartphones, tablets, and other Internet-enabled devices capable of valorizing creativity and personal information through "two-way conversation" begets reconsideration of advertising, marketing, and branding. While new devices and premium subscription services challenge incumbent business models built on exposure or attention, they also allow marketers to come ever closer to the long-standing goals of targeting known consumers and verifying returns on advertising investments (Lotz 2007; Napoli 2011). Marketers celebrate the proliferation of "touchpoints," or opportunities to solicit and monitor consumers (Jaffe 2005). Relationship marketing techniques range from using database and surveillance technologies to assess the value of individual customers (Turow

2005), to converting people into brand ambassadors in a process of exploitation euphemistically called "participatory marketing." Some marketing scholars recommend using interactive media to provide "social ecosystems that enable consumers to truly *live the brand*" (Martin and Todorov 2010, 64; emphasis added). Smythe's ideas provide a useful starting point for revisiting audience information systems, market research, and branding strategies.

Critique of "Technology"

Smythe was at pains to temper uncritical faith in technological progress. He regarded technology as part and parcel of capital, as a value-laden process of social development (1981, 217–222). Too much research to catalogue exhaustively deals with technology and new media; but it is worth noting some examples in immediate proximity to Smythe's work. Robert Babe's *Telecommunications in Canada* (1990) is a seminal treatise on technology and public policy. Babe attributes notable inspiration for the book to a lecture Smythe delivered under the title "Technology as Myth." In *The Digital Sublime*, Vincent Mosco (2004) unravels celebratory mythologies of technological determinism. James Carey (2009) has also written at length on myths furnishing the American imagination of technology.

Dan Schiller (1999a) historicizes the corporate seizure of computing networks that form the infrastructure for a global market system. "At stake in this unprecedented transition to neoliberal or market-driven telecommunications," Schiller argues, "are nothing less than the production base and control structure of an emerging digital capitalism" (1999a, 37). Darin Barney (2000) critiques utopian visions of techno-democracy in his book, *Prometheus Wired*—a magisterial treatment of the political, economic, and ontological dimensions of technology. Barney's thesis resembles the kernel of Smythe's project: "in so far as it combines productive activity with the gathering of significance, technology—especially information and communications technology—says something about what human beings are; what they wish to be; and how they live, or *might* live, together" (Barney 2000, 54). Nick Dyer-Witheford (1999) sets ICTs in a more explicitly theoretical (and specifically Marxist) context. Taken together, books by Schiller, Barney, and Dyer-Witheford triangulate the historical, philosophical, and theoretical vectors of digital technology in capitalism.

Materialist historians have studied television infrastructure, with emphases on national terrestrial broadcasting in the U.S. (Sterne 1999), cable in Canada (Babe 1975), direct broadcast satellite in North America (Comor

1998), and digital and interactive TV (Kim 2001; Boddy 2004; Turow 2006; Castañeda 2007). The institutional histories and implications of ICTs and real-time communication have been considered by numerous scholars (e.g., Mattelart 2000; Clement and Shade 2000; DeNardis 2012). Both Robin Mansell and William Melody have contributed indispensably to this literature, and they remain active instructors at an annual colloquium on the political economy of ICTs, which welcomes graduate students from around the world to study in Denmark. An especially rich resource, *Networking Knowledge for Information Societies*, edited by Robin Mansell, Rohan Samarajiva, and Amy Mahan (2002), pays tribute to Melody's research on telecommunications policy and economics.

Yuezhi Zhao (2007) discusses the "capitalist road" of technological development in China, in an article entitled, "After Mobile Phones, What?"—an homage to Smythe's (1973/1994) "Bicycles" report. Zhao contextualizes the social contradictions endemic to China's "digital revolution," and she reiterates Smythe's concerns about idealist philosophies and policies that treat "technology" as abstracted from social (economic/political/cultural) processes and conditions. Joseph Turow (1997a) also challenges belief in the neutrality of media, arguing that media technologies and the relatively civic or industrial logic of their institutional organizations are biased toward either building community or segmenting individuals into targeted markets. Turow warns against an increasing imbalance toward segment-making media—a shift effected by the advertising industry—and his analysis proceeds from a premise evocative of Smythe's: "The aim [of marketing] is to package individuals, or groups of people, in ways that make them useful targets for the advertisers of certain products through certain types of media" (1997a, 1). Media convergence owes in part to the advertising industry's efforts "to maximize the entire system's potential for selling" (ibid, 2).

Others add to a critique of the political economy of financial information (Lee 2012, 2013). Ubiquitous connectivity and the instantaneity with which information is circulated and brought to bear in the mediation of global markets have deepened human dependency on ICTs. High-frequency securities trading, for example, relies on computer algorithms to automate stock market transactions in which fortunes are conjured or effaced in fractions of a second. Ongoing technological developments owe to the vagaries of investment capital as well as many "new economy" policies stimulating market-directed innovation (Manzerolle 2013). Along with brand managers and market researchers, we should count technologists and finance pundits among agents of the Consciousness Industry. Acceleration and financialization of capital processes have been addressed in various critical analyses (e.g., Graham 2000; Manzerolle and Kjøsen 2012).

Critical Research on Communication Policy and Cultural Policy

During his career, Smythe played an important role "in almost every major policy development in broadcasting and telecommunication in the United States and Canada" (Melody 1994, 1). As Chief Economist of the FCC, Smythe contributed to the agency's "Blue Book," meant to impose a mandate preserving educational and public interest criteria for issuance of broadcast licenses and allocation of electromagnetic spectrum. The spectrum was a lightning rod for Smythe's deliberation on public and private/commercial media systems (Smythe 1954, 1981, Appendix), and it remains a foremost concern as ever more interests compete to use the resource for mobile networks (Manzerolle 2010; O'Dwyer 2013).[17] The qualitative nuances of information technologies and digital media have prompted reflection on the normative foundations of telecommunications policy (McChesney 1996; Melody 1996; Bar and Sandvig 2008) as well as a revisiting of the history of policy-making in the twentieth century (Streeter 1996; Schiller 1998; 2007; Pickard 2010, 2011).

Consistent with his emphasis on practical action (Wasko 1993), Smythe was forthcoming with prescriptions for research agendas. Despite his radicalism, he willingly engaged with the mainstream of communications research (1981, Chapter 11). Writing with Tran Van Dihn, Smythe avers that administrative research can be an important resource for critical theory and policy analysis (1983, 118–120). Trade publications, for example, provide a beachhead for exposing notoriously mysterious relations among advertisers, market researchers, and media firms (e.g., Turow 1997a; Schiller 1999a; Napoli 2003a). Smythe's work remains a starting point for debates about critical research into cultural industries (Havens, Lotz, and Tinic 2009), media audiences (Nightingale 2011), and the political economy of media/communication (Meehan 2000; Wasko 2004; Mosco 2009; Babe 2009; Wasko, Murdock, and Sousa 2011; Nixon 2012; Meehan and Wasko 2013).

Conclusion as Introduction

While this review harbors biases and blindspots, diligent students will find that the studies referenced herein are not of uniform opinion or approach. Rather than presenting a cohesive argument, I have intended this chapter to be a guidepost: I hope students interested in the topics probed by Smythe and others might use this introduction to locate relevant research that is more thorough and thoughtful than what I can provide.

As promised, Smythe has provided dissertation topics to generations of graduate students. His ideas continue to provoke debate among critical scholars of media, culture, technology, and political economy. I will conclude with a prescription for students:

> At all levels of analysis [...] the objective of research/action should be the demystification of 'technology' and science; the necessary relationship between theories and practice; the decentralization of control of communications; the democratization of communications institutions and practices; mass mobilization for organization and action; and the paramount significance of communications for peace. (Smythe and Van Dihn 1983, 127)

As serious readers already know, Smythe was not driven to secure remuneration for the work of audiences. His ambition was to usher in a social configuration, with communications at the heart, that would place human values above commodification and commerce—a system in which alienation and exploitation are replaced by *fully human life*.

Notes

1 Many thanks to the following people for commenting on drafts of this chapter: Robert Babe, Eileen Meehan, Vincent Mosco, Victor Pickard and Vincent Manzerolle.

2 Elsewhere, Babe is gracious in acknowledging his indebtedness to Smythe (see Babe 1990, xiv; 1996).

3 Reportedly, however, Smythe had begun formulating the idea as early as 1951 (Mosco 2009, 84).

4 See instead, Babe (2000, Chapter Five); Mosco (2009, especially 82–87); Mosco, Wasko, and Pendakur (1993); Smythe and Guback (1994); Artz (2008); Fuchs (2012a).

5 This review is biased toward work in the English language and in North America. This owes to the limits of my language skills and the texture of my own education. I foreground scholarship with connections to Canada in part because Smythe's thinking was quintessentially Canadian—dialectical, holistic, ontological, and historical (Babe 2000). These characteristics flavor much of the research referenced herein.

6 Harold Innis (1894–1952) was one of Canada's preeminent economic historians. His meticulous studies of how empires and civilizations have been formed and maintained in relation to communication and media (understood broadly to include language, bureaucracy, transportation infrastructure, and so on) are standard references for students of media theory and cultural history (see Innis 1950, 1964). Innis observed that control of knowledge is instrumental to the durability of empires, and that media are relatively biased toward conveying information over space or preserving it over time. Imbalances in the ecology of media engender monopolies of knowledge.

7 Smythe characterized television content as a "free lunch" that whets the appetite; McLuhan described content as a "juicy piece of meat carried by the burglar to distract the watchdog of the mind" (1964, 18).

8 "Cultural Workers Organize," http://culturalworkersorganize.org/

9 Interest in prosumption has occasioned special issues of such periodicals as the *Journal of Consumer Culture* (2010) and *American Behavioral Scientist* (Ritzer et al. 2012).

10 More recently, Magder (2009) seems to accept Smythe's formulation. Describing the business of commercial television, he states "the attention of viewers…is sold to advertisers," and "[n]ow, throughout most of the world, TV relies heavily on the buying and selling of audiences" (2009, 145–46).

11 A fundamental contradiction pursued throughout Smythe's work is that between the human relations of people and the commodified relations of capital (see Babe 2000, 125–126). The dialectical tension between labour and capital is at the crux of Autonomist Marxism. Among the most sustained and insightful treatises on class struggle under conditions of ICT-mediated capitalism is Nick Dyer-Witheford's (1999) book, *Cyber-Marx*. Dyer-Witheford relates that Smythe intimated to him in a personal conversation toward the end of Smythe's life an affinity for Autonomist Marxism (1999, 271).

12 For works usefully comparing/contrasting Marx and Innis, see Parker (1981), Jhally (1993), and Comor (1994).

13 Dwayne Winseck, professor of communication at Carleton University in Ottawa, leads the Canadian Media Concentration Research Project, which takes up a range of issues related to telecommunication and broadcasting policy (http://www.cmcrp.org/).

14 Illustrating the centrality of commodity audiences to media economics, one fond commentator called the $30 billion Comcast-NBC acquisition "a calculated move to seize the reins in shaping the future of TV-viewer behavior and a bid to assume the lead in figuring out how to advertise to the new-media consumer" (Steinberg 2009).

15 *Surviving Progress*, a 2011 documentary produced by the National Film Board of Canada, confronts these issues with input from such stalwart Canadian ecologists asDavid Suzuki and Margaret Atwood.

16 On this topic, it is worth referring to Raymond Williams's book, *Television: Technology and Cultural Form* (1974). At nearly the same time, Smythe (1973/1994) and Williams separately considered the implications of interactive television. While singling out Williams for coming "closer than many Marxists to a realistic treatment of communications," Smythe rebuked the former's book about television for "mystifying" technology (Smythe 1981, 25, n2). It is puzzling that Smythe was so critical of Williams and others with whom he shares affinities. As Manzerolle notes in this volume, the Smythe-Innis comparison is particularly interesting (see also Babe 2000, 135–136). Consider one example: Smythe contributed a chapter to a book honouring Innis (Melody et al. 1981); in reiterating his "blindspot" argument, Smythe refers to Innis only in passing.

17 Canadian researchers have assembled a comprehensive educational resource on the history of and current issues in spectrum management policy (http://canadianspectrumpolicyresearch.org/).

PART ONE

FOUNDATIONAL TEXTS

Audiences, Commodities and Market Relations: An Introduction to the Audience Commodity Thesis

William H. Melody
Center for Communication, Media and IT (CMI)
Copenhagen Institute of Technology, Aalborg University, Denmark

Washington Connections

Washington was an exciting place in the late 1960s when the civil rights and anti-Vietnam war movements peaked, serenaded by Woodstock and flower power, creating an atmosphere promising opportunities for fundamental institutional reforms. As a young economist at the Federal Communications Commission (FCC) I was part of a small group engaged in the first investigation in a quarter century of AT&T's performance in meeting its public interest obligations under the Communications Act of 1934.

At the time AT&T had a monopoly over most telecommunication R&D, equipment manufacture and service provision in the U.S. It was the largest company in the world; it would have ranked 9th economically as a nation; it was the largest private employer in the U.S. and the largest contributor to political campaigns.

Part of the FCC investigation was receiving written and oral evidence from interested parties in special "hearings." AT&T produced an impressive team of "expert witnesses" consisting of leading economists and financial analysts from prestigious universities, consulting firms and Wall Street, documenting AT&T's performance as the best in the world and exemplary in every way. One day a veteran FCC colleague excitedly told me, "Dallas Smythe is coming back!"

Dallas was the first FCC Chief Economist in the 1940s. The major concerns then were trying to extend telephone service to poor rural areas to fulfill the universal service mandate of the Communications Act, and employment discrimination, primarily racial in the Southern states. He returned a generation later as an academic presenting expert evidence on behalf of the Communication Workers of America (CWA) which represented AT&T's vast labor force.

I had read Smythe's earlier publications on the radio spectrum, and knew of his central role in the commercial broadcast regulatory reforms in the 1950s, but that was about all. In true Smythe fashion his evidence was not about any of the normal economic indicators of performance—prices, quality, coverage, productivity, returns, etc.—but about the effects of AT&T's exploitation of its enormous economic and political power. I still chuckle whenever I think of his caricature of AT&T as more like a nation-state than a private corporation. It had its own flag and orchestra, as well as a standing army of lobbyists and captive experts to ensure its monopoly objectives were met in all of its relations, including labor relations.

Although his evidence was ruled beyond the scope of the FCC hearing by the judge, it was the beginning of a running debate between us about markets and political power in shaping economic, social and human development that continued until his death. He tended to approach issues from a macro critique of capitalism, whereas I tended to focus directly on the specific institutional arrangements.

What made our debates productive was that we resisted the temptation to see either markets or governments as the preferred agents for human development. Both are highly imperfect with a variety of strengths and weaknesses that can vary depending on circumstances; each can provide major benefits or fail miserably. Unchecked monopoly power is a major failing in both markets and government. Some of Smythe's socialist colleagues had difficulty reconciling his support for increased market competition to support socialist objectives in some circumstances.

Our views undoubtedly were shaped by our respective experiences in government attempting to implement the economic and social objectives of U.S. telecom and broadcast policies and regulations. Government regulation was necessary to address market failures (e.g., monopoly) and implement extra-market social objectives. But sometimes private markets and competition were necessary to address the failures of government policy and regulation.

The AT&T investigation of the late 1960s produced evidence showing that AT&T wasn't fulfilling its universal service mandate and that FCC regulation requiring it had failed. AT&T didn't even serve the rural areas;

rather 10,000 local co-operatives, municipal operations and small private companies did. The basic public telephone service wasn't being beneficently subsidized as claimed by AT&T, rather it was subsidizing advanced services to big business. Companies with new ideas for telecom equipment (e.g., mobile handsets, computer communication) and services (e.g., data, video) were being prevented by policy and regulation from entering the market.

The FCC agreed that it had not been very successful in trying to regulate AT&T, and concluded that economic efficiency, innovation and public interest social objectives all would be better served if the monopoly policy was changed and market competition supported. At about the same time, I was a member of a research team advising the State of Alaska which concluded that achieving universal access would require the state to enter the market as a competitive operator to compete with the private monopoly. The state did enter the market, providing a stimulus for network expansion and the eventual achievement of universal access.

Audiences: Commodification During Consumption

During the early 1970s, I would see Dallas periodically at conferences and we would pursue our debates on a variety of market and policy issues focusing particularly on possible implications of the new communication technologies of the time—satellites, cable TV, data and mobile—on the form and structure of personal and mass media communication.

As a father of young children, I took a keen interest in Children's TV programs, and was incensed by the manipulative exploitation of the innocence of young children evident in both advertising and program content on the commercial networks—especially sugar-only cereals and dangerous toys. When I joined the Annenberg School at the University of Pennsylvania in 1971, I took up the issue in association with *Action for Children's Television (ACT)*. I focused on this market arrangement as a market failure requiring both strong regulation to limit commercial exploitation and subsidies for the production of independent programs targeting children as the audience, not the commodity to be sold to advertisers. My book *Children's Television: The Economics of Exploitation* (1973) explained the market failure of commercial broadcasting in theory and practice, the egregious exploitation of innocent audiences that were sold to advertisers, and made the case for specific market and policy reforms.

In 1974, although an emeritus professor by then, he was invited to become Interim Chairman of a small Communications Department at Simon Fraser University while a search was undertaken for a permanent Chair. I took up the Chair in 1976 with a commitment that Smythe would stay on and new faculty positions could be filled as a foundation for building a strong department. Bob Babe was the first of a diverse group of critical scholars who were hired. We were also fortunate in attracting a number of superb PhD students, a number of whom are active contributors to the critical communication literature today. This helped create a lively environment for research and debate, perfect for testing Smythe's provocative ideas.

Dallas wasn't content with the conventional set of market failures that economists have identified and analyzed—monopoly, externalities, information deficiencies, communication biases, income mal-distribution, public and merit goods, tragedy of the commons, depleting resources, etc. He took a broader perspective from Marx's claim that under capitalism the march of markets will eventually penetrate all aspects of life, not only extending monopoly power and exploiting labor, but also eroding other values. Marx focused on capital and labor. Smythe was more concerned about the wider effects of commercial mass media markets that had enabled the mass consumer markets of the twentieth century. He sought an understanding of commercial media in the march of markets and the erosion of human values.

Dallas found my formulation of audiences as commodities useful but unduly constrained by the conventional economic conception of markets as simply supply and demand interactions. The broader implications needed to be developed. If audiences are "the commodities" in commercial broadcasting, they are very special and unique commodities, not simply passive and lifeless "eyesballs" (the industry metric of measurement) that are gathered to view advertisements. There is much more to audience commodities than just receiving information as a precondition for purchases as research on active audiences would show. These "commodities" have lives; they have memories; they engage in a variety of social behaviors that influence their understanding of the world; they are society. What was happening with media audiences was for Smythe a further step in the increasing commodification of society.

Although an overarching analytical framework for investigating the audience commodity was provided by Marx, Marxist scholars had not seriously examined it as a central element in the development of twentieth century capitalism. Smythe set out to open up this issue by stimulating research and debate that explored its broader implications. He observed that audience commodities were being activated by the mass media. They were being put to work stimulating demand on behalf of advertisers, and sustaining commercial broadcasting as a capitalist institution that was conditioning and directing other aspects of people's lives. Smythe thought this formulation of the issue could accommodate a wide range of research and new knowledge development.

My research on Children's TV provided evidence supporting Smythe's thesis. As the children couldn't afford the advertised products, the obvious intent of the advertising—and often the program content as well—was to turn the children into active lobbyists with adults. Pioneering psychological experimentation and research on cognitive stimulation was being undertaken on children watching TV and financed by advertisers. If this worked on children, why not adults? It seemed pretty clear that Smythe had identified a major issue in need of further exploration.

Audience Commodities in Information Societies

Although Smythe's exposure of the audience commodity issue and its potential ramifications did stimulate significant research attention, he had hoped to attract a wider set of scholars to it. Some people have attributed the relatively limited attention his thesis has received until recently to a recognition that the age of industrial capitalism with mass consumer markets generated by commercial mass media is passing. Mass markets and the reach of the mass media are already fragmenting. The increasingly rapid transition to information economies in information societies requires a fresh interpretation of the next stage of capitalist development and the role of markets and audiences in this new environment.

Dallas would smile and suggest that one should also look a little deeper. Mass markets are growing to global proportions, not declining. But the products and services are being differentiated and personalized in response to what marketeers present as people's desire for consumer individuality. This is requiring the surrender of personal details that violate traditional notions of privacy, and even security. The more details one surrenders, the better one can be served. The gathering and massaging of the "big data" banks made possible by the new ICTs allows those in the mass persuasion business to pinpoint more precisely the enthusiasts, the readily persuadables, the opinion leaders and the skeptics.

This is far more effective in manipulating audiences than the old single message scattergun approach of the past. As Smythe put it, "the shotgun is being superseded by rifles." The innovative techniques developed by the Obama team to unbundle the mass voter market and pinpoint those who could be persuaded to vote for the "Yes we can!" man are being employed in the service of commercial markets, including Las Vegas casinos to protect and enhance their client base.

Smythe might suggest that the audience commodity model is changing in some ways in the new information/communication economy, but the basic model is more relevant than ever. It has become the dominant market model on the Internet. Moreover the "work" of the audience, as Smythe has conceptualized it, is expanding significantly into providing program content, ranging from Big Brother to Facebook to the necessary acceptance of "cookies" on our PCs and iPods. One reason the terms "information economy" and "information society" are used interchangeably is because the former represents such a large portion of the latter. The audience commodity is a far more significant matter in twenty-first century economies and societies than it was in the twentieth. It has moved to center stage as a basic market model shaping economic and social development. Dallas would be gratified to know that a new generation of scholars is examining the commodification process at the heart of new forms of media.

CHAPTER THREE

Communications: Blindspot of Western Marxism

Dallas W. Smythe[1]
Simon Fraser University

The argument presented here—that western Marxist analyses have neglected the economic and political significance of mass communications systems—is an attempt to start a debate, not to conclude one. Frequently, Marxists and those radical social critics who use Marxist terminology locate the significance of mass communications systems in their capacity to produce "ideology" which is held to act as a sort of invisible glue that holds together the capitalist system. This subjective substance, divorced from historical materialist, is similar to such previous concepts as "ether"; that is to say, the proof of its existence is found by such writers to be the necessity for it to exist so that certain other phenomena may be explained. It is thus an idealist, *pre*-scientific rather than a *non*-scientific explanation.

But for Marxists, such an explanatory notion should be unsatisfactory. The first question that historical materialists should ask about mass communications systems is *what economic function for capital do they serve*, attempting to understand their role in the reproduction of capitalist relations of production. This article, then, poses this question and attempts to frame some answers to it. Much of what follows is contentious because it raises questions not only about changes in capitalism since Marx's death but also, in some cases, about the adequacy of certain generally accepted Marxist categories to account properly for these developments. However, as Lenin remarked in a different context, one cannot make an omelette without breaking the eggs.

The mass media of communications and related institutions concerned with advertising, market research, public relations and product and package design represent a blindspot in Marxist theory in the European and Atlantic basin cultures. The activities of these institutions are intimately connected with consumer consciousness, needs, leisure time use, commodity fetishism, work and alienation. As we will see, when these institutions are examined from a materialist point of view, the labour theory of value, the expenses of circulation, the value of the "peculiar commodity" (labour power), the form of the proletariat and the class struggle under monopoly capitalist conditions are also deeply involved. The literature of Marxism is conspicuously lacking in materialist analysis of the functions of the complex of institutions called the "consciousness industry."[2]

The blockage in recognizing the role of the consciousness industry traces back to a failure to take a materialist approach to communications. Both economic goods in general and communications goods in particular existed long before capitalism and monopoly capitalism. While *specialized* institutions for the mass production of communications (i.e. newspapers and magazines) appeared in capitalism in the eighteenth century, these institutions did not reach their mature form until monopoly capitalism shifted their principal economic base to advertising in the late nineteenth century. By a grave cultural lag, Marxist theory has not taken account of mass communications. This lag in considering the product of the mass media is more understandable in European (including Eastern European) countries than in North America. There the rise to ascendancy of advertising in dominating the policy of newspapers and periodicals was delayed by custom and by law. Even in the radio-TV broadcast media, the role of the state (through ORTF, BBC, ITV, East European state monopolies, etc.) has been resistant to the inroads of monopoly capitalism—as compared with the United States and Canada. But the evidence accumulates (recent developments in British, French, West German and Italian mass media, for example) that such traditional resistance is giving way under the onslaught to pressures from the centre of the monopoly capitalist system. Europeans reading this essay should try to perceive it as reflecting the North American scene today, and perhaps theirs soon.

At the root of a Marxist view of capitalism is the necessity to seek an objective reality which means in this case an objective definition of the *commodity* produced by capitalism. What is the commodity form of mass-produces, advertiser-supported communications? This is the threshold question. The bourgeois idealist view of the reality of the communication commodity is "messages," "information," "images," "meaning," "entertainment," "orientation," "education," and "manipulation." All of these concepts are subjective mental entities and all deal with *superficial* appearances. Nowhere do the theorists who adopt this worldview deal with the commodity form of mass communications under monopoly capitalism on which exist parasitically a host of sub-markets dealing with cultural industry, e.g., the markets for "news" and "entertainment." Tacitly, this idealist theory of the communications commodity appears to have been held by most western Marxists after Marx as well as by bourgeois theorists: Lenin,[3] Veblen, Marcuse, Adorno, Baran and Sweezy, for example, as well as Galbraith and orthodox economists. So too for those who take a more or less Marxist view of communications (Nordenstreng, Enzensberger, Hamelink, Schiller,[4] Murdock and Golding[5] and me until now) as well as the conventional writers exemplified in the Sage *Annual Review of Communications Research.*[6] Also included in the idealist camp are those apologists who dissolve the reality of communications under the *appearance* of the "medium," such as Marshall McLuhan.[7] No wonder, as Livant says, that "the field of communications is a jungle of idealism."[8]

I submit that the materialist answer to the question—What is the commodity form of mass-produced, advertiser-supported communications under monopoly capitalism?—is audiences and readerships (hereafter referred to for simplicity as audiences). The material reality under monopoly capitalism is that all non-sleeping time of most of the population is work time. This work time is devoted to the production of commodities-in-general (both where people get paid for their work and as members of audiences) and in the production and reproductions of labour power (the pay for which is subsumed in their income). Of the off-the-job work time, the largest single block is time of the audiences which is sold to advertisers. It is not sold by workers but by the mass media of communications. Who produces this commodity? The mass media of communications do by the mix of explicit and hidden advertising and "programme" material, the markets for which preoccupy the bourgeois communications theorists.[9] But although the mass media play the leading role on the production side of the consciousness industry, the people in the audiences pay directly much more for the privilege of being in those audiences than do the mass media. In Canada in 1975 audience members bore directly about three times as large a cost as did the broadcasters and cable TV operators, combined.[10]

In "their" time which is sold to advertisers workers (a) perform essential marketing functions for the producers of consumers' goods, and (b) work at the production and reproduction of labour power. This joint process, as shall be noted, embodies a principal contradiction. If this analytical sketch is valid, serious problems for Marxist theory emerge. Among them is the apparent fact that while the superstructure is not ordinarily thought of as being itself engaged in infrastructural productive activity, the mass media of communications are *simultaneously* in the superstructure *and* engaged indispensably in the last stage of infrastructural production where demand is produced and satisfied by purchases of consumer goods. Chairman Mao Tse-Tung provided the Marxist theoretical basis for such a development as that which created the contemporary capitalist mass media when he said:

> When the superstructure (politics, culture, etc.) obstructs the development of the economic base, political and cultural changes become *principal and decisive*.[11]

The basis entry to the analysis of the commodity form of communications is acceptance of the significance of the concept of monopoly in monopoly capitalism. Baran and Sweezy's *Monopoly Capitalism*[12] demonstrated how monopoly rather than competition rules contemporary capitalism, and it may be taken as the reference point from which to address this issue. Like J.K. Galbraith,[13] Baran and Sweezy emphasize the role of management of demand by the oligopolies which dominate monopoly capitalism. Both civilian and military demand are managed to provide the consumption and investment outlets required for the realization of a rising surplus. The process of demand management begins and ends with the market for the commodity—first as "test markets," and, when product and package production have been suitably designed and executed, as mass advertising-marketing. But Baran and Sweezy fail to pursue in an historical materialist way the obvious issues which are raised by demand-management-via-advertising under monopoly capitalism.

What happens when a monopoly capitalist system advertises? Baran and Sweezy answer, as does Galbraith, *psychological* manipulation. They cite Chamberlin as providing in 1931 the authoritative definition of contemporary advertising.[14] Moreover, they somewhat prematurely foreclose further investigation by stating flatly: "The immediate commercial purposes and effects of advertising have been thoroughly analyzed in economic literature and are readily grasped."[15] The mass media of communications possess no black box from which the magic of psychological manipulation is dispensed. Neither bourgeois nor Marxist economists have considered it worthwhile to ask the following questions which an historical materialist approach would seem to indicate:

(a) What do advertisers buy with their advertising expenditures? As hard-nosed businessmen they are not paying for advertising for nothing, nor from altruism. I suggest that what they buy are the services of audiences with predictable specifications who will pay attention in predictable numbers and at particular times to particular means of communication (TV, radio, newspapers, magazines, billboards, and third-class mail).[16] As collectivities these audiences are commodities. As commodities they are dealt with in markets by producers and buyers (the latter being advertisers). Such markets establish prices in the familiar mode of monopoly capitalism. Both these markets and the audience commodities traded in are specialized. The audience commodities bear specifications known in the business as "the demographics." The specifications for the audience commodities include age, sex, income level, family composition, urban or rural locations, ethnic character, ownership of home, automobile, credit card status, social class and, in the case of hobby and fan magazines, a dedication to photography, model electric trains, sports cars, philately, do-it-yourself crafts, foreign travel, kinky sex, etc.

(b) How are advertisers assured that they are getting what they pay for when they buy audiences? A sub-industry sector of the consciousness industry checks to determine. The socio-economic characteristics of the delivered audiences/readership *and* its size are the business of A.C. Nielsen and a host of competitors who specialize in rapid assessment of the delivered audience commodity. The behaviour of the members of the audience product under the impact of advertising and the "editorial" content is the object of market research by a large number of independent market research agencies as well as by similar staffs located in advertising agencies, the advertising corporation and in media enterprises.[17]

(c) What institutions produce the commodity which advertisers buy with their advertising expenditures? The owners of TV and radio stations and networks, newspapers, magazines and enterprises which specialize in providing billboards and third class advertising are the principal producers. This array of producers is interlocked in many ways with advertising agencies, talent agencies, package programme producers, film producers, new "services" (e.g., AP, UPI, Reuters), "syndicators" of news "columns," writers' agents, book publishers, motion picture producers and distributors. Last but by no means least in the array of institutions which produce the audience commodity is the family. The most important *resource* employed in producing the audience commodity are the individuals and families in the nations which permit advertising.

(d) What is the nature of the content of the mass media in economic terms under monopoly capitalism? The information, entertainment and "educational" material transmitted to the audience is an inducement (gift, bribe or "free lunch") to recruit potential members of the audience and to maintain their loyal attention. The appropriateness of the analogy to the free lunch in the old-time saloon or cocktail bar is manifest: the free lunch consists of materials which whet the prospective audience members' appetites and thus (1) attract and keep them attending to the programme, newspaper or magazine, and (2) cultivate a mood conducive to favourable reaction to the explicit and implicit advertisers' messages.[18] To say this is not to obscure the agenda-setting function of the "editorial" content and advertising for the populations which depend on the mass media to find out what is happening in the world, nor is it to denigrate the technical virtuosity with which the free lunch is prepared and served. Great skill, talent and much expense goes into such production, though less per unit of content than in the production of overt advertisements. Only a monstrous misdirection of attention obscures the real nature of the commodities involved. Thus with no reference to the "Sales Effort," Baran and Sweezy can say:

> There is not only serious questions as to the value of artistic offerings carried by the mass communications media and serving directly or indirectly as vehicles of advertising; it is beyond dispute that all of them could be provided at a cost to consumers incomparably lower than they are forced to pay through commercial advertising.[19]

Under monopoly capitalism TV-radio programs are provided "free" and the newspapers and magazines are provided at prices which cover delivery (but not production) costs to the media enterprise. In the case of newspapers and some magazines, some readers characteristically buy the media product *because* they want the advertisements. This is especially the practice with classifies advertisements and display advertising of products and prices by local merchants in newspapers and with product information in advertisements in certain magazines (e.g., hobby magazines). Regardless of these variations, the central purpose of the information, entertainment and "educational" material (including that in the advertisements themselves) transmitted to the audience is to ensure attention to the products and services being advertised. Competition among media enterprises produces intricate strategies governing the placement of programmes in terms of types of products advertised and types of "free lunch" provided in different time segments of the week (e.g., children's hours, daytime housewives' hours, etc.): all this in order to optimize the "flow" of particular types of audiences to one programme from its immediate predecessors and to its immediate successors with regard to the strategies of rival networks.[20]

(e) What is the nature of the service performed for the advertiser by the members of the purchased audience? In economic terms, the audience commodity is a non-durable producers' good which is bought and used in the marketing of the advertiser's product. The work which audience members perform for the advertisers to whom they have been sold is to learn to buy particular "brands" of consumer goods, and to spend their income accordingly. In short, they work to create the demand for advertised goods which is the purpose of the monopoly capitalist advertisers. While doing this, audience members are simultaneously reproducing their own labour power. In this regard, it is appropriate to avoid the trap of a manipulation-explanation by noting that if such labour power is, in fact, loyally attached to the monopoly capitalist system, this would be welcome to the advertisers whose existence depends on the maintenance of that system. But in reproducing their labour power workers respond to other realistic conditions which may on occasion surprise and disappoint the advertisers. It seems, however, that when workers under monopoly capitalist conditions serve advertisers to complete the production process of consumer goods by performing the ultimate marketing service for them, these workers are making decisive material decision which will affect how they will produce and reproduce their labour power. As the Chinese emphasized during the Cultural Revolution, if people are spending their time catering to their individual interests and sensitivities, they cannot be using the *same* time to also overthrow capitalist influence and to build socialism.

(f) How does demand-management by monopoly capitalism, by means of advertising, related to the labour theory of value, to "leisure" and to "free time"? As William Livant puts it, the power of the concept of surplus value "...rests wholly on the way Marx solved the great value problem of classical political economy, by *splitting the notion of labour in two*, into labour in productive use and labour power (the capacity to labour)."[21] Labour in productive use in the production of commodities-in-general was Marx's concern in the three volumes of *Capital*, except for Vol. 1, chapter 6 and scattered passages in the *Grundrisse*. It is clear from these passages that Marx assumed that labour power is produced by the labourer and by his or her immediate family, i.e., under the conditions of handicraft production. In a word, labour power was "home-made" in the absence of dominant brand-name commodities, mass advertising, and the mass media (which had not yet been invented by monopoly capitalism). In Marx's period and in his analysis, the principal aspect of capitalist production was the alienation of workers from the means of producing commodities-in-general. Now the principal aspect of capitalist production has become the alienation of workers from the means of producing and reproducing themselves. The prevailing western Marxist view today still holds the incorrect assumption that the labourer is an independent commodity producer of labour power which is his to sell. Livant says it well:

> What often escapes attention is that just because the labourer sells it (his or her labour power) does not mean that he or she produces it. We are misled by fixating on the true fact that a human must eat and sleep into thinking that therefore the seller of labour power must also be the producer. Again the error of two combines into one.[22]

We need a dialectical materialist description of the production of labour power, of the capacity and incapacity to labour and of the relationship of the production of labour power to our ability to live as human beings.[23]

Am I correct in assuming that all non-sleeping time under capitalism is work time?[24] William Livant in commenting on a draft of this article, points out that the assumption should be plainly stated. As he puts it, a Marxist view

> … sees leisure time correctly as time of production, reproduction and repair of labour power. This production, reproduction and repair are activities. They are things people must do. As such, they also require labour power. To be sure, this latter labour power you do not have to sell directly to capital. But you do have to use it to produce labour power in the form you do have to sell.
>
> Why was this hard to see? I think we can find the answer if we look at 'non-work' time. Marx points out many times (e.g., *Capital*, Vol. I, Ch. 6) that wage labour only becomes possible if your labour power becomes a *personal possession*, which it is possible for you to sell. You can do what you 'want' with it… Non-work time is labour power which is *yours not-to-sell*. Hence it seems to be doubly your personal possession…
>
> When we see this, we can fit it within what Marx called the 'false appearance' of wage labour (citing *Wages, Prices and Profit*, Peking, 1973, pp. 50–1)… I think this false appearance has its other side. Just as it appears, at work, that you *are* paid for all the labour time you *do* sell, so it appears, off-work, that the labour time *you are not paid for is not sold…*
>
> Work and non-work time bear interesting relations that need examination, to see beneath the false appearances. They in fact *divide* the whole world of commodities in *two*. For at work it is principally commodities-in-general that are made and distributed. Those who make and distribute these commodities do not sell them. But off-work, we find something else. What is being produced there is primarily the peculiar commodity, labour power. And off-work, those who make this commodity, also do not sell it. But it is sold, as surely as commodities-in-general made at the workplace.[25]

It should be clear that for at least several generations labour power in advanced monopoly capitalist countries has been produced primarily by institutions other than the individual and his/her family. The mass media of communications and advertising play a large and probably dominant role through the process of consumption (by guiding the making of the shopping list) as well as through the ideological teaching which permeates both the advertising and ostensibly non-advertising material with which they produce the audience commodity.[26] When cosmetic counters in department stores display "Boxed Ego" (Vancouver, December, 1975), the dialectical relation of the material and consciousness aspects of the production of labour power should be evident.

What has happened to the time available to workers and the way it is used in the past century? In 1850 under conditions of cottage industry, i.e. unbranded consumer goods, the average work week was about 70 hours per week (and the work force was predominantly male).[27] At about the time when Marx was writing the *Grundrisse*, workers' savings, under the most favourable conditions of exploitation, could make possible

> ... the worker's participation in the higher, even cultural satisfactions, the agitation of his own interests, newspaper subscriptions, attending lectures, educating his children, developing his taste, etc., his only share of civilization which distinguishes him from the slave...[28]

In that simple stage of capitalist development, Marx could see that the relentless accumulative process would proliferate commodities:

> Capital's ceaseless striving towards the general form of wealth drives labour beyond the limits of its natural paltriness (*Naturbedurftigkeit*), and thus creates the material elements for the development of the rich individuality which is as all-sided in its production as in its consumption...[29]

Many other references may be cited from the *Grundrisse* to similar effect. But all this assumed that consumer goods were not monopolized by brand names and that workers could dispose of their non-work time subject only to class and customary (i.e. traditional) considerations. In 1850, the average American worker could devote about 42 hours per week (168 hours minus 70 hours on the job and 56 hours of sleep) to such "cottage industry" type of production of labour power.

By 1960, the average time spent on the job was about 39.5 hours per week—an apparent reduction in work time of almost 30 hours per week (to which should be added 2.5 hours as a generous estimate of the weekly equivalent of annual vacations). Capitalist apologists equated this ostensible reduction in work time a corresponding increase in "free" or "leisure" time. The reality was quite different. Two transformations were being effected by monopoly capitalism in the nature of work, leisure and consumer behaviour. On the one hand, huge chunks of workers' time were being removed from their discretion by the phenomenon of metropolitan sprawl and by the nature of unpaid work which workers were obligated to perform. For example, in the contemporary period travel time to and from the job can be estimated by 8.5 hours per week; "moonlighting" employment at a minimum of one hour per week; repair work around the home, at another five hours per week; and men's work on household chores and shopping at another 2.3 hours per week. A total of 16.8 hours per week of the roughly 32 hours of time supposedly "free" as a result of capitalist industrialization is thus anything but "free." A further seven hours of the 32 hours of "freed" time disappears when the correction for part-time female employment is made in the reported hours-per-week.[30] Three-fourths of the so-called "freed" time has thus vanished.

The second transformation involves the pressure placed by the system on the remaining hours of the week. If sleeping is estimated at eight hours a day, the remainder of the 168 hours in the week after subtracting sleeping and the unfree work time thus far identified was 42 hours in 1850 and 49 hours in 1960. We lack systemic information about the use of this "free time" for both dates. We do know that certain types of activities were common to both dates: personal care, making love, visiting with relative and friends, preparing and eating meals, attending union, church and other associative institutions, including saloons. We also know that in 1960 (but not in 1850) there was a vast array of *branded* consumer goods and services pressed on the workers through advertising, point-of-sale displays, and peer group influence. Attendance at spectator sports and participation in such activities as bowling, camping, and "pleasure driving" of the automobile or snowmobile—all promoted for the sake of equipment sales by the consciousness industry—now take time that was devoted to non-commercial activities in 1850. In-house time must now be devoted to deciding whether or not to buy and then to use (by whom, where, under what conditions, and why) an endless proliferation of goods for personal care, household furnishing, clothing, music reproduction equipment, etc. Guiding the worker today in all income and time expenditures are the mass media—through the blend of advertisements and programme content.

How do Baran and Sweezy deal with the use made of this illusory increase in free time? Deploying Veblen's concept of conspicuous consumption and thereby emphasizing the status-seeking character of workers' consumption decisions, they treat leisure time (without quotation marks) in psychoanalytic terms as time spent willfully in passivity and idleness:

> This propensity to do nothing has had a decisive part in determining the kinds of entertainment which are supplied to fill the leisure hours—in the evening, on weekends and holiday, during vacations. The basic principle is that whatever is presented—reading matter, movies, radio and TV programs—must not make undue demands on the intellectual and emotional resources of the recipients: the purpose is to provide 'fun', 'relaxation', a 'good time'—in short, passively absorbable amusement.[31]

What is wrong with this partial truth is: (1) it ignored the relationship of monopoly capitalism's Sales Effort, particularly advertising, to the problem; and (2) it substitutes casual bourgeois observations[32] for an historical materialist attack on the problem.

As against the seven hours per week of apparent "non-work" time gained by the average worker between 1850 and 1960, how much time does he not spend as part of the audience product of the mass media—time sold to the advertisers? Here the audience-measurement sub-industry gives us some information. David Black, economist for the Columbia Broadcasting System, in 1970 found that the average person watched TV for 3.3 hours per day (23 hours *per week*) on an annual basis, listened to radio for 2.5 hours per day (18 hours *per week*), and read newspapers and magazines one hour per day (7 hours *per week*).[33] If we look at the audience product in terms of families rather than individuals, we find that in 1973, advertisers in the U.S. purchased TV audiences for an average of a little more than 43 hours per home per week.[34] By industry usage, this lumps together specialized audience commodities sold independently as "housewives," "children" and "families." In the "prime time" evening hours (7:00 to 11:00 p.m.), the TV audience commodity consisted of a daily average of 83.8 million people, with an average of two persons viewing per home. Women were a significantly larger proportion of this prime time audience than men (42 percent as against 32 percent, while children were 16 percent and teenagers, 10 percent).

We do not know even approximately how the worker's exposure to the mass media articulates with the other components in his/her use of "free time." It is relatively easy to determine how much radio listening and newspaper and magazine reading takes place while travelling to and from work. But much TV and radio programming s attended to incidentally while engaged in other activities such as performing household chores, visiting with friends, reading, and now even while attending to spectator sports.[35]

This is the context in which we may pursue the question, how demand management by means of advertising in monopoly capitalism relates to the labour theory of value, to "leisure" and to "free time." It should now be possible to obtain some clues to the nature of work which workers perform in relation to advertising. If freedom is the act of resisting necessity, what is the nature of the process by which workers react to advertising, and why is it profitable for advertisers to advertise? An advertising theorist, Professor T.N. Levitt, says, "Customers don't buy things. They buy tools to solve problems."[36] It appears that the purpose of advertising, from the perspective of the advertising corporation, is to establish in the worker's consciousness (1) the existence of a "problem" facing the worker (acne, security from burglars, sleeplessness), (2) the existence of a class of commodities which will solve that problem, and (3) the motivation to give to priority to purchasing brand X of that class of commodities in order to "solve" that "problem." Give this situation, the realistic process of audiences-members' work can be best understood in terms of the ever-increasing number of decisions forced on him/her by "new" commodities and by their related advertising. Unfortunately, while workers are faced with millions of possible comparative choices among thousands of "new" commodities, they lack scientifically objective bases on which to evaluate either the "problem" to be solved by buying the proffered "tool" or the efficacy of the "tool" as a solution to the "problem." In this situation, they constantly struggle to develop a rational shopping list out of an irrational situation.[37] As Linder puts it, the most important way by which consumers can cope with commodities and advertising is to limit the time spent in thinking about what to buy.

> Reduced time for reflection previous to a decision would apparently entail a growing irrationality. However, since it is extremely rational to consider less and less per decision there exists a rationale of irrationality.[38]

Monopoly capitalist marketing practice has a sort of seismic, systemic drift towards "impulse purchasing." Increasingly, the work done by audience members is cued towards impulse purchasing. Again, Linder is insightful:

> To begin with advertising is a means of making factual knowledge more accessible than otherwise. Second, it serves to provide quasi-information for people who lack time to acquire the genuine insights. They get the surrogate information they want to have, in order to feel that they are making the right decisions... The advertiser helps to close the information gap, at the same time exploiting the information gap that is bound to remain.[39]
>
> As the scarcity of time increases, the emphasis in advertising will be displaced in the direction of ersatz information. The object will be to provide a motive for an action for which no solid grounds exist... Brand loyalty must be built up among people who have no possibility of deciding how to act on objective grounds. As routine purchasing procedure gain in importance as a means of reducing decision-making time, it will become increasingly important to capture those who have not yet developed their routines.[40]

In this connection, the new and sophisticated interest of market researchers in the relationship of advertising to children is very significant. According to the publisher of one recent study:

> As the authors see it, consumption is a perfectly legitimate and unavoidable activity for children. Consequently they reject a strategy directed at protecting kids from marketing stimuli. What is necessary, then, is to acknowledge that children are going to watch television commercials and to prepare them to be selective consumers.
>
> *How Children Learn to Buy* provides evidence to confront existing theories in the emerging field of consumer socialization. The work is essential to everyone concerned with the effects of advertising: sponsors, ad agencies, the television industry, educators, governmental regulators, consumer researchers, and parents.[41]

Constrained by the ideology of monopoly capitalism, the bourgeois notion of free time and leisure is only available to those who have no disposable income (and for whom it is, of course, a bitter mockery) and to those who are so rich that, as Linder says, for them, "the ultimate luxury is to be liberated from the hardships of having to do one's own buying."[42] For everyone else, "free time" and "leisure" belong only in the monopoly capitalist lexicon alongside "free world," "free enterprise," "free elections," "free speech," and "free flow" of information.

What has happened to the time workers spend off-the-job while not sleeping is that enormous pressures on this time have been imposed by all consumer goods and service branches of monopoly capitalism. Individual, familial and other associative needs must be dealt with, but in a real context of producers and advertising which, taken together, make the task of the individual and family basically one of *coping* while being constantly on the verge of being overwhelmed by these pressures. In this context, the work of the audience members which advertisers find productive for them is one of learning cues which are used when the audience member makes up his/her mental shopping list and spends his/her income.

(g) Does the audience commodity perform an essential economic function? Baran and Sweezy state that "advertising constitutes as much an integral part of the system as the giant corporation itself"[43] and that "advertising has turned into an indispensable tool for a large sector of corporate business."[44] In this they go as far as Galbraith who said "... the marginal utility of present aggregate output, ex-advertising and salesmanship is zero."[45]

But is the production and consumption of the audience commodity for advertisers a "productive" activity in Marxian terms? Baran and Sweezy are contradictory in answering this question. They tell us that advertising expenses "...since they are manifestly unrelated to necessary costs of production—however broadly defined—(they) can only be counted as part of aggregate surplus."[46] But after some agonizing over whether finance, insurance and real estate (which account for about twice the volume of national income as represented by advertising) are productive, they abandon their theoretical footing for rejecting expenses of circulation as unproductive of surplus:

> Just as advertising, product differentiation, artificial obsolescence, model changing, and all the other devices of the sales effort do in fact promote and increase sales, and thus act as indispensable props to the level of income and employment, so the entire apparatus of 'finance, insurance, and real estate' is essential to the normal functioning of the corporate system and another no less indispensable prop to the level of income and employment. The prodigious volume of resources absorbed in all these activities does in fact constitute necessary costs of capitalist production.

What should be crystal clear is that an economic system in which *such* costs are socially necessary has long ceased to be a socially necessary system.[47]

I am aware that *Capital* can be and has been read frequently as denying the productivity of the expenses of middlemen in general. As I read the work, however, it seems to me that in *Capital* Marx was concerned to analyze the operation of capitalism under the then realistic conditions of competition and the organization of industry as being generally *unintegrated* from raw material processing through exchange to the consumption process.[48] Marx also clearly did not assume the predominance of branded commodities or the prevalence of advertising. If one turns to Marx's "Introduction to the Critique of Political Economy," however, it seems probable that his analysis of monopoly capitalism, had such been possible in his would, would have answered the question of the productivity of advertising differently. Indeed the following passage accommodates the phenomena of advertising, branded merchandise, and monopoly capitalism in managing demands.

Consumption produces production in a double way… because consumption creates the need for *new* production, that is it creates the ideal, internally impelling cause for production, which is its presupposition. Consumption creates the motive for production; it also creates the object which is active in production as its determinant aim… No production without a need. But consumption reproduces the need… Production not only supplies material for the need, but it also supplies a need for the material. As soon as consumption emerges from its initial state of natural crudity and immediacy—and, if it remained at that stage, this would be because production itself had been arrested there—it becomes itself mediated as a drive by the object. The need which consumption feels for the object is created by the perception of it. The object of art—like every other product—creates a public which is sensitive to art and enjoys beauty. Production thus not only creates an objects for the subject, but also a subject for the object. *Thus production produces consumption (1) by creating the material for it; (2) by determining the manner of consumption; and (3) by creating the products initially posited by it as objects, in the form of a need felt by the consumer. It thus produces the object of consumption, the manner of consumption and the motive of consumption. Consumption likewise produces the producer's inclination by beckoning to him as an aim-determined need.*[49]

It is clear, firstly, that the exchange of activities and abilities which takes place within production itself belongs directly to production and essentially constitutes it. The same hold, secondly, for the exchange of products, in so far as that exchange is the means of finishing the product and making it fit for direct consumption. To that extent, exchange is an act comprised with production itself. Thirdly, the so-called exchange between dealers and dealers is by its very organization entirely determined by production, as being itself a producing activity. Exchange appears as independent and indifferent to production only in the final phase where the product is exchanged directly for consumption.[50]

On such a footing it is possible to develop a Marxist theory of advertising and of branded commodities under monopoly capitalist conditions. When the president of Revlon corporation says: "We manufacture lipsticks. But we sell hope," he is referring to the creation of products initially posited by it as objects in the form of a need felt by the consumer—similarly with Contac-C, the proprietary cold remedy which so disturbed Baran and Sweezy.[51] The denial of the productivity of advertising is unnecessary and diversionary: a *cul de sac* derived from the pre-monopoly-capitalist attempt at reconciliation with *Capital*.

(h) Why have Marxist economists been indifferent to the historical process by which advertising, brand-name merchandise, and the mass media of communications have developed in monopoly capitalism over the past century? Why do they continue to regard the press, TV and radio media as having the prime function of producing news, entertainment and editorial opinion and not audiences for sale to advertisers? The evidence for the latter is all around us.

Baran and Sweezy do indeed indicate how much advertising has grown and when, i.e., by a factor of ten between 1890 and 1929.[52] *But not why, how and with what connections*.

In the first three quarters of the nineteenth century, newspapers and magazines in the countries going through the Industrial Revolution were characterized by: (a) diversity of support as between readers' payments, subsidies from political parties, and advertising (most of the latter being information about commodity availability and prices and not about branded merchandise); and (b) a cyclical process of technological improvement with consequent larger printing capacity, lower unit costs, lower unit prices of publications, larger profits, capital accumulation and reinvestment in new and more productive plants, etc.[53] In that period, marketing of consumer goods was characterized by: (a) predominance of unbranded merchandise; (b) unintegrated distribution of commodities with the middleman being the most powerful link in the production-to-consumer chain; and (c) consequently, lack of massive advertising as a means of managing demand.

In the second half of the nineteenth century, capitalism faced a crisis. The first stage of the development of the factory system under conditions of competition between relatively small capitalists had succeeded in mobilizing labour supply and exploiting it crudely under conditions documented so ably by Marx in *Capital*. The very success of the system bred grave threats to it. Politically conscious labour unions posed revolutionary threats to capitalism.[54] Moreover, capitalist manufacturers were vulnerable to the power of the workers because the highly skilled workers possessed more knowledge about the production process than did their employers.[55] Manufacturers were thus blocked from ready control of their work force and from innovating the new and increasingly sophisticated machine processes of mass production which the rapid progress in physical sciences and engineering made possible. When they looked at their marketing methods, manufacturers were also beset by chronic insecurities. The periodic business cycles in their crisis and liquidation phases forced manufacturers into cut-throat pricing (of unbranded merchandise, typically) because of the pressure of overhead costs. The result was a short life expectance for competitive industrialists.

In sum, a watershed in the development of capitalism had been reached. As M.M. Knight said, "Down to the last quarter of the nineteenth century, commerce dominated industry; after it industry dominated commerce."[56]

Capitalism's systemic solution to the contradiction between its enormous potential for expanding production of consumer goods (and the profits to be thus realized) and the systemic insecurities posed by people as workers and people as consumers was to move large scale rationalization of industrial organization (through vertical, horizontal, and conglomerate integration).[57] This conferred control over supplies and prices in the factor markets, and in the marketing of end-products. But to make such giant integrated corporations viable, their operations had to address directly the problem of people (1) as workers at the job where they were paid, and (2) as buyers of the end products of industry. The systemic solution was a textbook example of the transformation of a contradiction on the principle "one goes into two." This was an ideological task and it was solved by capitalizing on the deeply held ideological reverence for scientific rationality in the pursuit of possessive individualistic material goals.

After militant unions had been crushed by force between 1890 and 1910, scientific management was applied to people as workers. Knowledge about the work process was expropriated from skilled workers to management. The work process was reduced to "ladders" of dead-end "tasks" to complement whichever more sophisticated generations of mass production machines were innovated. And through varieties of "incentive" wage plans, linked with promotion-from-within on the basis of seniority, supported by company welfare plans (and later social insurance through government), the workplace where people got paid was transformed ideologically.[58] People learned there that work under monopoly capitalism involves competition between individuals whose possessive needs necessarily set them in conflict with each other rather than with the owners of the means of their (concealed) cooperative production. The carrot which systematically motivated them was the pursuit of commodities, which joined this half of the ideological exercise with the next.

Simultaneously the system dealt with its problem of people as buyers of end products. As on the job front, science was invoked. The objective was personal satisfaction, and the rationale was efficiency. The term "consumer" was invented to describe the desired object. Advertising and the creation of mass produced communications (press, radio and TV principally) were developed as the specialized means to this systemic end. Even if a seeming "over-production" of consumer goods threatened the profitability of an industry the ability of a company to distinguish its products from unbranded similar products allowed its sales and profits to grow in security. If studies are done—I have been able to locate none—of the history of brand names, it will be found that this was how brand name loyalty became an essential weapon in industry when the trusts which produced the present oligopolistic empires of monopoly capitalist industry became dominant features of the industrial landscape. Certainly the Baran and Sweezy thesis that monopoly capitalism manages demand through market control and advertising would seem to carry as its corollary the hypothesis that something like the *suction* of commodities from the material production line to the oligopolistic end-product markets has replace the atomistic *circulation* of commodities typical of Marx's time as the model of monopoly capitalist marketing. While historical scholarship in marketing seems conspicuously underdeveloped, fragmentary evidence from studies of marketing history tend to confirm the outline of the process here sketched.[59]

For example, Joseph Palamountain says, "Great increases in the size of manufacturers or retailers have changed much of the distribution from a flow through a series of largely autonomous markets to a single movement dominated by either manufacturer or retailer."[60] Simultaneously, the newspaper and magazine industries found themselves in a position to vastly increase the productivity of the printing trades in the last quarter of the nineteenth century. Technical advances in typesetting, printing (including colour), photographic reproduction, etc., could be financed if someone would foot the bill. The newspaper and magazine entrepreneurs (the William Randolph Hearst and their rivals) invented the "yellow journalism" which took advantage of this situation. The cycle of capital expansion ensured in accelerated speed and scope. Production and circulation were multiplied, while prices paid by the readers were held constant or decreased. And the "mass media" characteristic of monopoly capitalism were created in the 1890's. It was these mass media, increasingly financed by advertising, that drew together the "melting pot" working class from diverse ethnic groups which were flooding in as migrants to the United States into saleable audiences for the advertisers.[61]

The advent of radio-telephony in the first two decades of this century made possible the use of the same principle which had been proven in the print media. And so commercial radio broadcasting became a systemic innovation of, by, and for monopoly capitalism. When the pent-up civilian demand at the end of World War II, and the generous capital subventions of a government intent on winning that war had provided electronics manufacturers with shell-loading and other war plants easily convertible into TV set manufacturing, and when a complaisant FCC could be manipulated into favouring TV over FM broadcasting,[62] TV was approved and largely financed out of capital accumulated from commercial radio broadcasting's profits.[63]

Why was this media complex rather than some other mode of marketing developed by monopoly capitalism to create and control "consumers"? Because it offered a cheaper, more efficient mode of demand management than the alternatives which could be devised. What alternatives? The obvious alternative was "more of the same" methods previously used in marketing: heavier reliance on travelling salesmen to push goods to retailers, heavier use of door-to-door salesmen. To calculate the opportunity cost with a hypothetical elaboration of a marketing system designed to sell branded commodities without advertising was and is a horrendous prospect. Moreover, it would be pointless because mass production of (branded) consumer goods and services under capitalism would not have happened, absent advertising. An indication of the efficiency of the audience commodity as a producers' good used in the production of consumer goods (and a clue to a possible measure of surplus value created by people working in audiences) is provided when we compare advertising expenditures with "value added" by retailing of consumer goods and services. In 1973 in the U.S. some $25 billion was spent in advertising while personal consumption expenditures were about $800 billion. Three percent of the sale price as the cost of creating and managing demand seems very cheap—and profitable. The system also accrued valuable side benefits. Institutional advertising and the merchandising of political candidates and ideological points of view in the guise of the free lunch and advertising messages were only appreciated and exploited systematically after World War I when propaganda and its associated public opinion polling were developed for war promotion purposes.

To summarize: the mass media institutions in monopoly capitalism developed the equipment, workers and organization to produce audiences for the purposes of the system between about 1875 and 1950. The prime purpose of the mass media complex is to produce people in audiences who work at learning the theory and practice of consumership for civilian goods and who support (with taxes and votes) the military demand management system. The second principal purpose is to produce audiences whose theory and practice confirms the ideology of monopoly capitalism (possessive individualism in an authoritarian political system). The third principal purpose is to produce public opinion supportive of the strategic and tactical policies of the state (e.g., presidential candidates, support of Indochinese military adventures, space race, détente with the Soviet Union, rapprochement with China and ethnic and youth dissent). Necessarily in the monopoly capitalist system, the fourth purpose of the mass media complex is to operate itself so profitably as to ensure unrivalled respect for its economic importance in the system. It has been quite successful in achieving all four purposes.

If we recognize the reality of monopoly capitalism buying audiences to complete the mass marketing of mass-produced consumer goods and services much further analysis is needed of the implications of this "principal and decisive" integration of superstructure and base which reality presents. First, the contradictions produced within the audience commodity should be understood more clearly. I refer to the contradiction as between audience members serving as producers' goods in the marketing of mass produced consumer goods and their work in producing and reproducing labour power. I think that the consciousness industry through advertising-supported mass media produces three kinds of alienation for the members of the audience commodity: (1) alienation from the result of their work "on the job"; (2) alienation from the commodities-in-general which they participate in marketing to themselves; and (3) alienation from the labour power they produce and reproduce in themselves and their children. It would seem that the theory of work needs reconsideration.

Then connections to other areas need to be examined. Among such connections there come to mind those to Marxist theory about social consciousness (and false consciousness), to theory about the nature of class struggle, the nature of the proletariat under monopoly capitalism and sex chauvinism, and to theories of the state. The last of these seem obvious if this analysis is considered in connection with the recent articles by Gold, Lo, and Wright.[64] The role of the mass media and the consciousness industry in producing the audience commodity both as commodity-in-general and peculiar commodity might provide the real sinews to the structural-Marxist model of the state of Poulantzas and to the theoretical initiatives of Claus Offe in seeking the processes within the state which "guarantee" its class character. The connection to the work of de Bord[65] regarding consciousness is proximate. The relation of industrially produced images to the "real" world of nutrition, clothing, housing, birth and death is dialectical. The mass media are the focus of production of images of popular culture under monopoly capitalism, both through the explicit advertising and the "free lunch" which hook and hold people in audiences. Because the consciousness industry produces consumable, saleable spectacles, its product treats both past and future like the present—as blended in the eternal present of a system which was never created and will never end. The society of the spectacle, however, cannot be abstractly contrasted with the "real" world of actual people and things. The two interact. The spectacle inverts the real and is itself produced and is real. Hence, as de Bord says, objective reality is present on both sides. But because the society of the spectacle is a system which stands the world really on its head, the truth in it is a moment of the false. Because the spectacle monopolizes the power to make mass appearance, it demands and gets passive acceptance by the "real" world. And because it is undeniably real (as well as false) it has the persuasive power of the most effective propaganda.[66]

Finally, another example of necessary connections is that to the theory of imperialism and socialism in the present stage of monopoly capitalism. There are many ways by which a theory of commodity production through mass communications would strengthen the analysis, for example, of Samir Amin.[67] The cocacolonisation of the dependent and peripheral countries cannot be grounded in Marxist theory without attention to the production of audience commodities in the interest of multi-national corporations. It would link Amin's theory to Herbert Schiller's work on the relation of the mass media to the American empire.[68] And, when linked with analysis of the ideological aspects of science and "technology," it could strengthen the development of a non-economistic, non-positive, non-Eurocentered Marxism. Analysis of such connections is inviting but beyond the scope of the present essay.

Notes

1 Dallas Smythe, "Communications: Blindspot of Western Marxism" was first published in the *Canadian Journal of Political and Social Theory,* Vol. 1 No. 3, 1977. Published with the permission of the editors, Arthur and Marilouise Kroker.

2 To demonstrate this in detail would require a lengthy analysis which would deflect the present article from its affirmative purpose. Gramsci, the Frankfurt School writers (Adorno, Horkheimer, Marcuse, Lowenthal), Raymond Williams, Poulantzas, Althusser, and Marxists concerned with the problems of developing nations (e.g. Samir Amin, Clive Y. Thomas)—none of them address the consciousness industry from the standpoint of its historical materialist role in making monopoly capitalist imperialism function through demand management (concretely through the economic processes of advertising and mass communications). This is precisely the blindspot of recent Western Marxism. In the developing debate it would be useful to have studies bearing on whether and why such

writers have or have not dealt with this aspect of monopoly capitalism. Reality imposes a burden of proof on them as well as on me.

3 Lenin held a manipulative theory of the mass media and admitted naiveté in this respect. "What was the fate of the decree establishing a state monopoly of private advertising issues in the first weeks of the Soviet government? ... It is amusing to think how naïve we were ... The enemy i.e., the capitalist class, retaliated to this decree of the state power by completely repudiating that state power." "Report on the New Economic Policy," Seventh Moscow Gubernia Conference of the Russian Communist Party, October 21, 1921, in *Lenin About the Press*, Prague, International Organization of Journalists, 1972, 203. Lenin's Imperialism is devoid of recognition of the relation of advertising to monopoly capitalism and imperialism.

4 *The Mind Managers*, Boston, Beacon Press, 1973.

5 "For a Political Economy of Mass Communications," *The Socialist Register*, 1973.

6 Sage Publications, Beverly Hills.

7 Cf. Finkelstein, Sidney, *Sense and Nonsense of McLuhan*, N.Y. International Publishers, 1968; Theall, Donald, *The Medium is the Rear View Mirror*, Montreal, McGill/Queen's University Press, 1971; and my review of the latter in Queen's Quarterly, Summer, 1971.

8 I am indebted to Professor William Livant, University of Regina, for much hard criticism which he formulated in a critique of a draft of this paper in December 1975.

9 The objective reality is that the ostensible advertisements and the material which comes between them, whether in the print or electronic media, have a common purpose of producing the audience. It is an interesting consequence of the idealist perspective that in most liberal analysis the "advertising" is considered to be separate from the "new", "entertainment," "educational material" which is interlarded between the advertisements.

10 The annual cost to audience members of providing their own broadcast receivers (and paying for Cable TV), consisting of depreciation, interest on investment, maintenance and electric power, amounted to slightly more than $1.8 billion, while the over-the-air broadcasters' (Canadian Broadcasting Corporation plus private broadcasters) and Cable TV operators' costs were about $631 million.

11 "On Contradictions," *Selected Works of Mao Tse-Tung*, Vol. 1, Peking, Foreign Languages Press, 1967, p. 336. Emphasis added.

12 New York, Monthly Review Press, 1966.

13 *The New Industrial State*, Boston, Houghton Mifflin, 1967.

14 Chamberlin, E.H., *The Theory of Monopolistic Competition*, Cambridge, Mass., 1931.

15 *Monopoly Capitalism*, p. 116.

16 It is argued by one of my critics that a better term for what advertisers buy would be "attention." At our present naïve stage concerning the matter, it does seem as if attention is indeed what is bought. But where people are paid for working on the job, should Marxists say that what the employer buys is "labour power" or "the manual dexterity and attention necessary for tending machines"? Where I refer to audiences as being produced, purchased and used, let it be understood that I mean "audience-power"; however it may turn out upon further realistic analysis to be exercised.

17 The pages of *Variety* report on cases where the ostensibly non-advertising matter in the media, which I call the "free lunch," attracted an audience which had propensities incongruous with the particular product or service being advertised; in such cases the program is cancelled and the audience discarded.

18 The "free lunch" concept of the mass media was first stated by Liebling A.J., *The Press*, N.Y., Ballantine, 1961.
19 Loc. cit. 121. Or for elaborate obfuscation, see Machlup, Fritz, *The Production and Distribution of Knowledge in the United States*, Princeton, Princeton University Press, 1962.
20 See Brown, Les, *Television: The Business Behind the Box*, N.Y. Harcourt Brace Jovanovich, 1971.
21 Livant, William, "Notes on the Development of the Production of Labour Power," 22 March, 1975 (dittoed).
22 Livant, William, "More on the Production of Damaged Labour Power," 1 April, 1975 (dittoed), p. 2.
23 In arguing that all non-sleeping time under capitalism is work time, I go beyond Samir Amin who says "Social time is split into non-working time and working time. But here too the former exists only to serve the latter. It is not leisure time, as it is called in the false consciousness of alienated men, but recuperation time. It is functional recuperation that is socially organized and not left up to the individual despite certain appearances", ("In Praise of Socialism," *Monthly Review*, September, 1974a, p. 8). Amin also has the blind spot which does not recognize the audience commodity which mass media have produced.
24 I am perhaps wrong to exclude sleeping time from work. The dividing line between recreation of the ability to work while awake and sleeping may be illusory. It may be that the head coach of the Washington, D.C. "Redskin" professional football team, George Allen, is closer to the mark than most economists when he tells his players, "Nobody should work all the time. Leisure time is the five or six hours you sleep at night. You can combine two good things at once, sleep and leisure." Quoted in Terkel, Louis, *Working*, N.Y., Pantheon, 1974, p. 389.
25 Livant, William, "The Communications Commodity" (Xerox, University of Regina, 25 December, 1975), p. 7.
26 For present purposes I ignore the ancillary and interactive processes which contribute to the production of labour power involving also the educational institutions, the churches, labour unions, and a host of voluntary associations (e.g. YMCA, Girl Scouts).
27 The following analysis draws on the de Grazia, Sebastian, *Of Time, Work and Leisure*, N.Y., Anchor, 1964.
28 Marx, Karl, *Grundrisse*, London, Pelican, 1973, p. 287.
29 Ibid., p. 325.
30 Part-time workers (probably more female than male) amounted in 1960 to nineteenspercent of the employed labour force in the United States and worked an average of 19 hours weekly. If we exclude such workers in order to get a figure comparable to the 70 hours in 1850, we consider the weekly hours worked by the average American male who worked at least 35 hours per week and find that they averaged 46.4 (as against 39.5 for all workers). For the sake of brevity, I omit the counterpart calculation of "free time" for women. No sexist implications are intended.
31 Loc. cit., p. 346.
32 "...the manufacturers of paper and ink and TV sets whose products are used to control and poison the minds of the people..." (Ibid., p. 344).

33 Blank, David M., "Pleasurable Pursuits—The Changing Structure of Leisure-Time Spectator Activities." National Association of Business Economists, Annual Meeting, September, 1970 (ditto).

34 *Broadcasting Yearbook*, 1974, p. 69.

35 For many years patrons at professional baseball and football games have been listening to portable radios broadcasting the same game. In 1975 I observed that patrons at professional football games are beginning to watch the same game on portable TV sets for the "instant replays."

36 Levitt, T.N., "The Industrialization of Service," *Harvard Business Review*, September, 1976, pp. 63–75, 73.

37 I use the term "rational" here in the common sense usage, that the result should be one which can be "lived with," is "the right decision," which "makes sense." I imply no Benthamist calculus of utilities or pleasure or pain.

38 Linder, Staffen B., *The Harried Leisure Class,* N.Y. Columbia University Press, 1970, p. 59.

39 Ibid., pp. 70–71.

40 Ibid., p. 71.

41 Publisher's blurb for Ward, Scott, Wackman, Daniel B., and Wartella, Ellen, *How Children Learn to Buy: The Development of Consumer Information-Processing Skills*, Beverly Hills, Sage Publications, 1977.

42 Linder, op. cit., p. 123.

43 Ibid., *Monopoly Capitalism*, p. 122.

44 Ibid., p. 119.

45 Galbraith, J.K., *The Affluent Society*, Boston, Houghton Mifflin, 1958, p. 160.

46 Ibid., *Monopoly Capitalism*, p. 125.

47 Ibid., p. 141.

48 At the outset of *Volume II, Capital*, Marx says: "It is therefore taken for granted here not only that the commodities are sold at their values but also that this takes place under the same conditions throughout. Likewise disregarded therefore are any changes of value which might occur during the movement in circuits." (Marx, Karl, *Capital, Vol. II, Book II*, p. 26. Moscow, Progress Publishers, 1967.)

49 Marx, Karl, *Grundrisse*. London: Pelican Books, 1973, p. 91–92. Emphasis added.

50 Ibid., p. 99.

51 Referring to a reported $13 million advertising budget which produced $16 million in drug store sales, expressed in wholesale prices, they say: "Allowing for a handsome profit margin, which of course is added to selling as well as production cost, it seems clear that the cost of production can hardly be more than a minute proportion of even the wholesale price" (Op. cit., p. 119).

52 Ibid., *Monopoly Capitalism*, p. 118.

53 Cf. Aspinall, A., *Politics and the Press*, 1949.

54 The young and conservative Richard T. Ely, writing in 1886 presented a sympathetic and respectful account of the radicalism of the exiting union movement in the United States. Ely, Richard T., *The Labour Movement in America*, N.Y., T.Y. Crowell & Co.1886.

55 Stone, Catherine, "The Origins of Job Structures in the Steel Industry," *Review of Radical Political Economy*, Summer, 1974, p. 113–173.

56 Knight, M.M., Barnes, H.E. and Flugel, F., *Economic History of Europe*, N.Y., Houghton Mifflin, 1928.

57 Brady, Robert A. *The Rationalization Movement in German Industry*. Berkley, Univ. of Calif. Press, 1933, and his *Business as a System of Power*, N.Y., Columbia University Press, 1943. Also, Veblen, T., *The Theory of Business Enterprise*, N.Y., Viking Press, 1964.

58 Jeremy Bentham and Charles Babbage had publicized the ideas; Taylor and his successor were the experts who applied them. See Braverman, Harry, *Labour and Monopoly Capital*. N.Y., Monthly Review Press, 1974; Stone, Catherine, "The Origins of Job Structures in the Steel Industry," op. cit., and Palmer, Bryan, "Class, Conceptions and Conflict: The Thrust for Efficiency, Managerial Views of Labour and the Working Class Rebellion, 1903–1922," *Review of Radical Political Economy*, Summer, 1974, p. 31–49.

59 Edwin H. Lewis argues that: "Prior to Civil War, in the United States, the wholesaler was typically the dominant factor in the channel. Small retailers and frequently small manufacturers as well, depended on the wholesaler to carry stocks and to give credit or financial support. Following the Civil War, large scale retailers became the dominant element in the distribution of convenience goods and certain shopping goods. As manufacturers have grown larger and as oligopolistic conditions have prevailed in many industries, the manufacturer has held a position of strength in the channel." Lewis, Edwin H., *Marketing Channels*, N.Y. McGraw Hill, 1968, p. 163.

According to Philip Kotler: "A change began in the 1890's with the growth of national firms and national advertising media. The growth of brand names has been so dramatic that today, in the United States, hardly anything is sold unbranded. Salt is packaged in distinctive manufacturers' containers, oranges are stamped, common nuts and bolts are packaged in cellophane with a distributor's label, and various parts of an automobile—spark plugs, tires, filters—bear visible brand names different from that of the automobile." Kotler, Philip. *Marketing Management*, 2nd ed., Englewood Cliffs, New Jersey, Prentice Hall, 1972, p. 446.

60 Palamountain, Joseph C. Jr., "Vertical Conflict" in Stern, Louis W., *Distribution Channels: Behavioral Dimensions*, N.Y., Houghton Mifflin, 1969, p. 138.

61 Stuart Ewen, in *Captains of Consciousness* (N.Y., McGraw-Hill, 1976) provides abundant documentation of the purposiveness with which monopoly capitalism used advertising and the infant mass media for this purpose in the period around and following World War I.

62 Lessing, L.P. *Man of High Fidelity: E.A. Armstrong*, 1956.

63 Why did the cinema, generally conceded to be part of the mass media, not become producers of audience products as part of the systemic bulge of the consciousness industry after 1875? To this, there are several obvious answers. The cinema requires an audience assembled outside the home. It is in the ancient traditional mode of the theatre, arena, assembly, etc. As such it had its own momentum and defined its prime product as the sale of a seat at a particular location and time in relation to the exhibited film. What the advertisers needed—and what capitalism developed as a specialized part of the process of mass producing and mass marketing consumer goods—was a method of mobilizing people to work at being consumers in their alienated separate homes. This advertising supported media made possible. The motion picture industry is not so isolated from the marketing process as this explanation might suggest. "Tie-ins" for consumer goods are a normal part of the planning and receipts (often unreported for tax purposes) of the producers, directors, writers and star performers in theatrical films.

64 Gold, David A., Lo, Clarence Y.H., and Wright, Erik Olin, "Recent Developments in Marxist Theories of the Capitalist State." *Monthly Review,* October 1975, p. 29–43.

65 de Bord, Guy, *The Society of the Spectacle*, Detroit, Black and Red, Box 9546, 1970.

66 Ibid., p. 6–9.

67 Amin, Samir, *Accumulation of a World Scale*, N.Y., Monthly Review Press, 1974b, and "Toward a Structural Crisis of World Capitalism," Socialist Revolution, April 1975, p. 9–44.

68 Schiller, Herbert I. *Mass Communications and American Empire*, N.Y., Augustus Kelley, 1971. Communication and Cultural Domination, White Plains, IASP, 1976.

CHAPTER FOUR

Introduction to The Blindspot Revisited

Graham Murdock
Loughborough University

When Dallas sent me his "Blindspot" article in draft to comment on, it intersected with major strands in my own work. In collaboration with my colleague at the Leicester Centre, Peter Golding, I had been developing a critical political economy of communication. We set out to sketch a framework that would have general applicability, but our empirical work paid particular attention to conditions prevailing in the U.K. and Europe. Much of this effort centred on a sustained interrogation of the increasing consolidation of corporate control over public communications and the consequences of this shift for the diversity of public debate on key issues. The conceptual and policy issues this analysis raised became even more urgent with the election of market fundamentalist governments led by Margaret Thatcher in Britain and Ronald Reagan in the U.S. The resurgence of neoliberal economics provided rationales for dismembering the public sector, dismantling the public interest basis of regulation, and installing market forms of enterprise and evaluation across every sector of social activity, including the cultural and media industries.

This process was already gathering momentum when Dallas drafted his paper, and had led me to explore the potential of tax-funded initiatives in providing a counter balance. This in turn, connected with my developing interests in the role of public communications in providing cultural resources for citizenship and the ways this effort was constantly undermined by the ceaseless celebration of consumerism in the commercialized media. I saw the ethos of citizenship, with its invitation to contribute to the quality of collective life, locked in permanent tension with consumerism's emphasis on personal fulfilment through market choices. Since, in Britain, and elsewhere in Europe, Public Service Broadcasting was the non-commercial cultural institution most comprehensively embedded in everyday life, I chose television as the primary site for exploring the contradictions between the identities of consumer and citizen and the social practices they encourage.

Both consumerism and public service broadcasting were blindspots in Dallas's presentation. His work at the FCC and after, confronting unrelenting attacks from the cheerleaders for marketization, had instilled a deep pessimism about the possibility of a comprehensive public service system being introduced in the United States. In contrast, in the U.K. and Europe public service broadcasting dominated the post-war television landscape but was under mounting pressure from the commercial lobby. The first significant break occurred in Britain when advertising supported terrestrial services were introduced in the mid-1950s, but it was the arrival of commercial cable and satellite services that introduced competition into markets across Europe more generally. Britain's pioneering role in installing advertising supported programming directed attention to the central role of commodity promotion and commercial speech in cementing a consumerist worldview. It coincided with rising popular aspirations as the country began to emerge from a period of post-war austerity and shortage, but it cut across the new contract of citizenship, based on a model of Welfare Capitalism, introduced by a reforming post-war Labour government. This prompted me to re-read writings on ideology in the Marxist tradition and to focus on the idea of commodity fetishism.

Dallas was resistant to this because his "Blindspot" intervention expressly set out to counter what he saw as the over attention paid to questions of ideology and culture in much British and European Marxism and reinstate the centrality of labor and exploitation. He saw himself resisting the attractions of idealism and retrieving the material basis of Marxism.

When I read the piece in draft I immediately recognized it as a powerful and immensely useful codification of the exploitation hiding behind the promise of something-for-nothing held out by commercial television. But I felt strongly that in redirecting attention so concertedly from the analysis of what appeared on the screen to the commodification of audience power, he had side-lined a key dimension of Marx's account of how capitalism reproduces its domination and covers its tracks. This difference was rooted in our contrasted intellectual formations. Dallas had studied economics at university and after graduating with a doctorate from the University of California at Berkeley, had worked for a time at the Department of Labor in Washington before moving to the FCC. After completing my first degree at the London School of Economics I moved to Sussex University to pursue post-graduate work in the sociology of art and literature, where I encountered debates in visual and textual analysis. The study of media offered the chance to combine my interests. My work since then has been animated *conceptually* by the search for more adequate ways of understanding the interactions between symbolic arenas, economic dynamics and social life, and *politically* by a commitment to building communications systems that promote values of equity, justice, and mutuality.

Everything that happened subsequently firmed up three convictions for me. Firstly, in the need to track the ever expanding reach of commodity culture and the meta-ideology of consumerism that it promotes and anchors in everyday representations and practices. Secondly, in the continuing struggle to redefine the cultural rights of citizenship in complexly stratified societies and to resist the conflation of citizens and consumers. Thirdly, in the pivotal role of states and governments in constructing the field of action on which battles over public culture are fought out and in determining the distribution of the essential resources groups and individuals bring to these struggles.

Some observers have been tempted to see the Internet's arrival as a mass utility rendering work on established media as increasingly irrelevant. This argument, based as it is on the crudest form of technological determinism, manages to ignore the concerted mobilisation of cyberspace in the service of capitalism.

Returning to Dallas's work on the audience commodity and user labor offers an essential starting point for anyone wishing to develop a comprehensive critical analysis of business models based on the sale of consumer information to advertisers and the incorporation of voluntary digital labor into corporate research and development strategies. At the same time, the commodities media systems promote are always symbolic as well as material. The particular pleasures they offer and the individual stories they tell are woven into the grander ideological narrative of consumerism that anchors capitalism in everyday imaginings and practices and legitimates its injustices. Understanding how these ties are secured remains a core task for critical analysis. The retreat from regulation of established commercial media and the failure to regulate the Internet has opened up unprecedented space for advertising and product promotion. Its increasingly ubiquitous and immersive presence is extended and reinforced by the progressive annexation of urban space by retail environments. Challenging this promotional enclosure requires us to reengage with struggles over the role of government and the state. The goal here is not to simply re-regulate market institutions and dynamics in the public interest, but to create the conditions for revivified public cultural institutions capable of defending and demonstrating a conception of citizenship that speaks to the internal complexities of societies and the shared problems that transcend national borders.

CHAPTER FIVE

Blindspots About Western Marxism: A Reply to Dallas Smythe

Graham Murdock[1]

Dallas Smythe's recent article, "Communications: Blindspot of Western Marxism" deserves serious attention from anyone interested in the possibilities for a viable materialist theory of mass communications. According to Smythe, not only do we not have such a theory at present—we do not even have a firm basis for its development. And this, he argues, is principally because western Marxism suffers from a fatal blindspot on the subject. It is not only that communications has been a relatively underdeveloped area of work with Marxism; it is also that the attempts at analysis made so far have been fundamentally misconceived. They have treated the mass media primarily as part of the ideological superstructure and ignored or underplayed their integration into the economic base. Smythe argues that we need to reverse this emphasis and return economics to the centre of Marxist cultural analysis. For him, "The first question that historical materialists should ask about mass communications systems is *what economic function for capital do they serve*, attempting to understand their role in the reproduction of capitalist relations of production" (p. 1 italics in the original). It is a bold polemic, delivered with panache, and embracing almost all of the currently fashionable variants of European Marxism. His list of the "blind" includes, Adorno and Horkheimer, Gramsci, together with leading contemporary writers such as Louis Althusser, Hans Enzensberger and Raymond Williams.

Smythe is undoubtedly right about the underdevelopment of economic analysis in western Marxist work on culture and communications. However, he is by no means alone in this perception. A number of European Marxists would wish to go a good deal of the way with him. Raymond Williams' recent writings, for example, are peppered with attacks on versions of Marxism which over-emphasize the ideological role of communications. As he put it in a recent article then, "the main error" is that they substitute the analysis of ideology "with its operative functions in segments, codes and texts" for the materialist analysis of the social relations of production and consumption.[2]

In his latest book, *Marxism and Literature*, he insists that "the insertion of economic determinations into cultural studies is the special contribution of Marxism, and there are times when its simple insertion is an evident advance."[3] Moreover, questions of economic determination have recently provided the focus for several Marxist and *Marxiant* analyses of the British mass media.[4] However, these studies part company with Smythe on the question of the value and relevance of the western Marxist tradition. Whereas for him it is an obstacle to be cleared away, for Williams, for myself and for many others in Britain and Europe it is a resource to be drawn upon. Certainly it needs to be rigorously reworked and the dross jettisoned, but I want to argue that a critical engagement with western Marxism is still indispensable to the development of a comprehensive and convincing Marxist analysis of mass communications. Not least, this is because the central topics of western Marxism are precisely those which were left underdeveloped in the work of Marx and of classical Marxism: the nature of the modern capitalist state; the role of ideology in reproducing class relations; the problematic position of intellectuals; and the formation of consciousness in conditions of mass consumption. Smythe acknowledges the continuing importance and centrality of these issues and itemises them as areas requiring further development at the very end of his essay. Yet paradoxically he turns his back on the rich sources of insight and conceptualisation offered by European Marxism. This wholesale rejection seems to me to be rooted in an over-simplified view both of the tradition itself and of the historical experience to which it speaks. This is Smythe's own blindspot. Before I elaborate on this point however, it is necessary to outline his argument a little more fully.

As noted above, Smythe is not alone in insisting that contemporary mass communications systems must be analysed as an integral part of the economic base as well as of the superstructure. At its simplest this is so because communications are now big business with mass media companies featuring among the largest corporations in the major western economies. Indeed, some commentators have argued that recent developments, particularly the general shift from manufacturing to service industries and the investment switch from armaments to communications, have made the information industries "one of the economic leading edges of developing multi-national capitalism."[5] Smythe's primary interest though, is not so much in the emerging structure of contemporary capitalism as in its underlying dynamics. For him the crucial question concerns the role of mass communications in reproducting capitalist relations of production. His answer centres on the part they play "in the last stage of infrastructural production where demand is produced and satisfied by purchases of consumer goods" (p. 3). In particular he focuses on their articulations with advertising and on the way that the mass media create "audiences with predictable specifications who will pay attention in predictable numbers and at particular times to particular means of communications" (p. 4). In order to generate these stable consumer blocs he argues, media entrepreneurs offer their audiences inducements in the form of news and entertainment material designed to keep their attention and induce a favourable response towards the products being advertised. Hence, while he recognises that mass media content plays an important role relaying and reproducing dominant ideologies, for Smythe this is less important than their prime task of creating audiences-as-commodities for sale to monopoly capitalist advertisers. Through their exposure to mass media audience members learn to buy particular goods advertised and acquire a general disposition to consume, thereby completing the circuit of production. Moreover, while they are doing this they are simultaneously reproducing their labour power through the relaxation and energy replacement entailed in consumption.

Despite the reservations which I will come to presently, Smythe deserves credit on at least two counts. Firstly, in contrast to most Marxist discussions of communications which start from Marx's more obvious statements about ideology, notably *The German Ideology* and the *1859 Preface*, his analysis is firmly grounded in the central economic works; *Capital* and the *Grundrisse*. This redirection of attention enables him to highlight a number of formulations which have been passed over previously and which deserve the attention of Marxists interested in communications. Secondly, Smythe's own attempt to apply these insights to the contemporary situation succeeds well in demonstrating their importance for a full understanding of the role of the mass media in capitalist societies. Unfortunately though, his argument suffers somewhat from overselling.

In part the problem stems from his treatment of the North American situation as paradigmatic. "Europeans reading this essay," he argues, "should try to perceive it as reflecting the North American scene today, and perhaps theirs soon" (p2). Today New York, L.A. and Vancouver, tomorrow London, Paris and the world. Of course there is a measure of truth in this assertion. North America does occupy a pivotal place in the world media system; as a source of ownership and investment, as an exporter of products, technologies and organisational styles, and as a key market for English language material. Certainly no analysis of the media systems of Britain and continental Europe would be complete without an analysis of their various links with the systems of North America. At the same time though, the European situation displays important differences which are reflected in the emphases and preoccupations of Marxist theorising. Smythe's failure to acknowledge and come to terms with these departures has produced his own blindspots about western Marxism. There are three particularly important omissions in his presentation.

1. He drastically underestimates the importance and centrality of the state in contemporary capitalism. True he refers in passing to the recent work of Nicos Poulantzas and Claus Offe, but only on his last page and then very much as an afterthought. Certainly the implications of their work are not explored in the body of the essay.

The continuing crisis of profitability has produced two contradictory movements within European capitalism. Firstly, a number of industries including mass communications have witnessed a marked concentration of ownership as the large firms have absorbed smaller concerns in a variety of sectors. In an effort to maintain profit rates these emerging multi-media conglomerates have sought out new markets, thereby further extending their reach and influence. Examples include: the institution of aggressive export policies, the opening up of new outlays such as commercial radio in Britain and so-called "free" television in Italy, and the incursion of competition and market criteria into hitherto public communications sectors such as French television. At the same time however, as the crisis has deepened, so the state has assumed a larger and larger role in formulating and directing economic activity and policy in order to guarantee the necessary conditions of existence for continued accumulation. The result is an indissoluble but contradictory relationship between the centralized capitalist state on the one hand and concentrated monopoly capital on the other. Consequently, as Bob Jessop has recently noted, "the analysis of the state…is an absolute precondition of adequate economic theorising today."[6] Indeed, the very notion of a materialist *political* economy presupposes the centrality of state-economy relations. How exactly these relations are best analysed remains the subject of pointed debate among European Marxists, but it is a debate which is missing from Smythe's presentation. Nor are the questions to be settled solely economic. The problematic relations between capital and the capitalist state have important social and cultural repercussions. They are mapped onto the ideological conflict between criteria of profit as against need and onto the political struggles between public and private ownership and control. In Britain at the moment, for example, protracted struggles are being waged around the allocation of resources for the fourth television channel, and for local community radio and cable TV. And to varying degrees this pattern is repeated in the rest of continental Europe.

But more is at stake than a better account of the contemporary situation in Europe. If Marxism is to go beyond the critical analysis of capitalism to develop a genuinely comparative analysis of social formations it urgently needs an adequate framework for conceptualising the complex and shifting relations between modes of production and forms of the state. There are signs that this difficult but necessary enterprise is now gathering momentum within Marxism, with the revival of investigations into European Fascism, the spate of post-mortems on the fate of Chile, and the growing interest in the nature of socialist states and the problems of transition. This last is particularly crucial, for as Tom Bottomore has emphasized, an adequate "Marxist sociology at the present time would have to be capable of providing not only a 'real' analysis of capitalist society, but also a 'real' analysis of those forms of society which have emerged from revolutions inspired by Marxism itself, but which display many features that are problematic from the standpoint of Marxist theory."[7] On this problem Smythe is entirely silent. His analysis applies only to advanced capitalist economies.

2. Smythe's preoccupation with the relations between communications and advertising leads him to underplay the independent role of media content in reproducing dominant ideologies. This is particularly clear in the case of those sectors with minimal dependence on advertising revenue—the cinema, the popular music industry, comic books, and popular fiction. True, they are still articulated to the marketing system through equipment sales (you need a record player to play records), through the use of film and pop stars to endorse consumer products, and through the manufacture of commodities based around film and comic book characters—Star Wars T-shirts, Mickey Mouse soap and so on. But selling audiences to advertisers is not the primary *raison d'être* of these media. Rather, they are in the business of selling explanations of social order and structured inequality and packaging hope and aspiration in legitimate bundles. In short, they work with and through ideology—selling the system.

These non-advertising based media are almost entirely passed over in Smythe's presentation in favour of the press and commercial television which are the exemplars *par excellence* of his thesis. Although secondary, the sectors he neglects are not exactly marginal. Certainly an adequate analysis would need to incorporate them, and here again western Marxism has much of offer. Pertinent work includes: Adorno's writing on the music industry; Gramsci's analyses of popular literature; Dieter Prokop's investigations of contemporary cinema; and Armand Mattelart's dissection of the ideology of Disney comics. Alongside these analyses of content and production others have worried away at the problem of understanding how ideologies become internalized and fixed in the consciousness of audiences. The various efforts to explore the relationships between Marxism and the ideas of Freud are probably the best-known. These range from the early work of Wilhelm Reich and of the Frankfurt School to the recent appropriations of Jacques Lacan.[8] While these attempts have not always been particularly successful or convincing, they have at least grappled with the crucial problems of mediation and reception, and tried to explain how exactly the ideas of the ruling class come to constitute the ruling ideas of the epoch. In his eagerness to purge every last trace of idealism from his analysis, Smythe has abolished the problem of ideological reproduction entirely.

This is a serious oversight. Materialist analysis needs to begin by recognising that although integrated into the economic base, mass communications systems are also part of the superstructure, and therefore they play a double role in reproducing capitalist relations of production. They complete the economic circuit on which these relations rest and they relay the ideologies which legitimate them. This second function is not reducible to the first. Indeed, as several recent commentators have emphasized, the successful reproduction of ideology is one of the key conditions for the continued existence of prevailing production relations.[9] Therefore it is not a question of choosing between theories of ideology and theories of political economy, but of finding ways of integrating the two into a more adequate and complete account. To quote Tom Bottomore again: "This phenomenon, the maintenance of capitalist society through the reproduction of bourgeois culture" still "needs to be investigated in detail."[10]

3. Smythe tends to present the operations of mass communications systems as relatively smooth and unproblematic. Not only is this somewhat surprising theoretically, given Marxism's stress on contradiction and struggle; it does not fit the facts of the present situation. As mentioned earlier, the British media system is currently the site of protracted struggles over questions of use and control. There are demands for the extension of nationalisation and municipal ownership, for greater decentralisation and regionalisation, for various forms of worker control, and for greater public participation in planning and production. Similar demands are also being made across the rest of western Europe. Moreover, there struggles are mapped onto the broader patterns of class conflict: between different factions of capital, between owners and production personnel, between intellectual and technical workers within media organisations, and between producers and consumers. Smythe acknowledges the problem of class struggle as an important area requiring examination, but he gives no indication of how it might be accommodated within his framework. Once again however, some of the most fruitful pointers have come from western Marxism, particularly from the work of Gramsci.

Given that these issues of state economy relations, of ideological reproduction and of class struggle, appear to be central to an adequate materialist theory of mass communications, why does Smythe give them such short shrift? The obvious reason is lack of space. Clearly it is unreasonable to expect a single article to offer a fully comprehensive framework. However, it is reasonable to expect a degree of balance between the important elements. Unfortunately Smythe's presentation is clearly unbalanced. In his eagerness to jettison western Marxism he reverses its priorities and treats its preoccupations as peripheral. Partly this is polemics, but I think it is also symptomatic of a real failure on Smythe's part to come to grips with the tradition. He doesn't settle accounts, he simply refuses to pay. What then is western Marxism and what does it have to offer?

At its broadest, the term "Western Marxism" covers all the variants of Marxism which developed in western Europe after 1918. Hence it stands in contrast to the other great current—Soviet Marxism. Although serviceable, this distinction has a tendency to blur around the edges. For example, one of the most influential western Marxists, Georg Lukács, spent long periods in the Soviet Union, an experience which is reflected in his writings. Conversely, Trotsky is often co-opted as a sort of honourary western Marxist. But even if we leave these ambiguous cases to one side, western Marxism still presents a remarkably complex and varigated intellectual tradition.

While it is broadly true that western Marxists have tended to concentrate their attentions on ideology and culture (for reasons we shall come to presently) there has always been a vigorous though subsidiary current of work on economics. Indeed, we are only just now beginning to explore this legacy; to come to terms with Austro-Marxism and particularly Hilferding,[11] to work through the implications of Pieree Sraffa's writings,[12] and to recognise the important contributions of neglected figures such as Sohn-Rethel.[13] Smythe is however correct in suggesting that the insights generated by Marxist economists have never been systematically applied to the analysis of mass communications. Among those mainly interested in culture and ideology however, other important divisions are evident, most notably the split between those who involved themselves in political activity and those who remained disengaged. Where the first group found their main base and audience within the left parties and the workers' movement, the latter found theirs primarily within the universities and the literary establishment. Hence the break was roughly between the activists and the academics. The first group includes Gramsci, Brecht and a number of lesser figures like the Trotskyist Franz Jakubowski,[14] while the second includes Adorno, Goldmann, Althusser, and Raymond Williams (in his later phase). From the list Smythe offers (p22) it is clear that it is this academic group that he has most in mind as representatives of western Marxism. Once again however, this distinction is not as firm as it first appears. Louis Althusser for example, is usually counted among the more theoreticist of contemporary western Marxists, yet he is also an influential member of the French Communist Party whose works are permeated albeit surreptitiously, by party polemics. Nevertheless, Smythe's assertion that "professional Marxists" have been preoccupied with questions of philosophy, ideology and culture is broadly correct.[15] He poses the question of why this should be so, and expresses the hope that others will try to suggest an answer. An even halfway adequate reply would take at least a book, but for the moment some sketchy suggestions will have to serve.

To understand the blindspots and *idées fixes* of western Marxism we need to place its development in the context of the history which formed it. As a beginning it is useful to distinguish three broad phases: the interwar years, the period from 1945 to the end of the 1960s and the years since. Although certain themes are common to all three, each has inflected them in a distinctive way.

The central problematic of the interwar years was formed by the failure of the revolutionary initiatives in the advanced western economies. One after another, promising advances were turned back and crushed. Later, with capitalism facing an unprecedented crisis, instead of a resurgence of socialism, fascism took root and flowered in the very places where revolution had seemed most possible: in Germany, in Austria and in Italy. Not surprisingly, explaining this spectacular reversal became a major priority among western Marxists. Since economic crisis had clearly failed to fuel the revolution, attention turned to the forces maintaining cohesions and domination. Some, like Trotsky, Franz Neumann and Sohn-Rethel,[16] focused on the new fascist forms of the capitalist state and their coercive apparatuses. Others, notably Gramsci, Adorno and Horkheimer highlighted the part played by communications and culture in engineering the consent of the governed. This second line of inquiry was given added impetus by the massive expansion of the communications industries. These were the years that saw the rise of radio as a mass medium, the introduction of talking "pictures," the sophisticated deployment of photo journalism, and their wholesale co-option into the ideological apparatuses of the fascist states. Against this background of escalating propaganda management, censorship and repression the question of the mass media's economic and commercial role seemed relatively unimportant.

Once ideology was viewed as a key weapon in the arsenal of class domination, critical intellectual work on culture could be regarded as a crucial contribution to the general struggle against fascism and the capitalist system which supported it. For Horkheimer and Adorno this meant preserving the gap between the actual and the possible; for Gramsci it meant constant educational work to build a radical counter culture among the dominated. This emphasis on the importance of critical intellectual work and cultural practice provided a convenient occupational rationale for Marxist intellectuals. For, as Pierre Bourdieu has wryly pointed out, nobody believed more fervently in the transforming power of ideas than the professional intelligentsia who owe their class position to their intellectual skills.[17] In a number of cases this occupation ideology was further reinforced by biographical experience. Adorno for example, came from a milieu where cultural activity and accomplishment was a central value. He had dabbled at composing and music criticism. Similarly it was not particularly surprising that Gramsci should value educational activity given than it had provided his own escape route from poverty and his entry ticket into the radical intelligentsia.

After the initial period of post-war reconstruction, the advanced capitalist economies of western Europe entered a cycle of boom which generated a rapid expansion in the consumption of leisure and entertainment goods. Many of these developments were dominated by American style products and organisations, and were firmly articulated to the advertising and marketing system which Smythe describes. Why then did western Marxists generally pay less attention to these aspects than to the problems of cultural form and ideological transmission?

Part of the answer has to do with classification of Soviet Marxism. The concentration on culture among western Marxists at this time can be seen as an over-reaction to the economism of the official Soviet line and to the Stalinist political practice that stemmed from it. Against the Soviet tendency to reduce cultural forms to reflections of class position and class interest, western Marxists stressed the relative autonomy of ideological production and the complexity of its internal dynamics. Raymond Williams, for example, left the British Communist Party in the late 1940s to begin a long interrogation of the British socialist tradition in an effort to find non-reductionist ways of conceptualising culture-society relations. Elsewhere, others were independently engaged on the same task. In France, for example, Sartre was struggling to marry his existentialism with his growing commitment to Marxism; Lucien Goldmann was exploring the possibilities offered by Lukács' work, and Roland Barthes was trying to integrate Sassurian linguistics with a Marxist account of popular culture. Once again, this general intellectual project was underpinned by occupational and biographical considerations. It is certainly no accident, for example, that a number of prominent European Marxists of the post-war period began either as professional philosophers (Goldmann and Althusser), or as writers and literary critics (Sartre and Williams).

Another part of the explanation however, lies in the changing texture of social conflict. The expansion of consumerism was accompanied by a dampening down of industrial conflict and class struggle. The contradiction between Capital and Labour receded from the centre of attention and its place was taken by conflicts grounded in age, in gender, in nationality, in race, and above all in the yawning gap between the developed and underdeveloped worlds, between the colonisers and the colonized. Moreover, these conflicts appeared primarily as political and cultural struggles for self determination, political liberation and cultural autonomy. To many observers on the left it seemed that culture was not just one important site of struggle among others, but perhaps the *most* important. This misreading of history reached its height during 1967–1968, when for a brief moment it seemed that the construction of a radical counter culture coupled with the control of key institutions of transmission would bring about a bloodless transformation of capitalism.

The seventies have provided a sharp corrective to this utopianism, and as the economic crisis has deepened so the intellectual pendulum has begun to swing back, and questions of economic dynamics and determinations have re-emerged at the centre of Marxist debate. The reappropriation of Marx's mature economic works, the renewed attention to the core problems of crisis and the falling rate of profit, and the revival of interest in figures such as Sraffa, all indicate a resurgence of Marxist political economy. This development in turn has opened up new issues in the other key areas of contemporary debate; the structure and role of the state in contemporary capitalism, the dynamics of class structuration and class struggle, and the nature of legitimation processes. At the present time then, Marxism in Europe is at a point of transition. It is simultaneously assimilating the *culturalist* legacy of western Marxism and confronting the implications of the emerging political economy. Certainly a choice has to be made, but it is not as Smythe would have it, a choice between a theory of economic processes on the one hand and a theory of ideology on the other. Rather it is a choice between different ways of conceptualising the complex relations *between* the economic, ideological and political dimensions of modern capitalism.

Western Marxism still has an indispensable role to play in this enterprise. Firstly, it speaks to real theoretical silences within classical Marxism, silences which cannot be adequately filled by Smythe's schema. Secondly, because it is grounded in historical processes which are still unfolding themselves it provides points of entry into the analysis of contemporary experience. The problem of understanding the resurgence of neo-Fascism within Europe is one obvious example of its continuing relevance.

To react to western Marxism's over-emphasis on culture and ideology as Smythe does by jettisoning it completely and calling for a new improved "non-Eurocentred Marxism" (p 21) seems to me to be an over-reaction which substitutes one set of biases and blindspots for another. Rather than rejecting the European tradition *tout court*, we need to critically rework it, to confront the theoretical problems and possibilities that it opens up, to sort out the concepts and insights that remain viable, and to consign the rest to the history of ideas. There is no doubt at all that Marxism needs to be overhauled if it is to produce convincing analyses of contemporary mass communications systems. As part of this task we shall certainly need to develop the fertile line of analysis sketched by Smythe, but we shall also need to assimilate and build on the contributions of Gramsci, Althusser, Williams and others. For without them, the Marxism of the 1980s will be very much the poorer.

Notes

1 Graham Murdock, "Blindspots About Western Marxism: A Reply to Dallas Smythe" was first published in the *Canadian Journal of Political and Social Theory*, Vol. 2, No. 2 1978. Published with the permission of the editors, Arthur and Marilouise Kroker.

2 Raymond Williams, "Notes on Marxism in Britain since 1945," *New Left Review*, No. 100, January/February 1977, p. 90.

3 Raymond Williams, *Marxism and Literature*, London: Oxford University Press, 1977, p. 138.

4 See Graham Murdock and Peter Golding, "Capitalism, Communications and Class Relations," in James Curran et al. (eds.) *Mass Communication and Society*, London: Edward Arnold, 1977. See also John Westergaard's contribution to the same volume.

5 Nicholas Garnham, "Towards a Political Economy of Culture," *New University Quarterly*, Summer 1977, pp. 341–42.

6 Bob Jessop, "Remarks on Some Recent Theories of the Capitalist State." Unpublished paper, University of Cambridge, 1977, p. 40.

7 Tom Bottomore, *Marxist Sociology*, London: Macmillan 1975, p. 22.

8 See for example, Rosalind Coward and John Ellis, *Language and Materialism*. London: Routledge and Kegan Paul, 1977.

9 This argument is put with particular force in Anthony Cutler, Barry Hindess, Paul Hirst and Athar Hussain, *Marx's Capital and Capital Today*, London: Rouledge and Kegan Paul, 1977.

10 Tom Bottomore, op. cit., p. 30.

11 See Anthony Cutler et al, op. cit., Part 1.

12 See Ian Steedman, *Marx After Sraffa*, London: New Left Books, 1977.

13 A work by Sohn-Rethel on one of his major preoccupations ("Intellectual and Manual Labour") is due for publication by Macmillan later this year.

14 Jakubowski's major work, *Ideology and Superstructure in Historical Materialism*, was published in English in 1976 by Alison and Busby.

15 This argument is also central to Parry Anderson's recent analysis and critique of the western Marxist tradition, *Considerations on Western Marxism*, London: New Left Books, 1976.

16 Franz Neumann's *Behemoth: The Structure and Practice of National Socialism* has been available in English since the early 1940's. Trotsky's writings on the fascist state were only published in English in 1971 under the title: *The Struggle Against Fascism in Germany*. Sohn-Rethel's major work, *Okonomie and Klassenstruktur des Deutschen Faschismus*, remains as yet unavailable in translation.

17 Pierre Bourdieu, "Cultural Reproduction and Social Reproduction," in Richard Brown (ed.) *Knowledge, Education and Cultural Change*, London: Tavistock Publications, 1973.

CHAPTER SIX

Introduction to "Ratings and the Institutional Approach"

Eileen R. Meehan[1]
University of Southern Illinois Carbondale

I've been asked to provide some context for this essay and my response is made with Anthony Giddens' theories on agency and structuration in mind. As I look back at this essay, three contexts immediately spring to mind.

The first context was intellectual: a long-term interest in critical theory and research that drove my graduate studies and, in 1978, got me to the conference of the International Association for Mass Communication (IAMCR) in Warsaw, Poland, where Dallas Smythe and Graham Murdock enacted the Blindspot Debate based on their printed exchange (Smythe 1977 and 1978; Murdock 1978). Smythe's claims about the audience commodity resonated with me but seemed too broad. Murdock's call for greater specificity and for a more nuanced approach also struck a chord. I was then searching for a dissertation topic and the audience commodity and ratings seemed ripe for exploration. Subsequent reading in Thomas Guback's class on political economy confirmed that impression.

A second context was formed by collegial relationships. I was one of Guback's students at the Institute for Communications Research (ICR) at the University of Illinois, Urbana Champaign. By the early 1980s, two dissertations on audience measurement were underway. Donald Hurwitz, a student of James Carey and Clifford Christians, pursued an historical-culturalist project. I dug through Congressional hearings, industry publications, and trade press coverage to trace continuity and discontinuity in demand across the relevant markets. While I took no classes from Christians, I respected his work and so regarded him as a very senior colleague.

The third context was institutional, i.e., shifts in the field of communications research in the 1970s and 1980s, and responses to them. For me, these shifts suggested that political economy and materialist cultural studies were on the move. That was evident at IAMCR in 1978 when conferees established the Political Economy Section, thus creating an institutional base for the pre-existing network of political economists inside this professional association. Subsequently, ICR provided institutional support for an international newsletter focused on critical communications (*Communication Perspectives*, 1978–1985) founded by Janet Wasko, Jennifer Daryl Slack, Fred Fejes, Martin Allor, Guback and me. Simultaneously, at Stanford University, Oscar Gandy, Noreen Janus, Tim Haight, and other students organized conferences on critical communications. By 1981, these and other efforts combined, creating the Union for Democratic Communication (UDC). The UDC's inaugural conference was supported by Temple University and George Gerbner, Dean of the Annenberg School (University of Pennsylvania) and long-term editor of the *Journal of Communications (JoC)*, in which he had published articles by scholars working in 'fringe' areas like critical research or feminist scholarship.

In 1983, Gerbner published a special *JoC* issue, "Ferment in the Field—Communication Scholars Address Critical Issues and Research Tasks of the Discipline." The 41 contributors represented multiple paradigms, including administrative and critical research. Gerbner's epilogue concluded that the "critical task of a discipline is to address the terms of discourse and the structure of knowledge and power in its domain and thus to make its contribution to human and social development. Those who search and struggle towards that end are critical scholars in the best and basic sense of the word" (Gerbner 1983, 362). By encouraging everyone to be critical "in one's own fashion," Gerbner loosened the term's connection to Marxism. In 1984, the Speech Communication Association (SCA) launched *Critical Studies in Mass Communication* (*CSMC*) as "a forum for cross-disciplinary research (welcoming) a wide range of theoretical orientations and methodological approaches" where "each contribution will, in its own way, make a critical statement" (SCA 1984). *CSMC*'s editorial board was comprised by scholars from a wide variety of research traditions including critical cultural scholars like Stuart Hall and political economists like Vincent Mosco. Overall, institutionalization may have diluted the connotation of "critical" but it also provided scholarly venues supportive of truly critical research.

In *CSMC*, my intellectual, collegial, and institutional contexts converged. The founding editor of *CSMC*'s "Review and Criticism" section was Christians. For the second issue, he assembled a "symposium-in-writing" with three essays on "Broadcast Ratings," each addressing a key issue and reflecting a particular paradigm (Christians 1984, 194). Christians invited William S. Rubens, NBC's vice president for research, Hurwitz, and me. I know of no connection between Christians and Rubens but Christians,

Hurwitz, and I were connected by our shared time at ICR. By inviting his student and me, an ICR alumna, to contribute, Christians supported two junior colleagues—one who could explicate ratings from a culturalist position, the other from an institutional perspective. In his roles as editor and senior colleague, Christians had the institutional power to publish administrative, culturalist, and political economic research as equally worthy of space in SCA's bold, new journal.

To me, the resulting symposium remains meaningful. Each essay provides some insights and encapsulates the author's paradigm. After briefly calling for more "public criticism" of ratings (Rubens 1984, 195), Rubens concentrated on methodological issues whose resolution would better serve advertisers and broadcasters. Hurwitz traced how that methodological focus made ratings "a mechanism to rationalize and legitimate the assumptions and arrangements of an industry growing increasingly remote from its origins" (Hurwitz 1984, 212). Using an institutional approach, I explored the commodity form of television's messages, audiences, and ratings, and their interlinkages. I hope you find the essay useful.

Notes

1 Long after this essay was published, International Association for Mass Communication Research became International Association for Media and Communication Research; the Speech Communication Association became the National Communication Association; and *Critical Studies in Mass Communication* became *Critical Studies in Media and Communication.*

CHAPTER SEVEN

Ratings and the Institutional Approach: A Third Answer to the Commodity Question

Eileen R. Meehan[1]

Political economists have long approached the study of mass media and communications industries from an institutional perspective. That is, their research has consistently focused on large-scale, impersonal relationships between relevant corporations, governmental entities, and trade associations in an attempt to trace flows of influences and pressure across political and economic structures. In this approach, researchers have sought to uncover the hidden economics of mass media industries within capitalism. Primary issues, then, have included questions of financial control and ownership (e.g., Guback 1979; Wasko 1982), effects of oligopoly or monopoly (e.g., Danielian 1939; Yorke 1931), the role of the State (e.g., Benda 1979; Mosco and Herman 1981; Smythe 1957), pluralist ideology and hegemonic content (e.g., Bennett 1982a and 1982b; Gitlin 1980; Hall 1982; Skornia 1965), cultural imperialism and one-way flows (e.g., Schiller 1973, 1976, 1977; Nordenstreng and Varis 1974), and so on across a spectrum of both issues and authors in the Americas and in Europe.

One question that has excited particular debate—and the question more directly relevant to ratings—has been the deceptively simple question: what commodity is produced by mass communications industries? The various answers posed each identifies a rather different role for the mass media, privileges different topics for inquiry, and establishes different analytic frameworks. In studies of broadcasting, two answers stand out to the commodity question. The first is that the broadcasting industry produces messages; the second, that the industry produces an audience. There is, of course, a third answer, which was posed but never seriously pursued by *Variety*'s industry pundit (Brown 1971): the commodity produced, bought, and sold is constituted solely by the ratings.

I will review the major positions within an institutional perspective that develop out of each answer. But, beyond this, I will argue that the most basic answer and the one most useful as a focus for research and for re-interpretation of earlier research is indeed the least obvious answer. The blindspot, I suggest, has been neither cultural production nor audience production, but rather the effects of the closed market for ratings that is formed by the intertwining of the advertising, ratings, and broadcasting industries. To get to that point, however, the positions that historically precede the third answer must be considered.

Cultural Production and the Commodity Message

The assumption that the message is the commodity underlies much critical research into the processes of cultural production. Claiming that the message is the commodity locates mass media between the economic base and the political/ideological superstructure, assigning to mass media the role of mediator. In this way, one can posit and account for both the economic and ideological roles played by mass media industries. The position calls attention to the duality inherent in the industrial production of symbols and thus to theories of ideology, representation, and hegemony. In the U.K., this is exemplified by the work of Raymond Williams, the Leicester Centre, and the Birmingham School; in the U.S., materialist ideological critique has been primarily confined to film theory (cf., MacBean 1975; Guback 1983).

In broadcasting studies, the situation has been somewhat different with industry organization taking precedence over textual analysis. Notions of internal production constraints, external commercial constraints, and of professionalism form the basis for explaining why cultural products reproduce hegemonic ideology. In other words, it explains why in a broadcasting system lacking direct State censorship or control, themes, ideas, claims, values, visions, and notions supportive of the capitalist system are promoted via news and entertainment programming, while others critical of the system are reconstructed and re-interpreted in such a way as to co-opt their critical content. Still others are simply excluded (e.g., Barnouw 1978; Carey 1969; Ewen 1976; Hall et al. 1978; Murdock 1982; Smith 1973). By tracing this selection process from the level of individual decision-makers to organizational constraints to industrial parameters, these researchers have demonstrated how economic institutions—specifically, networks, production companies, national advertisers, professional associations—combine to set the boundaries within which cultural production occurs. From economic processes emerge ideological products, which reinforce capitalist ideologies and presumably affect human consciousness. By examining the message as commodity, the impersonal relationships surrounding the production of such commodities become susceptible to study.

For our purposes here, however, one significant relationship seems curiously absent: how do ratings and the ratings industry fit into the production of the commodity message? As one would expect, researchers taking this position have examined ratings and the ratings industry from the perspective that only the message constituted a commodity. With the focus on the commodity message, researchers have treated ratings in much the same terms as broadcasters; that is, as imperfect but basically scientific measurements of audience size and type which could be influenced by manipulations of program schedules. The critique, then, has centered on issues of methodological adequacy and proper interpretation of ratings, with only a superficial treatment of the industrial structure and business history of ratings firms. Despite this oversight, this critique has become standard within both scholarly examinations of the broadcasting industry as well as in memoirs of that industry's employees. Indeed, the critiques offered in particular by Skornia (1965) and Shanks (1977) are so widely cited that their arguments seem quite representative of the general position on ratings that grows out of the commodity message perspective.

Both Skornia and Shanks begin their discussion of ratings by referring to the great number of firms producing ratings, but then quickly narrowing their focus to a single firm, the A.C. Nielsen Company (ACN). For Skornia, ACN is merely "one of the best known" of the over two hundred commercial firms that comprise the industry (1965, 127); for Shanks, ACN's ratings system is "the only one that finally counts" (1977, 245) despite the many systems available in the market. Both imply that the ratings industry per se and the more specific form of producing ratings for national television are open and competitive—and just happen to have a single most important firm. This lends the status quo within the ratings industry an aura of naturalness; the industry appears to "simply exist" abstracted from the processes of history and the imperative of capitalist economics. This supposition is significant not only because it prefaces Shanks' and Skornia's discussion of measurement practices, but also because it limits their analysis by bracketing our considerations of cost efficiency in ratings production, corporate struggle to safeguard and expand market share, cross-industrial pressures from either advertisers or networks, as well as corporate alliances across these tripartite industries. Instead, Shanks and Skornia limit their critique to how ratings firms implement scientific procedures and how broadcasters misuse the results—without asking why or how ratings firms select particular procedures in the first place.

Their critique of measurement practices, then, turns on problems of sample size and data collection with Skornia and Shanks producing similar litanies: the sample is too small; the sample is too unrepresentative; the act of measurement changes behaviors; the act of measurement changes reports of behaviors; too many meters break down; turning is treated as viewing; insufficient compensation is given to cooperating households; ratings firms are too secretive about sampling procedure and the life of their samples. To this, Skornia adds four related points: figures based on sets-in-use reflect only forced choices and not public taste; reporting of such figures in percentages misrepresents the actual size of viewership within already small samples; further, these percentages mask the "protest vote" of those who refuse to watch; finally, the intrusion inherent in even the most unobtrusive forms of measurement may bias the sample since "few intelligent people seem willing to have gadgets put on their sets..." (1965, 128). While these criticisms are relatively standard, they are valid only if ratings firms are indeed producing research rather than commodities, that is, items manufactured for sale within a business environment. The rules of survey research apply to the former activity; to the latter, apply such constraints as cost effectiveness, productivity, and profitability.

But corporate interests do not evade Skornia and Shanks completely. Indeed, when identifying the causes of these shortcomings in measurement and reporting, both place the blame directly on the broadcasting industry. While Skornia posits shared frames of reference as a possible cause of biased ratings, Shanks identifies the broadcasting industry's inflexible demand for inexpensive ratings as the source of the problem. But where Shanks stops after evoking budgetary constraints, Skornia goes further in his analysis to explain the broadcasting industry's—more specifically, the networks'—chronic underfunding of the ratings industry. Arguing that the networks' drive for profit has turned television into a purveyor of homogenizing and stultifying programming, Skornia charges that ratings are but a tool manipulated by the networks to justify their programming policies to relevant governmental bodies and to the public.

Deriding the networks' frequent characterizations of ratings as votes and television as a cultural democracy (cf. Nielsen 1966), Skornia details three manipulations used by networks to consolidate their control over television programming, all of which are relevant to the commodity message perspective. First, Skornia examines the networks' ability to manipulate schedules to eliminate programs and genres not congruent with the networks' definition of television as a mass medium. Beyond this, networks can manipulate the ratings reports by comparing the incomparable—for example, documentaries on social problems with quiz programs—thereby focusing on forced choice behaviors rather than issues of public taste and public service. Finally, Skornia argues that the networks support inadequate measurement techniques precisely because of those inadequacies—because the ratings measure audience size and not interest or enjoyment or learning. The ratings are poor, then, because the networks find such inadequacy useful. By extending his analysis beyond Shanks' evocation of the budget, Skornia uncovers network practices that alternately misshape and misconstrue the ratings so that programming can be kept within the boundaries of corporate profits and corporate preferences. The significance of Skornia's work for understanding the cancellation of controversial series such as *Lou Grant*, for the elimination of genres such as the dramatic omnibus format, and the "ghettoizing" of public service or education programs should be obvious.

From the perspective of the commodity message position, ratings are not only a powerful rationale available to members of the broadcasting industry to justify corporate decisions regarding cultural production, but a rationale that shifts responsibility from the decision-makers to the public. Just as the networks blame the public for poor programming, so too networks are blamed for poor ratings. This narrow focus on the message producing industry closes out the possibility that other influences and other interests might combine with those of the broadcasting industry to shape ratings production. In doing this, the position acquires two further assumptions—the first, that ratings production is an exercise in social science rather than a business venture; the second, that the audience is constituted by a naturally occurring viewership whose members consume the commodity message. In 1977, the problems posed by the second assumption sparked the "blindspot" debate in which the most basic conceptions of mass media and of the commodity message were questioned.

Blindspot in Western Marxism?

Launching the controversy was Smythe's (1977) claim that all mass media industries produced but a single commodity—the audience. In Smythe's formulation, the message was merely bait, just a "free lunch" designed to lure the audience to the point of sale. Once attracted, the audience then spent its "leisure time" with the mediated bait in such a way that media industries could organize that audience into salable categories for purchase by advertisers. Each person using mass media thus produced something of value—one's self as a member of a salable audience—which was surrendered to media corporations free of charge and was then sold by them to advertisers. Drawing his case rather dramatically, Smythe then argued that, in capitalism, all time was work time, that all media functioned in this way even when the audience directly paid for the "lunch" (as in the film or newspaper industry), that all mass media were thus part of the economic base. Any attempt to locate media in the superstructure, or in both the superstructure and the base, Smythe argued, was fundamentally wrong. The resulting exchange among critical scholars within the institutional approach was, to say the least, heated (cf. Smythe 1977; Murdock 1978; and Livant 1979; also publicly debated by Smythe, Murdock, and others in the 1978 meeting of the political economy interest group during the conference of the International Association for Mass Communication Research). But when the dust finally settled, and Smythe's claims were trimmed back to describe the media that they best suited—commercial newspapers, television, and radio—a new truism about television emerged, but in tandem with old truisms about the audience and the ratings industry.

Joined by Livant and others, Smythe argued for his analysis using television as his primary example. Here networks commissioned and designed programs and constructed schedules specifically to attract certain kinds of viewers in large groups and then sold those audiences to advertisers. Smythe noted that the economic transaction between advertisers and broadcasters did not depend on faith, but rather on the verification of network claims by a scientific, competitive ratings industry, which was subordinated to the unified demands of its clients:

> How are advertisers assured that they are getting what they pay for when they buy audiences? A subsector of the consciousness industry checks to determine. The socio-economic characteristics of the delivered audience/readership *and* its size are the business of A.C. Nielsen and a host of competitors who specialize in rapid assessments of the delivered audience commodity. (Smythe 1977, 5, emphasis in original)

Despite the differential market value of various audiences, Smythe and Livant go to some length to argue that "everyone is in the audience" (Livant 1979, 101)—a point that is generally accepted even by their most careful critics (cf., Murdock 1978, 109–11). Indeed, research has proceeded in this general area to examine the specifics of advertiser subsidies for particular kinds of messages geared to attract particularly desirable audiences (Gandy 1983a), but again working from the notion that the audience is a kind of raw material occurring in nature. Advertisers want audiences; networks capture audiences; raters measure audiences; advertisers buy audiences from networks. Yet, the audience not only fails to receive compensation for its participation in these transactions, but actually pays for much of the production process by investing capital in television sets, by absorbing advertising costs hidden in the prices of advertised brands, and by paying for the electricity to run the machinery necessary to make one available for sale. Thus, the second perspective emerges, illuminating the hidden economics of audience production and uncovering the motive force directly behind message production.

As Smythe noted, many of these ideas were not new; still, as Murdock pointed out, Smythe's re-articulation of them not only called attention to some central ideas argued by Marx in *Capital* and the *Grundrisse*, but also "succeed[ed] well in demonstrating their importance for a full understanding of the role of mass media in capitalist societies" (Murdock 1978, 111). This worked directly into the ongoing discussion over the relationship between political economy and materialist cultural studies. While most researchers recognized the need for integration, few were clear on how to achieve it, although some tried via research (e.g., Hall et al. 1978) and others through yearbooks (e.g., Mosco and Wasko 1983), some through critique (e.g., Williams 1977a) and others through readers (e.g., Gurevitch et al. 1982). The upshot of this debate has been a serious rethinking of the theoretical frameworks around the issue of economic processes and ideological effects especially in terms of corporate structure, audience production, and hegemonic/oppositional readings (cf. Guback and Douglas 1983; Fejes and Schwoch 1987; Slack 1983; Gandy 1983b; and Meehan 1983b).

But while this has been significant for the institutional approach and the materialist cultural approach, there still remains a crucial assumption requiring close scrutiny. While most communication scholars within the materialist paradigm now agree that audiences are indeed produced, that some audiences are more salable than others, and that audiences are the primary commodity in broadcasting, most still regard the audience commodity as substantially untouched by the manufacturing process that makes the commodity visible. The process of translating viewers into a verified audience—that is, into ratings—remains peculiarly aloof from the constraints and pressures that shape all other forms of commodity production in capitalism. But can political economists afford to assume that? Given the terms in which I cast the question, the answer is rather obviously "no." And this brings us to the third answer to the commodity question.

Commodity Ratings

Here the focus widens to include both ratings firms and their products as objects for analysis. No longer can the industry be dismissed as a mere slave of either the broadcasting industry (Skornia 1965; Shanks 1977) or the advertising industry (Barnouw 1978) or those industries' unified demand for audience verification (Smythe 1977). Instead, the internal economics of the ratings industry as well as relationships with client industries and client firms must come under the most careful scrutiny. Similarly, we must probe behind the image of ratings firms as technicians simply applying scientific procedures in order to discover the relationships between such procedures and actual production practices. Finally, ratings per se must no longer be treated as reports of human behavior, but rather as products—as commodities shaped by business exigencies and corporate strategies. In another context (Meehan 1983a) I have pursued these issues in terms of State inquiry and intervention. In this context, I will briefly sketch the demand structure that constitutes the closed market of these three intertwined industries. To do this, I return to the truism of the commodity audience.

While AT&T's invention of toll broadcasting resolved funding for early broadcasters, it created instrumental problems that required solutions in order to facilitate transactions between advertisers and broadcasters. The most obvious was to demonstrate the existence of an audience. In economic terms, this meant some measure of productivity. But besides adequate numbers of listeners, a measure was needed to demonstrate that the right product was being produced, that the audience for which advertisers expressed a demand was in fact being delivered by broadcasters. Only after agreeing on a basic method for producing measures of productivity and quality could broadcasters and advertisers move to the real business at hand—the buying and selling of audiences according to a rational price structure.

Inherent in the situation, then, is both continuity and discontinuity in the interests of the advertising and broadcasting industries. While continuity rests in the need for an official description of the audience, discontinuity arises from the connection between that description and pricing. Ways of generating that official description which might slightly inflate or deflate the numbers could have significant consequences in the cost per thousand that broadcasters could demand and that advertisers would pay. Thus, neither industry could trust the other to measure productivity and quality, yet both needed some measure in order to transact sales. In this way, despite unified demand for measures of productivity and quality, the situation presented contradictory interests and thereby the possibility for independence for any ratings producer that could successfully manipulate those differences. Further, the opportunity to obtain a monopoly and hence greater control was unmistakable, given the clients' need for a common ground, for *a* set of numbers. Both of these possibilities were used by the various firms attracted to the most lucrative form of ratings production, that is, resellable reports of national audiences for network programming over the primary form of broadcasting, otherwise called national syndicated reports.

But these ratings firms also faced a problem when approaching entry into national production and attempting to gain dominance in that market. Given the continuities in demand, how would one firm differentiate its reports from those of the dominant firm? Here again the ways of producing ratings figure significantly as firms used methological variations to differentiate similar products and then promoted these differences as being more scientific—hence more accurate and more objective—than procedures used by the dominant firm. This has become the primary strategy used by ratings firms from the time that the Cooperative Analysis of Broadcasting was unseated by the C.E. Hooper Company in the early 1930s to the replacement of that company by the A.C. Nielsen Company in the early 1950s through the various challenges to Nielsen up through the early 1960s to present day. First, develop a new procedure or a variation in a procedure; then, shape that innovation to manipulate differences in demand; finally, promote that innovation as another step forward in the progress of science while building cross-industrial alliances with those firms whose particular interests are best served by the innovation. The last significant contributions to these tactics were made by the A.C. Nielsen Company: develop an innovation capable of patent protection and then use that protection to safeguard market control by barring the entry of possible competitors or limiting their operations. From this brief sketch, it should be clear that, although continuities in demand do constrain ratings production, those continuities do not directly determine ratings production. A ratings challenger has a number of standard tactics available within the industry that can be incorporated into the particular corporate strategy originating in the firm.

With regard to defining the audience, one cannot overemphasize the effect of the Nielsen Company's contributions. The costs of running a metered operation—including the meters themselves, their installation and maintenance, data reporting systems whether over telephone lines or by film cartridge, and so on—militate against frequent turnover of the sample as do the processes of designing the sample and designation households (both of which are tied to the U.S. Census), as does the problem of securing cooperation (which encourages metered operations to be cost effective by carrying over some cooperating households from one sample to the next). Further, constraints in continuities of demand for some demographic categories over others, means that the commodity audience is a priori different from the viewing public. All viewers are not equally in demand and hence not equally profitable to either broadcasters or raters. As a result, a ratings firm that acts rationally within this market structure will not measure the public; to do so would be to produce an unsalable commodity. Similarly, it would be irrational for broadcasters to program for any viewership other than the fixed and semi-predictable sample; to do so would be to refuse to produce the commodity audience. For all practical purposes, then, the fixed sample, shaped and limited by economic constraints within the ratings industry and across these three intertwined industries, becomes the commodity audience. Thus, the structural effects of this closed market for ratings begin to become visible by applying an institutional perspective at the level of cross-industrial relationships.

Towards an Integration of Commodities

So what commodity *does* broadcasting produce? By now, the difficulties of answering that simple question should be clear. While none of the three answers considered here is wrong, each is incomplete without the other two. Yet, if we are to understand the structural context of message and audience production, we must begin by privileging the ratings commodity as setting the broad parameters that bound the other two forms of production. At the level of cross-industrial analysis, neither messages nor audiences are exchanged; only ratings. And those ratings are produced at a particular juncture by a single company that—like any other company in any industry at any point in time—seeks to maximize its profit and minimize its cost, to safeguard its market position and expand its sphere of independence, to manipulate discontinuities in demand and satisfy continuities. Institutional researchers can no longer afford to overlook the political economy of ratings production. Exploring the least obvious answer builds an interpretive framework that integrates these three answers.

Notes

1 Eileen Meehan, "Ratings and the Institutional Approach: A Third Answer to the Commodity Question" was first published in *Critical Studies in Mass Communication*, Vol. 1, No. 2, 1984. Reprinted with permission of the author.

CHAPTER EIGHT

Introduction to "Watching as Working"

Sut Jhally[1]
University of Massachusetts-Amherst

> **Morpheus**: The Matrix is everywhere, it is all around us, even now in this very room. You can see it when you look out your window, or you turn on your television. You can feel it when you go to work, when you go to church, when you pay your taxes. It is the world that has been pulled over your eyes to blind you from the truth.
>
> **Neo:** What truth?
>
> **Morpheus:** That you are a slave, Neo. Like everyone else, you were born into bondage…born into a prison that you cannot smell or taste or touch. A prison for your mind.
>
> *The Matrix*

Given the massive changes we have seen in the media landscape over the last 25 years, a legitimate question to ask of an article written in 1986 is whether it is at all relevant to the situation we find ourselves in today? While I have not actively engaged in any extended research and writing on this topic since that time, the themes of the article have continued as the basis for much of my speaking and teaching, and I remain convinced that the basic arguments we advanced have not only remained relevant, but perhaps have never been more so, offering insights into recent changes and the logics underlying them. Among the most important of these developments is the spread of advertising and commercial speech into every nook and cranny of the culture, (reflecting the continuing problems of the circulation of commodities that we argued commercial media were an initial response to). Just as Marx thought that nineteenth century capitalism presented itself as "an immense accumulation of commodities," twenty-first century capitalism presents itself as being dominated by the *promotion* of the immense accumulation of commodities. The structure of this cultural field, the matrix, given shape and substance by advertising and the sale of audience-time as a commodity is what interested us, and the article was a first attempt at an answer. The history of media in the last 25 years has done nothing but confirm the main tenets of our argument.

Although we used highly specialized, technical and concrete concepts (e.g., necessary/surplus watching, absolute/relative surplus value, etc.) our analysis was really pitched at another level of conceptualization and abstraction—the need to identify the broader contours of the *logic* that would shape the emerging media terrain. In that sense there was much that we could remain agnostic about regarding some of the responses to the article (for example, that the labor of audiences was not really productive as it was drawing upon the real surplus-value produced elsewhere in the economy and thus should be thought of in terms of rent rather than labor; that it was not exploited labor in the Marxist sense as there was no labor contract and thus no compulsion, etc.) as they were comments directed at another level of concreteness and specificity. As my own history and background was British cultural studies (with its main concerns of subjectivity, consciousness, and identity) I was interested in a political-economic approach to the extent that it helped my understanding of these domains, which in the contemporary context could not be thought of outside the logic of the commercial media that occupied so much of the time of audiences. "Watching as Working" helped me to identify the structures of contemporary culture.

That concern was also the basis for our claim that to properly understand contemporary culture we had to break with messages as our starting point, as that led to a focus on content that was *there*, and not on the broader *context* within which meaning is only ever made in the interaction of audiences and messages. Within the framework of advanced capitalism that context is one where the logic of commodity production not only dominates the domain of culture, but where it is its actual structuring logic—the move that Marx identifies as *the shift from the formal to the real subsumption of labor*. Starting with messages (who made them, under what conditions, how did audiences make meaning of them, etc.) is to mistake the *appearance* of media for its essence. In the same way that Marx, when searching for "the secret of profit-making," argued that we have to move from the visible sphere of the market and descend into "the hidden abode of production" of the factory (Marx 1976, 270–71), we similarly thought that the place to find an answer to the question of the secret of profit-making in the media was not in the Hollywood production process, but in the factory in the living room, and especially in the productive activity of audiences.

We were particularly concerned with countering the idea of passive audiences. As we argued, without audiences watching, there is literally nothing to sell and our watching is therefore a productive activity. We are not couch potatoes where messages are being pumped into a passive consciousness; we are active creators of meaning. However that human capacity to watch and make meaning is being organized and sold, and in the process, massive private profits are being derived from our alienated activity. I hope the reversal we attempted at the very end of the article, when we wrote that the "mass media are not characterized primarily by what they put into the audience (messages) but by what they take out (value)" is one its major contributions. I believe it explains the spectacularly successful integration of the Internet and social media into the existing "matrix" where active (and loyal to brands) audiences are key to its operations and its massive profits. (An example I often use for students to illustrate this idea of media and audiences are the creatures from the *Harry Potter* novels and films, the Dementor's, whose deadly kiss sucks the soul out of individuals. In contemporary culture, the commercial media are highly organized value-extracting dementors!)

Additionally, from the starting point of messages, there is no way to systematically identify *absences*. We have to have a way of understanding not only what is there but also to be able to ask *what is not there?* What is systematically excluded? Which values are systematically underplayed? Which media forms never see the light of day? Which ideologies do we never get to interact with? We can only get to absences from a study of the *logic of the system*. Only our understanding of the context can do that, and that context is that the media system is first and foremost a vital cog in the commodity-marketing system whereby audience-consciousness is being delivered to advertisers.

This notion was also behind our discussion of whether the idea of watching as working was a *real* process or a *metaphor*? At the time of writing the article as part of my dissertation work, I was deeply immersed in a study of Marx's general methodology, as well as the specifics of *Capital*. We know that Marx spent many years searching for the correct starting point for his masterwork, and that he chose the commodity because it was *both* concrete and abstract, that is both real and a metaphor. Not only was the commodity the way in which capitalism presented itself, the way in which it appeared, in its abstract form it also contained information of where it was produced and under what conditions. By unpeeling the outer layers of the commodity-form and penetrating to its inner-core, we can unveil the secret of profit-making, of the logic that governs the matrix. That is why we referred to the media as a *fishbowl*, which if properly understood, could reveal the workings not only of the media system but of the entire society itself. Indeed, since that time I have used the concepts we developed in "Watching as

Working" in my teaching as a way to describe what would otherwise be difficult Marxist economic concepts, but in a form that is more understandable to beginners. In this we were following in Marx's methodological footsteps.

Interestingly, when I have presented this argument in my talks and speeches over the years the response has been a kind of detached and amused admiration at the elegance of the argument, but without a really sustained understanding of its political and moral implications. To get to the level of necessary outrage requires the application of our main arguments to another audience—children—and if I had to revise our original paper I would include this discussion.

There is rightfully a lot of concern with child labor in the global economy and the role it plays in the production of the products that become the basis of our identities. Within the western economies, for very good humanitarian and moral reasons (and as a result of class struggle), there are generally child labor laws restricting who can work in factories. But there are no child labor laws when it comes to the factory of the home! And just as cynical factory owners in export-processing zones around the world have no interest in the health and well-being of their child workers, so the commercial media have no interest in the health and well-being of their child laborers. All commercial media want to do is wring out the last drop of value from the watching time of children, before discarding them onto the scrap heap. Their activity (like that of adults) is being organized in a system of industrial, factory production and is the reason why children's programming (where there is no pretense at separating commercial content from non-commercial content) is even worse than other kinds of programs. If the logic of programming was that it had to deal with where children were as children, that they are at vulnerable stages of their pyschic and cultural development, and that programming should be centrally concerned with their health and well being, it would look very different. Instead, in the words of media scholar Robert McChesney, our children are being "carpet bombed" by commercial messages in a way that no previous seneration had ever experienced (McChesney 2004, 10).

There is a lot of talk of television as "the electronic babysitter," and for very good reasons (increasing work hours for harried parents, the growth of single-parent families, the erosion of extended family structures, etc.) that is the social role that has been thrust upon it. And indeed, we should be able to trust is as a babysitter! Instead we have turned our children over to a private industrial institution that cares not one iota for their well-being, but is interested only in exploiting their value-producing watching activity.

That in itself is bad enough, but having produced children as a commodity, the media then delivers them to advertisers who want to turn them into lobbyists for products against their parents! This is not the model of the classical market where there are sellers and buyers, and consumers need information about how to spend their money. That model just does not hold in the sphere of children's media culture. What advertisers are doing is trying to convince vulnerable kids to pester their parents, to lobby them for particular kinds of products. The commercial media in their totality are *child abusers*, with their abuse being systemic.

Further, there is evidence that children are suffering serious industrial diseases by being immersed in this commercial culture. Juliet Schor in her book *Born to Buy* has found that consumer involvement undermines children's well-being, concluding "high consumer involvement is a significant cause for depression, anxiety, low self-esteem, and psychosomatic complaints" (Schor 2004, 167). More specifically about commercial television, she writes that "it induces discontent with what one has, it creates an orientation to possessions and money, and it causes children to care more about brands, products and consumer values." Perhaps most worryingly for parents, "higher levels of consumer involvement result in worse relations with parents" (Schor 2004, 170).

How to respond to this from the perspective of "watching as working"? There is a lot of talk about turning TV off, and while that may work as a one-day demonstration strategy, it is not a general strategy of opposition. We should instead, using an old-fashioned term, *collectivize*. Our aim should not be simply to turn TV off, but to take the factory back into democratic and collective hands, and in the process change what it produces, from a commodified consciousness to a diverse and contradictory cultural environment that is the pre-requisite for democracy.

Notes

1 Acknowledgment: In our acknowledgments to the original article in the *Journal of Communication* we referred to Dallas Smythe as "a materialist in the jungle paradise of idealism." That is even truer for my co-author, Bill Livant, who passed away in 2008. Bill was the "truest" Marxist I ever knew, not in a dogmatic form, but in the way its philosophic, moral and analytical components were the very fiber of his being as an individual and theoretician. He was the smartest person I have ever met.

CHAPTER NINE

Watching as Working: The Valorization of Audience Consciousness

Sut Jhally[1]
University of Massachusetts-Amherst
with Bill Livant

Does the audience work at watching television? Is the notion a real economic process, or does it serve as a metaphor? Our short answer is that it is both. It is a metaphor because it is a real economic process, specific to the commercial media that produces value. How this process occurs is the argument of our article.

The metaphorical power of watching as working arises from the particular relationship of the media and the economy as a whole. In the media, the whole economy exists as an image, an object of watching—more precisely, an object of the activity of watching. At the same time, the media exist as a reflection of the whole economy of which they are a part. The media, therefore, are at once a real part of the economy and a real reflection of it. This is why we say that watching as working is both a real economic process, a value-creating process, and a metaphor, a reflection of value creation in the economy as a whole.

Metaphor and Reality: The Production of Watching Extra

Let us begin with the advertising-supported commercial media as part of the whole economy. How do they make a profit? A short answer would be that the media speed up the selling of commodities, their circulation from production to consumption. Hence, they speed the realization of value (the conversion of value into a money form) embodied in commodities produced everywhere in the economy. Through advertising, the rapid consumption of commodities cuts down on circulation and storage costs for industrial capital. Media capital (e.g., broadcasters) receives a portion of surplus value (profits) of industrial capital as a kind of rent paid for access to audiences. The differences between this rent and its costs of production (e.g., wages paid to media industry workers) constitute its profit.

But what is it precisely that industrial capital rents? Media capital sells something, sells the use of something, to capital as a whole. If media capital could not sell this something, if this something, when used by the buyer—capital as a whole—did not speed the realization of value embodied in commodities in general, the media would receive no payment (rent). Thus, the production of this something is the central problem for commercial media.

This something has been fuzzily described in the communications literature. Is it attention? Is it access, and, if it is, access to what? To markets? To audiences? And what are these audiences? Are they materials, tools, conditions? What is fuzzy about these answers is that they describe the something from the point of view of the interest of the buyer: in terms of how capital as a whole proposes to use it. But they do not describe the something from the point of view of the seller: in terms of how it is produced. In short, such answers do not describe the something as a problem for the media, which is the clue to the development of their specific practices.

For the media, above all, this something is time. But whose time? What kind of time? What happens in it that takes time? It is clear that this time would be empty and unsellable if people didn't watch. It is therefore something about watching-time that is sold by the media to advertisers.

This watching is, first, a human capacity for activity. It is not a thing, not simply a product in which value is, in Marx's word, "congealed." It is a capacity for doing something. We use doing generically to include seeing and listening—in general, capacities of perception. Watching is human activity through which human beings relate to the external physical world and to each other.

Watching is guided by our attention, so we often see or hear less than there is to be seen or heard. But we also see or hear more than there is. Elementary psychology books are filled with examples in which watching completes a figure, makes a connection, fills in the blanks. Gibson (1979) has found that we actually see behind obstructing or occluding objects: We see hidden surfaces. Indeed, this seeing more is the basis of Williamson's (1978) analysis of decoding advertisements. All watching contains an element of what we call watching extra.

Precisely because watching is an unspecialized, general-purpose capacity, it is capable of being modified by its objects, by what we watch, how we watch, and under what conditions we watch. Watching has a historical character; this is especially true of watching extra. What is so striking about the modern commercial media is that for the first time this extra has a specific social form: it is a commodity. Recall that this something that media capital sells is a human capacity for a certain kind of activity, which can be put to use by the buyer. The trade literature offers many testaments to the problem of getting people to watch. As ex-sportscaster Howard Cosell noted, "You've got to deliver 40 million people. Do you know the strain of that? You've got to deliver them…If you don't, you're gone. The business chews you up" (*Sport*, February 1979). But when the activity of watching becomes subject to commodity production, the central problem for the media is not simply to get people to watch but to get them to watch extra. The problem for the commercial media is to maximize the production of this commodity and to attempt to minimize the costs of doing so.

What is the form in which these costs appear? We can answer this if we ask the following question: when we, the audience, watch TV, for whom do we watch? It is not hard to get people to tell you that some things they want to watch and some things they don't particularly want to watch but they do anyway. Indeed, as Jerry Mander's son Kai remarked, "I don't want to watch but I can't help it. It makes me watch it" (Mander 1978, 158). Although formally free to watch or not to watch, we are often practically compelled. The literature is full of the phenomenology of felt compulsion (see Mander 1978; Winn 1977). This phenomenology in itself does not directly describe the commodity form of watching, but it does point to the fact that somehow this extra watching is being extracted in our interaction with the media. But extra watching does exist as a commodity. Some of what appears on television is a cost of production to media capital; some of what appears is not a cost but revenue that media capital receives from those who will use it. The costs are incurred to produce what we will call necessary watching-time—necessary to reproduce our activity of watching. The revenues are received for the surplus watching-time that is extracted. The problem for commercial media is to extract the maximum surplus watching-time on the basis of the minimum necessary. The logic of the media is governed by the expanded reproduction of surplus watching-time.

We argue that necessary and surplus watching exist as real economic magnitudes, identifiable and measurable. They are conceptually defined on the basic generalized human capacity for watching, the fundamental activity that constitutes a population as an audience. Watching-time is the mode of expression of value. What we are exploring is the struggle over the valorization of the activity of watching.

We believe that this argument offers an understanding of important features in the modern history of the commercial media: the changes in its technology; its composition and segmentation of audiences; the development of the blended forms of messages, such as the advertorial and the infomercial; and, above all, the acceleration of time that pervades the media. The test of our conception, of course, will be the understanding that it offers of this history.

We have found the real meaning of watching as working by looking into the media, by "putting the audience into the tube" so that we can watch its watching. But having done so, we can now look out from the media into the whole economy; we can treat the media as a metaphor for the economy.

We know that virtually anything can appear on television and that, today, virtually everyone watches. In short, the media are potentially a reflection of everything. But the vast literature of what TV teaches has overlooked the possibility that it might show us, right there on the screen, the production and realization of surplus value; that all its devices might reflect the organization of human labor in the economy as a whole; that, through the relation of the watching populations to media capital, it might reflect the relation of the working population to capital as a whole.

In our view, the media are a great fishbowl. Every economic process, every movement of value, every step in the circuit of capital appears as a reflection, not simply through or in the media but as the media-audience process itself. The media economy is a fishbowl that reflects the whole thing of which it is a part. In particular, the struggles over the valorization of human activity in the media reflect these processes in the whole economy. The media are indeed a metaphor.

Watching as working? Really? Metaphorically? Again our answer is "both." It is metaphorical because it is real; it is real because it is a part of something real, which it reflects and for which it therefore can stand as a metaphor. We will look into the commercial media in order to see how the whole economy, embodied in it, is reflected there.

From Use-Value to Exchange-Value: Breaking with Message-Based Analysis

Most studies of the media (both in the mainstream and critical traditions) have focused on messages as their central unit of analysis. Despite the many differences within the field, there is a broad unstated agreement that the discipline of media communication is about the production, distribution, reception, interpretation, and effects of messages. From two-step theory, to uses and gratifications, to cultivation analysis, to agenda setting, to the study of ideology and texts, to the controversy over the New World Information [and Communication] Order, and even to the debate concerning the effects of the new information technologies, the focus has been on messages. More specifically, the concentration has been on how these messages are used, on what meanings are generated in the interaction between messages and people. The history of communication, then, has been a study of the use-values of messages, their meaning.

That this should be the focus of the study of the commercial broadcast media, in particular, is somewhat surprising when we view the industry in historical perspective, for messages have never been the central commodity that has been produced and traded. In the early years of broadcasting, as Williams reminds us, there was little attention given to the content of the new media: "Unlike all previous communications technologies, radio and television were systems primarily designed for transmission and reception as abstract processes, with little or no definition of preceding content . . . the means of communication preceded their content" (1974, 25). In the United States, the first role of the electronic media was to stimulate the sale of radio sets (Barnouw 1978). Later, as the commercial networks developed, the sale of audiences took precedence as the industry's major activity. Messages were integrated within this wider industrial production.

There has been increasing recognition in recent years, particularly among critical scholars, of a failure to penetrate to the core understanding of the role of media in advanced capitalism. The traditional concepts of base/superstructure, relative autonomy, ideology, and hegemony are not sufficient to explain the dynamic changes taking place in mass media. As Garnham wrote,

> So long as Marxist analysis concentrates on the ideological content of mass media it will be difficult to develop coherent political strategies for resisting the underlying dynamic of development in the cultural sphere in general which rests firmly and increasingly upon the logic of generalized commodity production. In order to understand the structure of our culture, its production, consumption and reproduction and of the role of the mass media in that process, we need to confront some of the central questions of political economy in general. (1979, 145)

Smythe (1977, 1981) also expressed explicit dissatisfaction with the existing state of critical media analysis. For Smythe, Marxism has had a blind spot about communications, concentrating on the concept of ideology instead of addressing the issue of the economic role of mass media in advanced capitalism. Smythe gave two original formulations to this problem. First, he argued that mass media produce audiences as commodities for sale to advertisers. The program content of mass media is merely the so-called free lunch that invites people to watch. It is the sale of their audience-power to advertisers, however, that is the key to the whole system of capitalist communications. Second, he claimed that advertisers put this audience-power to work by getting audiences to market commodities to themselves. Audiences thus labor for advertisers to ensure the distribution and consumption of commodities in general. While one cannot overestimate Smythe's contribution to a proper understanding of the political economy of communications, the stress on audience labor for the manufacturers of branded commodities has tended to deflect the specificity of the analysis away from communications to the ensuing consumption behavior of the audience. Ultimately, Smythe was concerned with drawing attention to the place of communications in the wider system of social reproduction and the reproduction of capital. We believe that the exploration of the blind spot needs to be located more firmly within the media industries rather than focusing on their wider role.

Broadly speaking, Garnham and Smythe were attempting to break with symbolism and meaning as the starting point of materialist analysis: They were seeking to break with message-based analysis and the study of use-values. We strongly support this attempt but wish to phrase it in slightly different terms in trying to establish a general framework for a critical materialist analysis. While all messages have a use-value, within the commercial media messages are part of not only a system of meaning but also a system of exchange. They form part of the process wherein media industries attempt to generate profit by producing and selling commodities in a market setting. Within the sphere of commercial mass media, messages have both a use-value and an exchange-value. More precisely, the use-values of messages are integrated within a system of exchange-value. To understand use-value, we have to adequately contextualize its relation to exchange-value. This means a switch in focus from the question of the use-value (meaning) of messages, not because the understanding of meaning is unimportant, but because we can understand it within its concrete specificity only once we fully understand the conditions created by exchange-value. The remainder of this article attempts to unravel the system of exchange-value that constitutes the system of advertising-supported media.

To properly comprehend the system of exchange-value within which the commercial media are based, we need to understand its economic logic and to answer three related questions: What is the commodity-form sold by commercial media? Who produces this media commodity and under what conditions? What is the source of value and surplus value in this process? Once we have answered these questions, we can then formulate an adequate context within which we can understand the role of messages.

What Is the Commodity-Form Sold by the Commercial Media?

At first glance, the answer to the question of what commodity-form is sold by the commercial media seems obvious and straightforward: Media sell audiences to advertisers. We need, however, to pin down specifically what about audiences is important for the mass media. For all his emphasis on communications, Smythe does not ask this question directly. For him, it is audience-power put to work for advertisers that is important. There is no doubt that this is what advertisers are interested in, but it does not mean that the media are interested in the same thing. What advertisers buy with their advertising dollars is audiences' watching-time, which is all the media have to sell. That advertising rates are determined by the size and demographics of the audience is ample confirmation of this. When media sell time to a sponsor, it is not abstract time that is being sold but the time of particular audiences. Furthermore, this is not (as Smythe contends) time spent in self-marketing and consuming advertisers' commodities, but rather time spent in watching and listening—communications-defined time. What the media sell (because they own the means of communication) is what they control—the watching-time of the audience.

Most critical analyses of advertising and media have been stalled at this point. The watching-time of the audience has been (quite correctly) characterized as the domination of so-called free time by capital to aid in realizing the value of commodities in general. For example, Ewen (1976) and Baran and Sweezy (1966) concentrate on this point, as does Smythe. No matter how much Smythe stresses the oppositional activities of audiences in constructing alternative lifestyles, however, he drifts back to the use-value of messages—meanings and their relationship to consumption. The discussion of audiences should not stop here. The audience as a market is the first form of organization of this commodity but not the last.

The recognition of watching-time as the media commodity is a vital step in the break with message-based definitions of media and audiences. It makes the problem an internal one to mass communications. The focus on watching-time is crucial to establishing an audience-centered theory of mass communications from a materialist perspective based on the analysis of exchange-value.

Who Produces This Media Commodity and under What Conditions?

Networks consider themselves the producers and sellers of audiences (see Bergreen 1980; Reel 1979, 4–5), and critical thinkers have tended to take this at face value, accepting the notion that, because networks exchange audiences, they also produce them. It is surprising to find this confusion in the writing of Marxist critics on the topic of communication, since they do not make this error in writing on, say, the auto industry or petrochemicals or, indeed, on communications hardware itself. But, when it comes to communications, the myth of the "productivity of capital" (Marx 1976) still befuddles us.

To avoid this trap we have to distinguish several common confusions. First, we must distinguish the production of messages from the production of audiences. The staff in a network newsroom produces news. The viewers watching it do not produce the news, but they do participate in producing the commodity of audience-time, as does the network staff. Networks could produce messages that no one might watch, in which case they would barely be able to give away that time, let alone sell it. The commodity audience-time is produced by both the networks and the audience.

Second, we therefore have to distinguish between the production of audiences and their exchange. There is a lot of talk in the industry about the media producing audiences, but they have not produced what they are selling. The networks merely sell the time that has been produced for them by others (by the audience). It is only because they own the means of communication that they have title to the commodity, which has been produced for them by others. Like Manchester manufacturers more than a century ago (see Engels 1891), networks suffer from the self-serving myth of the productivity of capital. Once we have sorted out these confusions we can see that the answer to this second question is that both audiences and the networks produce the commodity audiences' watching-time.

What Is the Source of Value and Surplus Value in This Process?

Through their own station licenses and those of their affiliates, and through their ownership of the means of communication, the networks have control of twenty-four hours a day of broadcasting time. How is this time that the networks control made valuable—how is it valorized?

The surface economics of commercial television seem quite simple. Network expenses can be defined as operating costs plus program costs. Their revenues are advertising dollars from advertisers who buy the time of the audiences that the programming has captured. Networks hope that revenues are more than expenses—more than an empty hope, of course, for it is almost impossible to lose money if one owns a network station. The average cost of a thirty-second network prime-time commercial in 1985 was $119,000. Based only on prime-time (8:00 p.m. to 11:00 p.m.) sales, each network collects $60 million per week from advertisers. We need, however, to dig beneath the seeming superficiality of commercial television economics and ask specifically how and by whom value and surplus value (profit) are produced. Let us trace through the process in detail. The networks buy (or license) programs from independent producers to entice the audience to watch. Networks then fill this empty time that they control by buying the watching-power of the audience. Having purchased this "raw material," they then process it and sell it to advertisers for more than they paid for it. As a concrete example, a network pays independent producers $400,000 per episode for a half-hour situation comedy. The program is in fact twenty-four minutes long; the other six minutes is advertising time. Let us presume that this six minutes is divided into twelve thirty-second spots that sell for $100,000 each. They thus yield $1,200,000 in income, which results in a surplus of $800,000 for thirty minutes of the broadcasting day.

If we keep in mind that it is the watching activity of the audience that is being bought and sold, we can see precisely where value and surplus value are produced. It is necessary for the audience to watch four of the twelve spots to produce value equal to the cost of programming. For four spots, the audience watches for itself; for the remaining eight spots, the audience is watching surplus-time (over and above the cost of programming). Here, the audience watches to produce surplus value for the owners of the means of communication, the networks or the local broadcasters.

Networks wish to make necessary watching-time as short as possible and surplus watching-time as long as possible. The struggle to increase surplus time and to decrease necessary time animates the mass media. One way in which this ratio can be manipulated is to make the advertising time longer. Program time is made into ad time, so that, in the previous example, two more thirty-second spots could be added by making the programming only twenty-three minutes long. In that case, the ratio between necessary and surplus time (presuming program costs remain the same) extends to 4:10 from 4:8, resulting in more surplus time. This, indeed, is what local stations do to syndicated shows. Portions of the program are cut out to make space for more ads. This strategy, based upon extending advertising in real time, can be labeled the extraction of absolute surplus value. In this scenario, there is a continual attempt to expand total advertising time.

However, at a certain point, there is a limit to the expansion of advertising time. Audiences will simply stop watching if there is too much advertising and not enough programming. The TV Code of the National Association of Broadcasters (NAB) limits nonprogramming time to 9.5 minutes per hour in prime time (although most stations violate this limit [see Ray and Webb 1978]). The networks in this situation must adopt new strategies to manipulate the ratio between necessary and surplus time. If the networks cannot make people watch advertising longer in absolute terms, they can make the time of watching advertising more intense—they can make the audience watch harder.

From Absolute to Relative Surplus Value

Since the late 1950s, as market research has grown in sophistication, and as advertisers are able to pinpoint quite precisely their target market, the media have found it profitable to deliver these segmented audiences to sponsors. Barnouw (1978) has given a powerful account of how the obsession with producing the right demographics has come to dominate the everyday practices of broadcasters (see also Gitlin 1983). Advertisers judge the effectiveness of various media in terms of their cost per thousand—how much it costs to reach one thousand people. However, the watching-time of all types of audiences is not the same; some market segments are more valuable because that is who advertisers wish to reach.

For instance, advertisers will pay more to buy time during sporting events, because the audience for sports includes a large proportion of adult men whom advertisers of high-price consumer articles (such as automobiles) are anxious to reach. As John DeLorean put it,

> The difference in paying $7 a thousand for sport and $4 a thousand for "bananas" [prime time] is well worth it. You know you're not getting Maudie Frickert. You're reaching men, the guys who make the decision to buy a car. There's almost no other way to be sure of getting your message out to them. (Johnson 1971, 224)

Now men certainly do watch prime time, but, in prime time, automobile advertisers are paying not only for the male audience, but also for the rest of the audience, many of whom are presumed to have no interest in purchasing cars. For every one thousand people whose time is bought by advertisers on prime time, then, there is much wasted watching by irrelevant viewers. Specification and fractionation of the audience leads to a form of concentrated viewing by the audience in which there is (from the point of view of advertisers) little wasted watching. Because that advertising time can be sold at a higher rate by the media, we can say that the audience organized in this manner watches harder and with more intensity and efficiency. In fact, because the value of the time goes up, necessary watching-time decreases, and surplus watching-time increases, thus leading to greater surplus value.

The other major way in which relative surplus value operates in the media is through the division of time. Whereas the concern with demographics reorganizes the watching population, the concern with time division reorganizes the watching process. This involves a redivision of the limited time available to increase the ratio between necessary and surplus time (Bergreen 1980, 289). The major way to accomplish this is to move toward shorter commercials, and, indeed, over the last twenty-five years, the number of nonprogram elements has dramatically skyrocketed, although the absolute amount of advertising time has increased only by 2.5 minutes per hour. In 1965, the three major networks showed an average of 1,839 ads per week. The figure rose to 2,200 in 1970, to 3,487 in 1975, to 4,636 in 1980, and to 4,997 in 1983 (*Television/Radio Age*, June 1985). Today, the thirty-second commercial predominates, although there are a great number of fifteen-second commercials also. They comprised 6.5 percent of all network ads in 1985, and it is estimated that, in 1986, this figure will climb to 18 percent (*Fortune*, December 23, 1985).

The basic economic logic works as follows: Assume there are five thirty-second commercials in a commercial break. If each sells for $100,000, income to the network is $500,000. To increase the revenue derived from this time, the network divides it into ten fifteen-second slots offered to advertisers for $60,000 each. If there were enough demand to sell these spots, the income to the network would be $600,000 instead of the previous $500,000.

But why would advertisers agree to this price hike? After all, they are now paying more per second, although less per spot. Advertisers, however, are not concerned about the value of time but about the frequency with which the market can he reached. The shorter spots give them twice the number of ads without raising the price by a proportionate amount. And, indeed, advertisers believe that a combination of thirty-second and fifteen-second versions of the same ad works well in conveying almost the same information. If the program price remains the same, viewers will have to watch for less necessary time to cover its cost. We must emphasize that the time of audiences is the key to the process by which networks valorize the time they control. It is also the limits of human perception (that is, the limits to watching) that guide the division of time. Advertisers may be able to construct beautifully crafted ten-second commercials, but these are useless if they do not work on the audience in that short time. "If we can demonstrate that the American consuming public can absorb and act on a 15-second unity, can the 7.5-second commercial be far behind?" (*Fortune*, December 23, 1975). Human watching, listening, perceiving, and learning activities act as a constraint to the system.

Watching and Labor

Our use so far of the familiar concepts of Marxian economic theory to analyze the valorization of time by the networks has been a pointed one. Central to the whole paradigm of Marxian economics is the notion that human labor—not capital or technology—is the basis of the productivity of societies. Similarly, in the analysis of broadcasting economics, it is audience watching that is vital to the whole process. In a very real sense, we can see that there are many similarities between industrial labor and watching activity. In fact, watching is a form of labor.

Again, this relationship should be seen as both metaphorical and real. Watching is a real extension of the logic of industrial labor, even if it is not the same as industrial labor. However, as metaphor, it illuminates the obscure workings of the economy in general. As Ricoeur writes, "metaphor is the rhetorical process by which discourse unleashes the power that certain fictions have to redescribe reality" (1977, 7).

Watching as metaphor reflects the dynamic of the capitalist economy. In Marx's analysis of the work day, the productivity of capitalism is based upon the purchase of one key commodity—labor-power. This is the only element in the means of production that produces more value than it takes to reproduce itself. Like all commodities, it has a value, a cost—the cost of its production (or reproduction). The cost of labor-power (the capacity to labor) is the cost of the socially determined level of the means of subsistence: that is, what it costs to ensure that the laborer can live and be fit for work the next day. The amount of labor-time that it takes to produce value equivalent to this minimum cost is labeled by Marx as socially necessary labor (necessary to reproduce labor-power). Socially necessary labor-time produces value that is equivalent to wages. The remaining labor-time is labeled as surplus labor-time, through which surplus value is generated. In the nonwork part of the day, workers spend wages (on shelter, food, children, etc.) that will ensure that they will be fit and healthy enough to go to work. During nonwork time, they thus reproduce their labor-power.

How is this process reflected as metaphor within the broadcast media? The network owns the means of production—communication—which makes possible the production of commodities and gives the network ownership of those commodities. While workers sell labor-power to capitalists, audiences sell watching-power to media owners; as the use-value of labor-power is labor, so the use-value of watching-power is watching, the capacity to watch. In addition, as the value of labor-power is fixed at the socially determined level of the means of subsistence (thus ensuring that labor-power will be reproduced), so the value of watching-power is the cost of its reproduction—the cost of programming, which ensures that viewers will watch and be in a position to watch extra (the time of advertising). In this formulation, it is only the time of advertising that comprises the so-called work day for the audience. The programming, the value of watching-power, is the wage of the audience, the variable capital of the communications industry. It is also time for the reproduction of watching-power, the time of consumption, the time of nonwork. As the work day is split into two, so the work part of the viewing day—advertising time—is split between socially necessary watching-time and surplus watching-time.

For instance, the early history of industrial capitalism is tied up with attempts by capital to extend the time of the working day in an absolute sense, thus manipulating the ratio between necessary time and surplus time. Within the development of the commercial media system, this phase is represented by broadcasting from the late 1920s to the early 1960s. In the first years of commercial broadcasting (extending into the 1930s), broadcasters struggled to persuade advertisers to sponsor shows. The more shows were sponsored, the more audiences could be sold to advertisers. This was an extension in the amount of time that people watched and listened for capital. It also has to be remembered that, until the introduction of spot selling in the 1960s, programs were advertising agency creations, with the sponsor's name and product appearing everywhere (not only in ads) (Bergreen 1980).

However, as Marx realized, this absolute extension of the working day cannot go on indefinitely. Unions and collective bargaining limited the length of the working day, forcing capital to increase the intensity of labor. The concept of relative surplus value initially meant the cheapening of consumer goods that reproduce labor-power, so that the amount of necessary time would be decreased. In the era of monopoly capitalism, two other major factors contribute to the extraction of relative surplus value—the reorganization of the workplace and the introduction of technologically efficient instruments of production As Marx writes, "The production of absolute surplus value turns exclusively upon the length of the working day: the production of relative surplus value revolutionizes out and out the technical processes of labor and the composition of society" (1976, 645). We have already referred to the stress on demographics (reorganization of the working population) and the redivision of time (reorganization of the work process). Watching and labor, then, display many historical similarities in the movement between absolute and relative surplus value.

From the Formal to the Real Subsumption of Watching

In a very important text published in English for the first time in 1976, Marx distinguishes between the formal and the real subsumption of labor (1976, 943–1085). As capitalist relations of production expand, they come into contact with other types of relations of production—for example, feudal relations in agriculture. Capitalism does not effect a change in these other relations but merely "tacks them on" to its own operations: "capital subsumes the labor process as it finds it, that is to say, it takes over an existing labor process, developed by different and more archaic modes of production" (1976, 1021). Thus, while capital subsumes the process, it does not establish specifically capitalist relations of production in that sphere; it does not need to. The old relations are used in ways that benefit capital without being organized under its relations of production. Marx argues that the formal subsumption of labor is based upon increasing the length of the working day: that is, on absolute surplus value.

In broadcasting, the formal subsumption of watching activity is linked to the period when advertisers had direct control of programming (when they wrote and produced it). Broadcasting did not develop initially as an advertising medium; its first purpose was to aid in the selling of radio sets. Only later was time on the airwaves sold by AT&T to bring in additional revenue. Even when advertising became prominent in the late 1920s and 1930s, networks did little more than lease facilities and sell air time to advertisers who had total control of broadcasting. Thus, capital (advertisers) took over more archaic modes of watching for their own ends. Advertisers were interested primarily in the activities of the audience as it related to the consumption of their products. Watching here was tacked on to specifically capitalist relations of production without being organized in the same manner.

But the two different relations of production cannot exist side by side indefinitely. Indeed, capitalism constantly works to wither away the other mode of production and to introduce capitalist relations of production into that domain. This is labeled as the real subsumption of labor. At this stage (which corresponds to the extraction of relative surplus value), "the entire real form of production is altered and a specifically capitalist form of production comes into being (at the technological level too)" (Marx 1976, 1024). The archaic forms of production are replaced with capitalist relations of production. The old realm is no longer directly subordinate to other domains but itself becomes a proper capitalist enterprise interested primarily in its own productivity rather than being peripheral (yet vital) to something else.

By the late 1950s, it was proving inefficient (for the network) to have the audience watch exclusively for one advertiser for thirty to sixty minutes. The media could generate more revenue for themselves if they could reorganize the time of watching by rationalizing their program schedule. The move to spot selling was an attempt to increase the ratio of necessary to surplus watching-time. There was a limit to how much one advertiser could pay for a thirty- or sixty-minute program. If the networks could control the programming and the advertising time within it, then they could generate more revenue (by selling spots) from multiple advertisers, all of whom individually paid less. Initially, advertisers resisted this rationalization and the subsumption of their individual interests under the general interests of media capital. In the end, however, rising program costs, legal objections to advertisers' control, and scandals drove the networks to move toward full control of their schedules (Barnouw 1978). This resulted in the double reorganization of the watching population and the watching process under specifically capitalist relations of production.

Alienated Watching

There is another dimension along which watching-labor shares characteristics with labor in the economy in general—both are viewed as unpleasant by the people who have to perform either activity. The history of working-class resistance to the process of wage labor and various sociological studies illustrate that, for many people in modern society, work is not an enjoyable activity. People, on the whole, work not because they like their jobs but because they have to work. Work has become a means to an end rather than an end in itself; labor is a form of alienated activity.

Similarly, consider the attitudes of the watching audience to the time of advertising. Despite the fact that huge amounts of money (much more than on programming) are spent on producing attractive commercials, people do all they can to avoid them. Data indicate that almost 30 percent of viewers simply leave the room or attend to alternative technologies during the commercial breaks (Fiber 1984). They also simply switch channels in the hope that they can find another program to watch rather than more ads. (Switching between the major networks is rather unproductive on this score, as they all tend to have their commercial breaks at the same time.) Indeed, a 1984 report by the J. Walter Thompson advertising company estimated that, by 1989, only 55 to 60 percent of television audiences will remain tuned in during the commercial break. Commercial viewing levels are decreasing. The remote channel changer is a major factor in this so-called zapping of commercials, as is the spread of video cassette recorders (VCRs). When programs are recorded to be watched at a later time, one can simply skip over the commercials by fast-forwarding through them. The owners of the means of communication are faced here with a curious problem—the audience could watch programs (get paid) without doing the work (watching commercials) that produces value and surplus value.

These findings have not been lost on the advertising or television industries, who have increasingly recognized that the traditional concept of a ratings point may no longer be valid. Ratings measure program watching rather than commercial watching. Indeed, it seems that there is much disparity between the two, and advertisers are starting to voice their discontent at having to pay for viewers who may not be watching their ads at all. This has led the ratings companies to experiment with new measures of the audience. The most intriguing development is the people meters, a device on the TV with a separate button for each participating household member. Individuals punch in when they start watching and punch out when they stop, providing advertisers and broadcasters with a more precise measure of the level of commercial viewing. There could be no clearer indication of the similarities between watching and labor. Just as workers in a factory punch in and punch out, so too will viewers be evaluated along similar lines.

It is instructive to note that no one would be worried if people were zapping the programs and watching ads in greater numbers; the industry would be undisturbed. But when the new technologies of cable and VCRs threaten the viewing patterns of commercial time, then the very foundations of the broadcasting industry begin to shake in anticipation of the consequences.

Although we have pointed out many similarities between watching and labor, we do not regard them as identical activities. For instance, watching has no formal contract for the exchange of watching-power, and there can be no enforcement of the informal contract. We have sought to identify the broad dynamic through which watching activity is brought into the realm of the economic and the manner in which watching activity, under the conditions of advanced capitalist production, reflects in a spectacular way the workings of the real.

Narrowcasting and Blurring

Since the continued spread of cable in the 1970s and 1980s, there has been a very dramatic shift in viewing patterns in the United States. In 1975 and 1976, according to Nielsen figures, the three major networks commanded among them 89 percent of the watching audience during prime time (*New York Times*, October 16, 1985). By 1985, that figure was down to 73 percent. This does not mean that people are watching less television; indeed, by 1984, the average family viewed an all-time high of almost fifty hours a week. People thus are watching more TV and less of the networks. The extra viewing has been diverted largely into offerings available on cable television. Those homes with access only to regular over-the-air broadcast television watched only forty-two hours and twenty-two minutes a week in 1985, while those with cable and subscription services watched almost fifty-eight hours a week. Clearly, cable television (based upon narrowcasting to specific audiences) increases the total amount of time that people watch television. While some of this extra watching goes to pay TV services (without advertising), much of it is still bound up with commercially sponsored programs. Narrowcasting, then, also increases absolute surplus value.

Up until now, we have made a rather strict distinction between programming and advertising. In the historical development of the commercial media system, however, the boundaries between the two were very often blurred. The function of programming is much more than merely capturing the watching activity of a specific demographic group of the market. Programming also has to provide the right environment for the advertising that will be inserted within it. Advertisers seek compatible programming vehicles that stress the lifestyles of consumption. Thus, in the 1950s, the very popular and critically acclaimed anthology series were dropped by the networks, because they focused on working-class settings and complex psychological states, neither of which was conducive to the advertisers' needs for glamorous consumer-oriented lifestyles and the instant and simple fixes offered by their commodities to the problems of modern living. The anthology series were replaced by programs that were much more suited to the selling needs of advertisers. Furthermore, actors and stars moved easily between programs and commercials. At a more explicit level, advertisers sought to have their products placed within the program itself. In all of these ways, we can see a blurring between the message content of the commercials and the message content of the programming.

Although many writers (see Barnouw 1978) have commented on this phenomenon, they have not noticed how this blurring is enormously intensified by the move to narrowcasting. In each portion of the fractionated audience, from the point of view of the message content the difference between the program content and the ad content constantly diminishes. Both ads and programs draw upon the specific audience to construct their message code. The drawback of the mass audience for broadcasting is usually thought to be that the program may attract a mass audience without necessarily attracting a mass market for certain commodities; hence, the importance of demographics for advertisers. But the problem has not usually been perceived within the sphere of watching itself. Broadcasting produces only a loose compatibility between programs and commercials. Broadcasting limits blurring while narrowcasting overcomes these limits.

Although there may appear to be a formal symmetry at work, within the sociomaterial conditions of the media the first aspect dominates the second. The commercial form is the dynamic element in the process. If part of the program is really an ad, then part of the program time is not really consumption-time: rather, it is labor-time, and the length of the working day has been extended. The program as the extension of the ad shows us the increase in the magnitude of labor-time of watching. It contributes to absolute surplus value. Within narrowcasting, the progressive fractionation of the audience intensifies both absolute and relative surplus value. From such a viewpoint, there appear to be two media revolutions. The first—broadcasting—converts non-work leisure time into the sphere of watching-time (both consumption watching-time and labor watching-time). The second—narrowcasting—converts consumption watching-time (programming) into labor watching-time (ads). Because there is a limit to advertising time, media have to gain more surplus from the existing time. Blurring accomplishes this by converting program into ad, by converting consumption watching-time into labor watching-time. While this process is observable within broadcasting, it is greatly intensified by narrowcasting.

Televised sports are one such example of blurring. Values of masculinity and fraternity are present in both ads and programs; sports personalities flit between the two. The sponsorship of televised sporting events (e.g., the Volvo Grand Prix of Tennis, the AT&T Championships) is also an attempt to convert program into ad.

Within the field of broadcasting itself, segmented programming leads to blurring. For instance, in 1983, Action for Children's Television petitioned the Federal Communications Commission to recognize certain Saturday morning children's programs—those that featured toys successfully sold in stores as their primary characters—for what they really were: thirty-minute-long commercials.

The formation of the MTV cable network highlights this movement most dramatically. On MTV, the entire twenty-four-hour viewing day is advertising time—between the ads for commodities in general are placed ads for record albums (rock videos—the so-called programming). Objectively, all time on MTV is commercial time. Subjectively, also, it is very difficult to distinguish between ad and program. The same directors, actors, dancers, artists, etc., move between videos and ads until the lines between the two blur and disappear (Jhally 1987).

The best example of the blurring under consideration is that of the commercial on the commercial. The Commercial Show, a cable program in Manhattan, "consists of old commercials; advertisers can buy time to put new commercials between the old ones" (*Wall Street Journal*, February 4, 1982). The blurring here is so complete that it shows dramatically the difference between consumption-time and labor-time. There is no better example of the fact that the same kind of message has two fundamentally different functions. One could hardly find a better reason to abandon a message-based definition of the messages themselves.

Conclusion

We believe that this theory explains in a unified way a number of recent developments: the move to shorter commercials, the stress on demographics, the evolution of narrowcasting, the creation of new ratings measures. Looking farther afield, this framework also lends an explanation to the movements in Western Europe toward commercial broadcasting in which the watching activities of the audience can be more fully integrated within a productive sphere.

Moreover, this framework allows us to understand the basic division in the message system between programs and ads. Many writers have commented that ads are much better constructed than the bulk of network programming. Barnouw (1978) believes that ads are a new American art form; it does indeed seem that the artistic talent of our society is concentrated there. On the average, ads cost eight times as much to produce as programs. Why should this be the case? Our theoretical framework provides an answer: The reason ads are technically so good and programs are generally so poor is that they have a different status within the communications commodity system. Programs are messages that have to be sold to consumers—they are, in fact, consumer goods. Like most consumer items in the modern marketplace, they are products of a mass production system based upon uniformity and are generally of a poor quality. Program messages, like consumer goods in general, are designed for instant, superficial gratification and long-term disappointment that ensures a return trip to the marketplace. They are produced as cheaply as possible for a mass audience.

In contrast, we can label commercials capital goods—they are used by the owners of the means of production to try to stimulate demand for particular branded commodities. Like machines in a factory (and unlike consumer goods), they are not meant to break down after a certain period of time. Although the objects of their attempted persuasion are consumers, commercials are not sold to consumers, and consumers do not buy them (as far as we know, people do not tune in to television to watch commercials as a first priority).

As with other objects used by capital, no expense is spared in producing the best possible good. Also like capital goods, commercials are tax deductible. During programming time (consumption watching-time), audiences create meaning for themselves. During commercial time (labor watching-time), audiences create meaning for capital. It is little wonder that commercials whose function is to communicate (and not just get attention) should be the "best things on television" and the only part of television to have realized the potentials of the medium.

The main contention of this article has been that the activity of watching through the commercial media system is subject to the same process of valorization as labor-time in the economy in general. This is not to suggest that they can be identified as exactly the same type of activity for, clearly, they produce different types of commodities. Factory labor produces a material object, whereas watching activity does not. However, the modern evolution of the mass media under capitalism is governed by the appropriation of surplus human activity. The development of this appropriation is a higher stage in the development of the value form of capital. Its logic reproduces the logic described by Marx for the earlier form, but its concrete form is, in fact, a new stage: the value form of human activity itself. The empirical reflection of this is that the process of consciousness becomes valorized. There is thus a partial truth in the label attached by Enzenberger and others to the modern mass media as consciousness industry—except that they have so far conceptualized it upside-down. Mass media are not characterized primarily by what they put into the audience (messages) but by what they take out (value).

Notes

1 Sut Jhally (with Bill Livant), "Watching as Working: The Valorization of Audience Consciousness," was published in *The Spectacle of Accumulation: Essays in Culture, Media, and Politics*, 2006, New York: Peter Lang. Reprinted with the permission of the author and Peter Lang.

PART TWO

SOCIAL MEDIA, AUDIENCE MANUFACTURE, AND THE WORK OF SELF-MARKETING

CHAPTER TEN

The Institutionally Effective Audience in Flux: Social Media and the Reassessment of the Audience Commodity

Philip M. Napoli [1]
Rutgers University

As Eileen Meehan (1984) demonstrated in her classic analysis of the political economy of the media, the pursuit of audiences is really the pursuit of ratings—those statistical representations of audience exposure to content (and advertisements) produced by third party audience measurement firms. And so, fully understanding the audience commodity means understanding how these ratings are produced; what assumptions about the audience are inherent in their production and usage; what dimensions of audience behavior they seek to capture; and what dimensions of audience behavior they neglect (see Ang 1991; Bermejo 2007; Napoli 2003a, 2011).

When we examine the audience commodity through this lens of the organizations and systems that quantify audience behavior so that it can be bought and sold in the audience marketplace, we must pay particular attention to the implications of any changes in, or alternative approaches to, how audience behavior is measured and monetized. Such changes can dramatically reconfigure the competitive dynamics and the allocation of advertising dollars, within and across media sectors; and as a result can dramatically affect cultural production in ways that can have meaningful social, political, and economic consequences (see, e.g., Napoli 2005, 2009).

Today, television audience measurement and valuation is in a state of flux. Audiences are becoming increasingly fragmented across a growing range of delivery platforms. Expansive bandwidth and diminished barriers to entry associated with many of these platforms mean that the range of content options available to the viewer has increased dramatically as well. In this new media environment, the traditional business model of selling audiences on the basis of their size and demographic characteristics has become more difficult to implement, particularly for those content providers residing deep in the lengthening (and difficult to measure) "long tail" (Anderson 2006) of the distribution of audience attention (Napoli 2011).

However, many of these new platforms that contribute to the fragmentation of television audiences also facilitate new ways of gathering information about them. The interactivity inherent in many new media platforms creates a backchannel of audience data that provides participants in the audience marketplace with an unprecedented flow of information about a wide range of dimensions of the audience, some of which are proving to be of value to advertisers.

One of the most prominent aspects of this period of disruption in television audience measurement and valuation is the emergence of social TV analytics. Social TV analytics assess and rank television programs on the basis of the volume and valence of audiences' comments on a wide array of social media platforms. This chapter explores the emergence of social TV analytics in the U.S. as an alternative representation of the audience commodity. It examines the institutional context in which this alternative analytical approach is emerging; as well as the nature of this new conceptualization of the television audience; how it compares with traditional approaches; and how it could impact our understanding of the audience as a commodity.

Institutionally Effective Audiences and their Evolution

Ettema and Whitney (1994, 5) have used the term "institutionally effective audiences" in reference to media audiences as reflected in the practices of media organizations, advertisers, and audience measurement firms. Institutionally effective audiences are those that can be efficiently integrated into the economics and strategy of media industries. Articulating this perspective within the specific context of audience measurement, Meehan (1984, 218) emphasized that commercial audience measurement should not be approached as producing research, but rather as producing "commodities...items manufactured for sale within a business environment." Moreover, as she and subsequent researchers on the audience commodity (see, e.g., Buzzard 2002; Napoli 2003a) have noted, this marketplace possesses a strong orientation towards monopoly, "given the clients' need for a common ground" (Meehan 1984, 222). As a result, there tends to be one uniform commodified audience representation that serves as the coin of exchange in the audience marketplace.

Another key characteristic of the institutionally effective audience is its potential for malleability. A common observation about institutionally effective audiences has been that they are largely a socially constructed phenomenon (see, e.g., Ang 1991). Peter Dahlgren (1998, 307) has gone farther, arguing that "'audiences' are always at least in part discursive constructs, shaped by specific institutional needs and discursive domains." Research has demonstrated how various technological and institutional factors can, over time, produce significant reconfigurations of established constructions of media audiences (see, e.g., Andrews and Napoli 2006; Balnaves and O'Regan 2010; Bermejo 2007, 2009; Bourdon and Meadel 2011; Buzzard 2002).

The point here is that, as institutional needs and interests evolve, and/or as certain aspects of the technological environment change, the institutionally effective audience can be adjusted to better serve the needs of stakeholders in the marketplace (though the interests of different stakeholder groups often conflict). Napoli (2011) has labeled this process audience evolution, arguing that such shifts can only take place when both: a) the dynamics of audiences' media usage are changing in ways that undermine traditional approaches to audience measurement and valuation; and b) affordable and acceptable new analytical approaches are available. The contemporary television industry appears to be illustrative of this set of conditions, and thus represents an interesting window into the institutional dynamics surrounding a potential reconfiguration of the audience commodity.

Methodology

In addressing this issue, this study analyzed data obtained via multiple methods. First, participant-observation was conducted at multiple industry convenings in which the issue of the measurement and valuation of media audiences was a central concern. These included the Advertising Research Foundation's Audience Measurement 6.0 conference (June 2011), the Online Media, Marketing, and Advertising Metrics and Research Conference (February 2012), the AdMonsters OPS TV Conference (July 2012), and the June 2012 meeting of the Collaborative Alliance, a consortium of media, advertising, and audience measurement organizations that explores and assesses the potential of new audience measurement systems. Other participant-observation events included private meetings between media and advertising industry executives and representatives of three of the leading new social TV analytics providers; two audience measurement and advertising industry association meetings in which the author was an invited participant; and three online training sessions on three of the leading social TV analytics platforms. These events (all of which took place in New York

City), and the presentations, discussions (and arguments) that took place at them provide a useful window into how the television audience is being conceptualized, reassessed, negotiated, and valued by various industry stakeholders.

Next, textual analysis was conducted on relevant industry trade publications, corporate marketing and promotional materials (including online videos), presentations, reports, and white papers that addressed the topic of evolving approaches to measuring and valuing television audiences. Trade publications incorporated into this analysis included *Advertising Age*, *AdWeek*, *MediaWeek*, *Broadcasting & Cable*, *Media Life Magazine*, *MediaPost*, and *paidContent*. Because this analysis focused on current developments, the gathering and analysis of this material was targeted at the time period from November 2010 through October 2012. These data are useful for both establishing the range of stakeholders and technological developments relevant for this analysis and for gaining insights into the key issues and controversies confronting the various industry sectors.

In addition, 21 semi-structured interviews were conducted with professionals in the television, advertising, and audience measurement industries. These interviews ranged from 30 to 90 minutes; and were conducted either in-person or over the phone. Interview subjects were obtained via snowball sampling that originated from the author's personal and professional network of contacts, and from initial outreach to key organizations relevant to the subject matter of this study. Interview subjects held positions in a wide range of areas, including program development, ad sales, research, and information technology. Interview subjects came from a variety of organizations, including broadcast networks, cable networks, measurement firms, advertising/media buying agencies, and industry associations. Years of experience ranged from two years to over 35 years, with titles ranging from Research Analyst to Executive Vice President. All interview subjects participated with the understanding that their identities would be kept anonymous in order to maximize the extent to which they could provide candid responses to the interview questions. Subjects were asked general questions about the state of audience measurement and the audience marketplace, as well as more targeted questions about emerging analytical approaches.

Finally, a limited amount of data was obtained from three of the leading social TV analytics providers (Bluefin Labs, Trendrr.tv, and General Sentiment). Specifically, these firms provided ranked lists of the top 25 regularly scheduled, primetime programs for a sample week (March 5–11, 2012),[2] in order to facilitate comparisons across service providers. Separate ranked lists were provided for both overall volume/quantity and for sentiment (i.e., ratio of positive to negative comments). The unit of analysis for these data was the individual program, as opposed to the individual telecast (given that some programs are broadcast multiple times in a given week).

Social TV Analytics in Context: The Evolving Television Audience Marketplace

The realm of television audience measurement was described by one interview subject as undergoing "more change in the past five years than has probably happened in the entire history of audience measurement dating back to the 1950s." The state of affairs today involves two parallel development tracks. The first involves a wide range of efforts to strengthen and preserve what will be referred to here as the traditional approach to measuring and valuing television audiences on the basis of their exposure to programs and advertisements. The second involves efforts to go beyond audiences' exposure and quantify (and valuate) some dimension of the audience's engagement with the content, under the presumption that there is additional, untapped value embedded in the extent to which audiences are engaged in the content they consume.

Rehabilitating the Status Quo: Efforts to Preserve the Measurement and Valuation of Exposure

The traditional approach to measuring and valuing television audience, in which a relatively small, (presumably) representative sample of television households has their exposure to television programs measured either via electronic People Meters attached to individual television sets, or via paper diaries, has come under assault from a variety of fronts, due to the perceived inability of this approach to keep pace with the increasingly complex, increasingly fragmented television viewing environment (see, e.g., Abzug 2011). However, industry stakeholders have implemented a variety of initiatives in an effort to preserve the traditional exposure model.

Consider, for instance, the impact that the DVR has had upon audiences' exposure to television advertisements. According to recent estimates, DVR usage is leading to roughly 12 percent of commercial exposures being skipped (Fitch 2011). Obviously, such behaviors have diminished the value of the traditional 30-second commercial spot and, more importantly, highlighted a shortcoming of traditional television ratings, which never distinguished between program and commercial viewership levels.

The television, advertising, and audience measurement industries responded to the disruptive effects of the DVR by transitioning in 2007 to the C3 currency. Under this approach, advertisers pay only for those viewers exposed to the program's commercials within the first three days of the program's initial airing. This approach was the compromise that arose from contentious negotiations between national programmers and advertisers over how to incorporate DVR usage and commercial avoidance into the valuation of television audiences (see Napoli 2011).

Perhaps even more important than the DVR is the ongoing process of the fragmentation of the television audience, both within and across viewing platforms. This process creates a range of new technological contexts in which audience behavior must be measured. It also segments the audience into ever-smaller, and thus more difficult to measure, groupings (for details, see Napoli 2011).

As a result, a variety of initiatives have been launched to try to capture "cross-platform" exposure to programming. Nielsen recently announced its new "cross-platform campaign ratings," which allow for the calculation of a single ratings number for a television program, derived from audience exposure on TV and online (Stelter 2012). NBC collaborated with Google, comScore (NBCUniversal 2012), and radio audience measurement firm Arbitron (Mandese 2012a) to measure cross-platform viewership for the summer, 2012 Olympics. And, most recently, comScore and Arbitron announced a partnership to launch a "five platform" ratings system that will provide unified audience ratings data across television, radio, computers, tablets, and smart phones (Mandese 2012b).

Another significant alternative methodological approach to preserving the measurement and valuation of audience exposure that is gaining traction is the use of digital cable and satellite set-top boxes to draw data from much larger numbers of television households than can be achieved via stand-alone systems such as the Nielsen People Meter (Brooks, Gray, and Dennison 2010). Through set-top boxes, it is potentially possible to obtain more accurate and reliable measurement of the networks and programs that extend further into the long tail of audience exposure than can be achieved via traditional People Meter samples. A number of firms, including Nielsen, are developing and offering performance metrics derived from set-top box data (see, e.g., Stilson 2011). The market leader is Rentrak, which offers local and national television ratings derived from a sample of 19 million digital set-top boxes (compared with a Nielsen national People Meter sample of 25,000 households). The set-top box approach is, however, limited in its ability to provide a generalizable representation of the entirety of the television viewing audience, given the persistence of "over-the-air" viewing in approximately 15 percent of TV households. This approach also does not account for the growth of alternative viewing platforms described above.

It is important to emphasize that these efforts all have one important element in common—they focus on providing measures of audiences' exposure to television programs; and in that regard are reflective of the basic dimension of audience behavior that has served as the primary criterion for audience value in the electronic media since the 1930s (Napoli 2011). These efforts have emerged in response to the fact that an established approach to measuring and valuing audiences on the basis of demographically-sorted exposure metrics derived from household samples has entered into a period of perhaps unprecedented uncertainty and criticism. However, as the next section will illustrate, this state of disarray has provided an opening for completely different analytical approaches to try to gain traction.

New Media, New Sources of Audience Value: The Rise of Social Media Analytics

The second major transition affecting the measurement and valuation of television audiences involves extending beyond basic audience exposure to programs and instead accounting for how viewers *feel* about the programs (and advertisements) they consume; and how they *respond* to them. The primary terminology that has emerged in the audience marketplace for these aspects of television viewing is "engagement." This is not to say that the marketplace is operating under a single, agreed-upon definition of engagement (indeed, the opposite is, to this point, the case; see Napoli 2011); only that many buyers, sellers, and measurers of audiences seem to agree that audience engagement is a relevant value criterion. And once again, this transition has been driven by the fragmentation that has undermined traditional approaches to measuring and valuing television audiences. As one digital media executive interviewed for this study noted, "engagement is much more of a focus today because of fragmentation." Essentially, as the value of the traditional audience commodity has been undermined, both buyers and sellers of audiences are searching for new sources of value.

Social media have emerged as the primary means by which audience engagement with television programs is being translated into performance metrics that can be used in the assessment and valuation of audiences. Like all sectors of the media industry, the television industry is in the midst of a potentially dramatic transformation at the hands of social media (see, e.g.,

Futurescape 2011a; Proulx and Shepatin 2012). To put the magnitude of this potential transformation in perspective, according to recent estimates, social media-related TV businesses will grow from their current level of $150 billion annually, to $256 billion by 2017 (Friedman 2012b). Social media are seen as a means to enhance the television industry on a variety of fronts, ranging from marketing and promotion (i.e., as a way to improve traditional ratings performance) (Friedman 2012c); to content development (i.e., as a way to innovate in terms of storytelling and to extend programming and associated content into multiple media platforms); as well in terms of providing new forms of audience feedback that can be used in program development and modification (i.e., social media as focus group; see "Twitter on TV" 2012; Marszalek 2012); and—most relevant to this analysis—in the valuation and selling of audiences to advertisers.

In regards to the latter dimension, some analysts have argued that the analytical opportunities provided by social media offer a much-needed corrective to the significant (and growing) limitations to measuring and valuing television audiences according to mere exposure (see, e.g., Abzug 2011; Leavitt 2011; Seles 2010). This perspective is well-reflected in this statement from a broadcast network executive: "We believe that the value of our brand is beyond just the C3 rating. We are now . . . able to qualify, in some way, that the show is much bigger than just the Nielsen measurement" (Tarrant, quoted in Proulx and Shepatin 2012, 115). These discussions reflect long-standing critiques of the limitations of traditional exposure-based approaches to audience measurement (Ang 1991; Lazarsfeld 1947; Meehan 1984).

Beyond Exposure: Exploring Social Media-Based Constructions of the Television Audience

The scenario at hand, then, is one in which traditional approaches to audience measurement and valuation are unusually vulnerable to alternative analytical approaches facilitated by recent technological developments (e.g., social media). This set of conditions raises the possibility of a rare, radical redefinition of the audience commodity.

Social media-based representations of the television audience bear very little resemblance to the television audience ratings that preceded them, both in terms of methodological approach and in terms of their output. They represent an entirely different conceptualization of the audience, one in which exposure represents the starting point, rather than the endpoint, of determining audience value. In the realm of social TV analytics, viewers are of no value unless they engage in some form of online expression about the program to which they were exposed.

Most social media analytics services utilize some form of "web scraping," in which the comments posted on a wide range of social media platforms are aggregated and classified via algorithms. Services such as General Sentiment, Trendrr.tv, Radian6, Crimson Hexagon, and Bluefin Labs operate in ways that are invisible to the social media user, scraping the social media space for relevant postings and activities that can be aggregated and incorporated into snapshots of the relative performance of individual television programs and networks.

It is important to note that there are a number of clear methodological differences that distinguish many of these social TV analytics services from one another. For instance, there are important differences in terms of the online platforms they draw upon to obtain their data. Some focus almost exclusively on Twitter. Others also incorporate platforms such as TV "check-in" services, blogs, and even online news media. Analytics services also differ in terms of the time windows that serve as the basis for their analysis. Thus, for instance, some services focus on analyzing social media activity on a weekly or monthly basis (e.g., Trendrr.tv, General Sentiment). Others seek to provide data on activities on individual days, or even on a minute-by-minute basis. Bluefin Labs, for instance, focuses on identifying online comments with specific moments in individual programs (or advertisements within these programs). SocialGuide aggregates and analyzes a program's mentions one hour before air time through two hours after, in an effort to focus on "linear TV" (Humphrey 2011). Some of these services now also seek to assess the presumed reach of social media comments via the incorporation of metrics such as the Klout scores of all commenters.

Other important differences cannot be so easily distinguished. There are, for instance, no doubt significant differences between the algorithms and coding criteria that the different services utilize to aggregate and categorize the data that they gather. However, as will be discussed in greater detail below, many significant methodological details are treated as proprietary by these measurement firms.

These services seek not only to quantify the volume of online discussion about television programs or networks, but also the valence of that discussion. General Sentiment, for instance, assigns a quantitative score (1–10) (determined by a language processing algorithm) to each comment it aggregates, with the score meant to reflect the extent of the positive or negative orientation of the comment.[3]

In many instances, the providers of these metrics present their analyses using concepts and vocabulary derived from the traditional approach to television audience measurement and valuation. Thus, for instance social media analytics firm Trendrr's television-specific product, Trendrr.tv (2012), provides a "share of voice" metric (echoing the traditional television audience share terminology), which represents the share of all television-focused social media activity that is attributed to a particular network or show for a specified monthly time period. Similarly, Bluefin Labs provides a "response share" metric that indicates an individual program's share of the online television conversation at the time the program aired (Lawler 2011).

In these ways, these services seek to maintain some conceptual connections with the established approaches to measuring and valuing television audiences. Such efforts are no doubt intended to reduce, to the extent possible, the perceived magnitude of the conceptual gulf between the old and new analytical approaches; and of course to maximize the extent to which the new analytical approaches appear compatible with established performance criteria and practices in the television and advertising industries.

Old and New Methods of Audience Representation (and Misrepresentation)

Nonetheless, the scraping of social media conversations to assess the performance of television programs represents a substantial methodological departure from the People Meters and paper viewing-diaries long-employed by Nielsen. Whereas traditional TV ratings require individuals to agree to be part of a relatively small measurement sample that is constructed in an effort to represent the population as a whole, social media metrics draw from the entirety of the online population's accessible expressions of their viewing habits, reactions, and opinions, with no opt-in (or associated incentive) required.

In the case of traditional TV ratings, whether representativeness was achieved was always a question; and as a result, the history of television ratings is rife with indications and accusations of failures to adequately represent the viewing habits of specific demographic groups (see, e.g., Ang 1991; Napoli 2003a, 2005). These issues of representation and misrepresentation re-emerge in the social TV context, though in this case generalizability to the television viewing audience as a whole is not something that these services are claiming—or striving—to achieve. Factors such as different levels of Internet access, general social media usage, and social media usage in relation to television programs across population groups, all definitively undermine any notions of social media conversation reflecting the television viewing population as a whole (see, Napoli 2012b).

Social TV analytics services are also limited in terms of their ability to capture the full scope of conversations happening online (Ampofo 2011). Many of the social TV analytics platforms rely heavily (in some cases, almost exclusively) on data from Twitter. Facebook, in contrast, does not provide access to all postings on its platform (only comments on public Facebook pages are accessible). Thus, as one network researcher noted, if the user base of Twitter does not represent your network's, or your program's, viewers, then the utility of many of these social analytics platforms for analyzing and monetizing your audience is dramatically diminished. One recent critique of social media analytics as a television audience measurement tool argues that the social media analytics firms face the same kinds of challenges in capturing data in an increasingly fragmented social media environment that have confronted Nielsen in its efforts to capture increasingly fragmented television viewing (Hussey 2012).

It is important to emphasize that the buyers and sellers of audiences interviewed for this study widely acknowledged and recognized the variety of ways in which social TV analytics do not provide a generalizable representation of the television viewing population as a whole; and as a result are limiting the extent to which these data are influencing their decision-making. However, even ongoing efforts to maintain the viability of traditional exposure metrics, such as the use of set-top boxes, are employing efforts that fall short of basic criteria for generalizability to the television viewing population as a whole (see above).

It seems reasonable to suggest that, at this point, generalizability (whether actual or claimed) is of diminishing importance to marketplace participants assessing representations of the media audience. The technological environment may simply be proving too complex for advertisers, media buyers, and content providers to continue to hold measurement firms to such a standard. At the same time, new technologies are facilitating the gathering of unprecedented quantities of data that are not necessarily generalizable to the population as a whole. The situation may be one of the perceived benefits of massive quantities of data eclipsing the perceived benefits of smaller quantities of data derived from more representative samples. Marketplace participants appear somewhat willing to operate from representations of the audience that clearly exclude certain audience segments.

The Many Social TV Audiences: Differentiation Across Competing Audience Representations

The current social TV analytics environment is one in which there are roughly half a dozen different social TV analytics firms offering competing services, though the predicted consolidation in this sector (see Napoli 2012a), has recently begun with Nielsen acquiring the measurement firm SocialGuide, and Twitter (already in a measurement partnership with Nielsen) acquiring Bluefin Labs and Trendrr (Friedman 2012a; Stelter 2013; Goel 2013). With so many providers of social media analytics serving the television industry either exclusively or as part of their broader business model, all of whom are employing different methodologies, algorithms, and analytical frameworks (see above), one industry observer has rightly asked: "Is there going to be one central hub for these experiences to play out or is it going to be sort of like a Tower of Babel that provides thousands of apps?" (Wallenstein, quoted in Proulx and Shepatin 2012, 100).

Reflecting these methodological differences, the social TV analytics data obtained for this study (described in the Methodology section) suggest that the different social TV analytics platforms are providing very different representations of the television audience. The following discussion reviews an exploratory analysis of the level of agreement across three of the major providers (and with Nielsen rankings) in terms of the most successful television programs, according to the primary performance metrics offered by these services. This analysis is fairly limited in scope, given the limited data access that was obtained. There are, however, some important implications in the patterns that were discovered.

Table 1 presents the percentage of overlapping programs in the top 25 program lists (in terms of total volume of online comments) between three of the leading social TV analytics providers and between Nielsen household and age 12–34 ratings for the week of March 5–11, 2012. As the table indicates, the degree of program overlap between the three social TV analytics providers ranges from 36 percent to 40 percent. The degree to which each of these top 25 lists overlaps with Nielsen's top 25 in terms of household ratings ranges from 16 percent to 32 percent. The degree to which the social TV analytics' top 25 lists overlap with Nielsen's 12–34 top 25 ranges from 32 percent to 40 percent. It is important to note that these computations do not take into account the degree of agreement across lists in terms of the placement of programs *within* the list (i.e., whether programs share the same rankings across lists); only whether individual programs are common to both lists, regardless of their placement.

Table 1: Top 25 Program Overlap (Volume of Social Media Conversation)

	Bluefin	Gen. Sent.	Trendrr	Nielsen HH	Nielsen 12–34
Bluefin	-	40%	40%	32%	32%
Gen. Sent.	-	-	36%	28%	40%
Trendrr	-	-	-	16%	32%

Table 2 presents the same comparisons, with the total volume of online comments replaced with the valence of online comments. These top 25 lists represent the programs with the highest proportion of positive comments. Comparing these rankings to Nielsen rankings, we see some strong differences between the providers. Bluefin's rankings have virtually no overlap with the Nielsen rankings. General Sentiment's overlap is at 12 (HH) and 16 (12–34) percent. The strongest overlap can be found with Trendrr, at 16 (HH) and 32 (12–34) percent. In terms of the comparisons across the three social TV analytics providers, two of the three have no overlap in their top 25 programs. In the third comparison (between General Sentiment and Trendrr), the level overlap is 20 percent.

Table 2: Top 25 Program Overlap (Valence of Social Media Conversation)

	Bluefin	Gen. Sent.	Trendrr	Nielsen HH	Nielsen 12–34
Bluefin	-	0%	0%	4%	0%
Gen. Sent.	-	-	20%	12%	16%
Trendrr	-	-	-	28%	36%

Interpreting these results is inherently subjective. However, the generally low levels of agreement between the social TV analytics providers and the Nielsen ratings would seem to confirm, to some extent, that audience exposure and audience engagement (at least as measured by social TV analytics) are markedly different phenomena. And, as the tables indicate, the degree of overlap between the most appreciated and the most popular programs is even less than the degree of overlap between the most discussed and the most popular programs. What constitutes a "success" in the social media space does not appear to correspond with what constitutes a success in traditional television ratings (though this is an issue that should, ideally, be investigated in more detail with larger data sets).

It also seems appropriate to interpret the level of agreement across the three social TV analytics' representations of the television audience as highly divergent (particularly in regards to analyzing/reporting valence, where agreement across the services was practically non-existent). This conclusion seems reasonable when we consider that this analysis focused on the 25 most popular/successful programs as measured via social media mentions, and that previous comparative analyses of similar audience measurement systems have shown that the general pattern tends to be one wherein the highest levels of agreement are found amongst the most successful content options (see, e.g., Lo and Sedhain 2006). This pattern reflects the fact that the most successful content options are often so successful that their positions are able to withstand the methodological differences across services; whereas with less popular content the methodological differences are likely to produce greater variability. This analysis focused exclusively on the "fat head" (i.e., the most successful programs) rather than the "long tail," and so essentially analyzed a small subset of programs with the greatest likelihood of overlap across services, and still the levels of agreement across services were relatively low. Consider also that, within these comparisons, there was no agreement across the three providers in terms of the #1 program of the week (in terms of either volume or valence).

In the end, when it comes to their depictions of the television audience, it appears that these social TV analytics services are almost as different from each other as they are from traditional Nielsen ratings. Despite the fact that these firms all are providing measures of audience engagement with television programs, as represented by the volume and valence of online conversation, the methodological differences across these providers are apparently so substantial that their outputs provide disparate representations of this phenomenon. That these disparities are particularly pronounced within the context of the valence of online comments is likely a reflection of the greater subjectivity and methodological complexity inherent in quantifying such a construct. And so, the audience commodity as represented by social TV analytics remains, at this point, highly fractured, and open to multiple, contradictory, interpretations.

These conflicting representations run counter to the very notion of the audience as a commodity. A true commodity is a product that is essentially interchangeable with other products of the same type; and as a result competition is based exclusively on price. One might argue, then, that the audience as represented by social TV analytics has not adhered around a true commoditized form; and given the methodological differences that characterize the competing providers, it is not likely to do so. Until there is some homogenization of methodological approaches across suppliers; or—more likely—the consolidation that will inevitably affect this industry sector, the television audience as represented by social TV analytics will remain a highly fractured and inconsistent construct.

Social TV Analytics Usage and Resistance

This final section addresses the issue of if and how social media-based representations of the television audience are being integrated into the audience marketplace. As with all aspects of social media today, it is vitally important to attempt to separate hype from reality. *Advertising Age* has gone so far as to ask, "Will social media be the new Nielsen for TV ad buyers?" (Patel 2011). Another recent trade press article asserted that "the amount of social media buzz that a show generates can be almost as important as whether it's gaining in ratings" (Vasquez 2011, 1). Reflecting this perspective, one recent industry report claimed that "Facebook and Twitter are TV's new power brokers" (Futurescape 2011b). And, in an interview with the New York Times, American Idol producer and star Simon Cowell described people who actively use Twitter and Facebook as "the only powerful people now on TV" (Stelter 2011, 1).

Resistance

The reality, not surprisingly, appears to be something not quite this extreme. The fieldwork at industry convenings noted a persistent skepticism amongst industry stakeholders toward social TV analytics, even (somewhat paradoxically) as these same stakeholders discussed integrating these services into their day-to-day activities. A number of interview subjects noted a lack of trust within their organizations in the findings that are emerging at this point from social TV analytics. As one television audience research executive noted, "People are afraid of what it's going to do."

Some of this skepticism emanates from uncertainty over how these social TV analytics firms generate their results. No doubt due in part to the current high level of competition in this area, many significant methodological details are withheld. Consequently, a number of interviewees described the new social TV analytics resources available to them as "black boxes" (for a discussion of the "black box" terminology, see MacKenzie 2005). One cable network executive described a social TV analytics tool as "a crock. They never tell you what's in the black box. There's no transparency with them." Another television executive, however, preferred to avoid the black box terminology, noting that the issue was less about transparency, but more about complexity, given that the nature of these new measurement systems requires training and expertise that do not overlap with the skill sets of traditional media research personnel. Of course, whether the cause is inadequate transparency or extreme complexity, the result is the same—an inability to fully understand the operation of a system/tool that is being utilized.

From this standpoint, it is perhaps not surprising that there are a number of ongoing efforts to broaden the marketplace's analytical focus beyond traditional exposure criteria that eschew social media analytics. These approaches instead rely on more traditional research methods (surveys, interviews, etc.) in an effort to tap into some of the same concepts (emotional attachment, word of mouth, etc.) that are being addressed by social media analytics (see, e.g., Crupi 2012; Friedman 2012d).

The emergent social TV analytics must also confront the inevitable entrenchment of established analytical approaches within the contemporary advertising, programming, and media buying sectors. As one television executive noted in reference to media buyers, the addition of new metrics "just makes their lives miserable," given their inadequate time and resources to "do things ten different ways." As one industry observer has noted, "The conundrum facing practitioners is that a single, established metric is easier to communicate, use, and manage; yet multiple measures (often new and not fully understood) are needed to accurately reflect engagement and persuasion of audiences" (Russell 2009, 50).

A related theme that emerged amongst the interviews conducted for this study was that the embracing of social TV analytics was being inhibited by the lack of standardization and/or consolidation. The quantitative data examined above highlighted the extent to which there is at this point relatively little commonality across the different services. As one network executive emphasized, "Someone needs to step up and say, 'This is the measure we're going to use.'"

Together, these conditions create strong incentives for stakeholders to maintain established, accepted practices. One audience measurement executive described ad agencies and media buyers as "large ships at sea," given that, in his view, it takes them a long time to change direction. A digital media executive elaborated on this perspective, noting that advertising agencies lack incentive to innovate. Rather, their energies are spent on keeping clients rather than trying to be innovative or creative. As one audience measurement executive has noted, "'No one was ever fired for buying adults 18–34'" (quoted in Abzug 2011, 20).

Organizational theorists have labeled such tendencies "cognitive inertia," and previous research has found evidence of this tendency in the media sector (see, e.g., Napoli 2003b). These perspectives are reflected in a recent consulting firm analysis, which concluded that, despite growing limitations, traditional exposure-based analytical approaches to television audiences (based on traditional criteria such as gross ratings points) are likely to persist in the television ecosystem, due to factors such as "the collective inertia on both sides of the table, the strong leverage TV buyers have over their digital counterparts, and the methodological rigor needed to standardize a new currency" (Glantz 2012, 6).

And, of course, for those programmers who are still faring well under the current model, there is little incentive to embrace alternative approaches to audience measurement and valuation. One cable network researcher noted that, during his time at a top-ranked broadcast network, the organizational interest in alternative performance metrics was minimal. As this researcher noted, "When you're the leader, you don't need to use alternatives."

Embracing

However, this skepticism, uncertainty, and resistance co-exist alongside ongoing efforts by many marketplace participants to integrate social TV analytics into their valuation of the audience commodity. One television sales executive described stakeholders in the marketplace as "oversubscribing" to the wide range of social TV data sources currently available, until they determined which sources were the most accurate, made the most sense, and were most likely to be adopted in the marketplace. Many individual media buying firms—and even television networks—have partnered with one or more of the measurement firms working in this space to develop specialized performance metrics that serve their particular needs (see Edgecliffe-Johnson 2010; Marich 2008; Patel 2011).

In the case of programs that do not perform well according to traditional criteria, social media data are being used to try to recast the programs' value in new ways. A very common expression amongst the television executives interviewed for this study was the value of social media metrics in helping them to tell an alternative "story" to advertisers in those circumstances when there was not an appealing ratings story to tell about a particular network or program. As one researcher at a poorly performing network noted, given his network's lack of "ratings stories to tell . . . we have to talk about audience quality," which in this case meant the levels of audience engagement and enthusiasm reflected in social media metrics. A cable network research executive noted that smaller networks will utilize alternative performance metrics (such as social analytics) because they "don't have the mass story."

Not surprisingly, marketplace participants that stand to benefit the most from a reconfiguration of the audience commodity appear most willing to look past any methodological shortcomings of social TV analytics. In the case of programmers, for instance, one cable network executive noted, somewhat cynically, that "if you have a good story you're going to believe the methodology." Another research executive emphasized the prominence of "research to show," as opposed to "research to know." It is important to recognize that this kind of pragmatism and opportunism are fundamental elements of the audience marketplace (see Meehan 1984), and can play an influential role in any potential reconfiguration of the audience commodity. These observations highlight the malleability of the institutionally effective audience. The greater the extent to which stakeholders are suffering under the established conceptualization of the media audience, the greater their willingness to embrace alternative conceptualizations.

It would appear, then, that social media metrics have begun to serve a fairly specific secondary role in the television audience marketplace. What seems to be emerging is a two-tiered audience marketplace, in which social TV measures are being used to supplement traditional exposure measures in specific situations, such as when: a) a point of differentiation is being sought between two programs that perform similarly according to traditional criteria; or b) the program's audience is too small to register in the traditional exposure-focused measurement system.

As one television executive noted, traditional ratings will now serve as "a piece of the pie, rather than the whole pie." Another television researcher described an evolving marketplace in which traditional ratings data would reside at the core, surrounded by a variety of supplementary measures that would vary according to the needs of advertisers and/or programmers. As one digital media executive declared, "nothing can be looked at on its own anymore. [. . .] I think in all aspects of media, there are many more ways to skin the cat, in terms of selling, buying, and measuring."

Regardless of whether these data sources are used as primary or secondary criteria for assessing audience value, the bottom line is that these metrics represent the opportunity for the introduction of a greater diversity of sources of value into the television audience marketplace than has been the case perhaps since the medium's invention. As one television executive noted, "Lots of people are trying to show lots of different types of success." And the idea that the audience marketplace could operate under multiple types of success—that is, multiple articulations of the audience commodity—represents a radical departure from how audience marketplaces traditionally have operated.

Conclusion

In the end, it would seem that we are in the midst of what could best be termed a "critical juncture" (see, e.g., Collier and Collier 2002; McChesney 2007) in the dominant conceptualization of the audience commodity. The emergence of such a critical juncture does not in any way guarantee that any radical changes will occur. It only means that the necessary conditions are in place to create the opportunity for such changes to occur. This chapter has attempted to illuminate the range of technological, methodological, and institutional forces that are simultaneously engaged in this potential reconfiguring of the audience commodity.

But, it would appear that the media environment has fragmented to such an extent that the traditionally unidimensional audience commodity now may be able—indeed may need—to be more multifaceted than has long been the case. If this is the case, then the fundamental premise of a single, monolithic representation of the audience (e.g., Nielsen ratings) that has informed a long tradition of research on the audience commodity may need to be abandoned. Going forward, the television audience marketplace appears likely to operate under a greater variety of institutionally effective audiences, which are not effective substitutes for one another. And so, within the context of the strict definition of a commodity, it would appear that the audience is in the process of, to some extent, being de-commodified.

Notes

1 This paper made possible by the Time Warner Cable Research Program on Digital Communications. The author wishes to thank Fernando Laguarda of Time Warner Cable for his support of this project, as well as Mitch Oscar of the Collaborative Alliance for his assistance in obtaining the data utilized for this study, and Fordham University graduate research assistants Abraham Kohn and Grace Zhang.

2 This week was selected in order to obtain a sample week with a minimum level of disruptions from regularly scheduled programming that were occurring during this time period (e.g., NCAA basketball, presidential primaries, etc.).

3 http://www.generalsentiment.com/tvaerindexdisplay.html (retrieved March 25, 2012).

CHAPTER ELEVEN

Extending the Audience: Social Media Marketing, Technologies and the Construction of Markets

Jason Pridmore
Erasmus University Rotterdam

Daniel Trottier
University of Westminster

This chapter explores the role of social media in the development of audiences—audiences that commercial brands intend to reach and influence. That these audiences are "produced" by particular marketing practices is a rather obvious and trivial statement. It suggests that different factors come together to make the relevance of a particular set of persons an optimal target for commercial interests. This has been the goal of marketing practices since their inception in the early twentieth century (see Pridmore and Zwick 2011). What we wish to suggest here is that these audiences are in fact 'performed' and 'enacted' through a configuration of practices that increasingly include and are reliant upon social media. This performative perspective acknowledges the productive capabilities of social media, but sees this production of audiences as a continual and ongoing affair for the purposes of profit, even though these profits may be fairly elusive. With few exceptions, social media were not developed as marketing tools, rather they are becoming and have become critical means for marketing, with the combination of marketing practices, consumer/user responsiveness and technologies that are drawn together to produce an audience. It is about the arrangement and ordering of marketing practices in ways that enable corporations to meet the "needs" and "desires" of consumers they "know" through social media. This chapter explores the continual enactment of brand audiences, looking at the people, practices and technologies of social media that facilitate this process and "extend" (swaths of) consumers as prime targets for branded messages.

In what follows, we describe the material practices of social media marketing in terms of how it enacts particular audiences. First, we discuss social media as a form of surveillance, enhancing the visibility of consumers for marketing practices. Second, we detail how this visibility allows for increased engagement by marketing professionals, in contradistinction to Dallas Smythe's notion of the audience commodity. This visibility has become part of relationship marketing strategies, for which social media has become particularly useful. Third, we look at how specific technical practices serve to extend brand audiences as they are integrated into marketing strategies focused on everyday forms of collaboration and participation. Last, we connect the enactment of social media audiences with critical examinations of consumption, noting how audiences have to be continually produced in order to remain valuable to corporations.

This chapter draws on both an analysis of the marketing use of one particular platform, that of Facebook, and an evaluation of the technical and commercial practices that surround social media Application Programming Interfaces (APIs). To produce a detailed account of Facebook's role as a consumer platform, we draw on thirteen in-depth, semi-structured interviews with employees and self-employed consultants engaged in promotional activities through Facebook. What started with a small group of readily identifiable employees working in this area has grown to the extent that social media are ubiquitous in the corporate realm. For this reason, these interviews were arranged and conducted in an exploratory manner. While the participants below come from diffuse backgrounds and perform different duties, roughly half of them do consulting work for clients, while the others are fully employed by a corporation. Our analysis of APIs is based on research into how the technical practices of social media facilitate increased consumer engagement, drawing on a number of resources to facilitate the organization and use of personal information from social media for marketing purposes. The use of APIs further engage and extend the participation of consumers in practices that make the enactment of the social media audience more consistent and predictable.

Social Media and Surveillance

The development of social media has made social life, even as it has changed our conception of what this "social life" might entail, increasingly more visible to businesses (see Trottier 2012). A broad set of organizational tasks—including market research, recruitment, and customer service—are augmented through a growing body of searchable personal information. Sites like Facebook have undergone a tremendous diffusion into the business world, the effects of which are only just becoming apparent (see Zarella and Zarella 2011). Add to this the integration of other social media such as Twitter and YouTube, and newer services such as Pinterest and Instagram, and you have a significant set of media that has become an intricate part of the everyday life of billions of people. These media rely on significant amounts of information, most of it personal. As such, the scale of information offered by these companies, coupled with their rapid spread to different social spheres, suggests new possibilities for market surveillance, including creating and monitoring market segments through transactional data.

We purposely describe these practices as surveillance, as this highlights the focused and sustained collection of personal information (Lyon 2001). Our discussions are also intended to further the conception that the use of social media constitutes a form of participatory (Albrechtslund 2008) or collaborative (Pridmore 2013) surveillance. Of course this connects with what has been described as a political economy of personal information (Gandy 1993; cf. Gandy 2009) or the personal information economy (Elmer 2004; Pridmore 2013). In addition, the use of social media for marketing purposes can also be seen as aligned with the study of audience labor (Smythe 1981) and political economic concerns in the age of new media (Dyer-Witheford 1999).

Descriptions of social media within the press and within marketing literature tend to highlight the revolutionary potential of new media. By treating the range of online services and mobile devices as a landscape of information exchange, authors like Shirky (2008) claim they enable "organizing without organizations." Yet, instead of being made redundant, organizations face new challenges/risks and significant opportunities in exploiting the use of social media platforms. Social media allows for an exponential increase in visibility for both individuals and organizations in ways that may be seen as positive or less than positive. Disgruntled clients and co-workers may broadcast compromising information on Facebook and Twitter, yet their own personal lives are also made transparent through their prolonged engagement with these sites. The risks and opportunities associated with social media cannot be decoupled (boyd and Hargittai 2010).

So-called experts, "gurus," and "rockstars" present social media as an invaluable resource for businesses. Indeed, most technologies are ushered into the public by promoters (Mosco 2004), and social media are no exception. Enthusiasts claim social media help businesses by making it easier for them to communicate with and listen to their markets and a number of corporations, and organizations are taking proactive measures to leverage the use of these sites. In recognition of the heightened visibility of their brands, media consultants and marketers actively search social media for conversations between users about their brand and its products. They promote the brand when and where they can but are also especially concerned about complaints and other damaging statements that may occur in the unfiltered social media environment. Corporations are also using these services to gain new insights about their market. The open-ended nature of social media allows for other possibilities for engagement and exploitation.

We wish to suggest that these developments resemble other kinds of institution-led surveillance (Trottier 2011), yet market surveillance marks a further shift toward the enactment of particular audiences through categorical searching and extended engagement, moving beyond the use of individual profiles (see Pridmore and Zwick 2011). For instance, the forms of surveillance on sites like Facebook extend from previous attempts by businesses to gather data on a large scale (Elmer 2004) and are fueled by a number of initiatives, including the use of geodemographic information to locate markets (Burrows and Gane 2006). Market surveillance on social media is intended to make connections between current and potential consumption with personal information, but also with information that is more relational in character—that is, held in common with friends, family and colleagues that are the basis of social media connections. These are sites where users socialize with others, even though they are privately owned and are increasingly optimized for market focused scrutinizing, searching and sorting. This is important when considering market surveillance more generally, as social media represents a central form of market "know-how" that allows businesses that own or purchase this data to know their "market" more intimately. Businesses can extract value from social media while they manage their own publicity (Winseck 2003). In the case of Facebook, the growth of market scrutiny is facilitated by their push towards relational searching, however the development of new tools and technologies allow for further connections to be made that are location and interest focused as well. These approaches pull categorical content out of social media information, even though the content is either initially bound to individual profiles and reputations or projected onto them.

Commoditizing the Audience, Engaging the Consumer

As marketing strategies increasingly depend on social media, scholars have considered the role of the user in this process (e.g., Scholz 2013). In that the contemporary user performs tasks that closely resemble those of the mass media audience, it bears returning to Dallas Smythe's notion of the audience as commodity as a point of departure. The audience commodity stands as an intervention to conventional understandings of the political economy of mass media, and media studies more generally. Departing from a view of mass communications as "subjective mental entities" and "superficial appearances, divorced from real life process" (Smythe 1981, 231), it shifts towards demand management in monopoly capitalism, and specifically the role of audience power in this process. For Smythe, audience power "is produced, sold, purchased and consumed," and as such "commands a price and is a commodity" (233). This claim implies that just because audiences do not directly and knowingly participate in the sale of their labor, their labor can still be sold to advertisers. Smythe notes that audiences contribute an often-unnoticed form of capital to advertising through the purchase and maintenance of television and radio receivers. As well, they contribute an immaterial form of labor "which audience members perform for the advertiser to whom they have been sold" by "learning to buy goods and to spend their income accordingly" (243). Learning in this context refers not only to knowledge of a broader range of products, but also of specific brands and their attributes.

On first pass, this activity may not resemble labor as it is commonly conceived (Jenkins et al. 2013). One obstacle to accepting Smythe's contribution to the political economy of mass media is the fact that audiences not only need to sustain themselves through the purchase of consumer goods, but that they occasionally enjoy this process. Smythe acknowledges the above, but insists upon a dialectical tension whereupon audiences "feel it necessary to cooperate within the monopoly-capitalist system in a variety of ways and for a variety of reasons; yet at the same time, as human beings they resist such cooperation in a variety of ways, for a variety of reasons" (1981, 246). Prior to any substantial analysis of the connections between audience labor and social media, this tension resembles the ambivalence that many social media users report about accepting the necessity, value and pleasure of participating on platforms like Facebook, but also their reservations that these very platforms mark an unprecedented and volatile commoditisation of interpersonal communication (Trottier 2012). Participating in mass media, social media or other staples of "so-called leisure time" is inevitably a commodity-producing task, which according to Smythe "creates the contradiction between oppressive liberating activity in time for which people are not paid" (Smythe 1981, 249). This contradiction marks an expansion of alienation to which political economists and media scholars ought to consider:

> In Marx's period and in his analysis, the principal aspect of capitalist production was the alienation of workers from the means of producing commodities-in-general. Today and for some time past, the principal aspect of capitalist production has been the alienation of workers from the means of producing and reproducing themselves. (250)

To be sure, mass media had an integral role in workers' reproduction at the time Smythe was writing. Arguably, platforms like Facebook not only occupy a more integral role in the reproduction of workers, but as such also demand a greater degree of engagement through both the work of watching and the work of being watched (cf. Jhally and Livant 1986; Andrejevic 2002).

Recent developments in information and communication technologies (ICTs) have shifted everyday use of the Internet from web based information sites towards both social media and mobile communication. Both of these are key aspects of how consumers are increasingly engaged by marketers. With social media platforms in particular, marketers employ three broad strategies to interact with consumers. In this process, they can be seen to first "produce" or "enact" an audience—a social media audience. First, they may treat sites like Facebook as broadcasting platforms, concerning themselves more with self-presentation than with audience engagement. Second, they may use these sites as opportunities to watch over users. In doing so, they take an active interest in user-generated content, but not the actual user. Third, marketers may take a conversational approach that engages users by soliciting input from them, and acting upon this content. It is these latter two approaches that begin to demonstrate the enactment of market audiences. The following sub-sections consider these strategies in more depth by looking specifically at corporate use of Facebook.

First, however, it is important to note the context in which these interviews took place. Businesses were partly concerned with managing their brands' reputations, but they were also eager to find new sources of relevant market information. These respondents experience some precariousness on Facebook as they do not have full control over their brand reputation, nor over the platform augmenting the visibility of that reputation. At the same time, they are taking advantage of user activity on Facebook and other social networking sites, repurposing interpersonal information for marketing strategies. In general they described their engagement with social media as a kind of personal information frontier.

Watching and Listening to Users

Marketing literature emphasizes the need for businesses to "know" their consumer. This has largely been translated into the concept of relationship marketing (Egan 2008), but a number of businesses choose to engage in a continued observation of consumer rather than develop the communicative reciprocal relationship implied by relationship marketing. This is evident in practices with social media, as a number of businesses choose to monitor users and change their marketing practices accordingly. Facebook and other social media are seen as a vast and accessible source of information where, according to Li and Bernoff, consumers "are leaving clues about their opinions, positive and negative, on a daily or hourly basis" (2008, 81). They recommend that businesses "listen" to social media in order to: find out how consumers interpret their brands; obtain a high-resolution and constantly updated understanding of the market; invest less capital for better feedback; identify key influencers in social media; effectively identify and manage public relation crises; and generate new product and marketing ideas (93–95).

Of course part of this process is to understand that users of social media are in fact "consumers." Though this may seem straightforward, consumption or interaction with consumer brands is rarely the primary reason for social media use. However, given the growing popularity of social media, businesses have been quick to develop Facebook "pages" that allow them to access user content directed towards their brand. These pages can be used to not only centralize relevant content for current and prospective customers, but also provide additional analytics on those that frequent the page (though increasingly this is connected to the use of APIs, discussed below). Facebook itself also offers detailed feedback through its advertising services (again, some of this connects the use of APIs detailed below). Finally, they may rely on external social media services like Radian6 or Reputation.com. Based on the access that any of these entry points offers to nearly one billion users, Marc, a sales representative for a software company, goes so far as to describe Facebook as a source of "unlimited information."

When asked why they chose to collect information through Facebook, respondents point to its methodological advantages. Jared, the director of new media at a radio station, notes that his station's audience uploads personal information on Facebook that does not appear elsewhere, for example disclosing that they are coping with a death in the family. While the immediate market value of this information is difficult to ascertain, it supplements a conventional way of understanding an audience in a risk-free manner. Other respondents collect information on Facebook because it is much cheaper than alternatives such as telephone polling or focus groups.

Conventional market research is seen as prohibitive and lengthy compared to social media. Joana, a marketing and communications manager for a major paint producer, notes the following:

> Market research is very, very expensive. So for us to get an Ipsos study done costs a lot of money and involves a lot of lengthy phone calls and interviews and a lot of compilation of data and this would be just another avenue of hearing consumers' comments and, you know, first hand, really.

Not only is market research riddled with logistical issues of cost and time, but the above quote also frames that process as burdensome compared to firsthand accounts. Joana goes on to suggest that Facebook leads to more authentic information, as it does not rely on leading questions. This assessment is indicative of how Facebook users begin to be "produced" or "enacted" as consumers, using the site to provide opinions and feedback in ways very unlike a survey or questionnaire. For Wade, who is developing a web-based application for a venture capital firm that relies on online communities, Facebook is not an explicit site of inquiry. He is fine with this, as he discredits users' ability to self-report:

> What I think even more important than where the criticism happens is being able to observe the actual behaviour. So, you know, it's often true that people will tell you that they want one thing, or they do something one way but they do it another. As subjects, humans are actually poor at self-reporting. [...] I think if you own the social network, or the social application, and you have good measures and analytics on the back end, you've really got this amazing tool set because you can look at not just what people say but what they do and where the delta is between those two things.

This suggests the real value of Facebook's data will come from emerging analytics used to process it. Facebook's potential for behavioural assessment is unclear. However, this has been a key ambition for market surveillance (Elmer 2004). Behavioural scrutiny sorts users based on their engagement with brands, their purchase histories, and other transactional data. By generating data about social ties as well as operating in a prolonged engagement with users, Facebook is regarded as offering better insight into consumer behaviour.

The reliance on user-generated content is also regarded as a way to sidestep methodological tools in order to reach user perspectives. Jared described how Facebook enabled them to access photos their listeners would take at events the radio station would host. These photos allowed the business to "understand what they're [the listener] looking at" and "allows us to see it from their eyes too." While Facebook is clearly a mediating link between businesses and users, the fact that it is not an explicit service for this end suggests it can provide a more intimate connection to a user base. That businesses can also collect information before engaging with users suggests it provides the former with a strategic advantage over the latter. Marc uses this advantage to gain leverage over potential clients:

> The more information you have, the more leverage you have when it comes to making your pitch. That's one thing I've realized. And that's what I like about the social networks, it's because it's a gateway for me to, kind of, look into their lives and figure out what they like and we have in common, you know, things I can bring up to build rapport when it comes to our phone conversation.

To look into the lives of consumers is to begin to create and develop an audience based on these marketers' understandings of these users. User-generated content is valued for how it allows marketers to understand markets and generate sales. This is not a new feature of marketplace surveillance, yet Facebook's ability to effortlessly locate a broad range of biographical and relational details about any single person is noteworthy. Indeed, it marks a more precise resolution, where marketers and sales staff can search both abstract categories and key individuals. This also suggests services like Facebook can be used to establish trust with clients in a way that may violate personal boundaries. This is not surprising, as the violation of such boundaries remains a pressing concern for social media users (Tate 2013).

While there are indications that polling and market research services are being made redundant by social media, Liane, a self-employed consultant, believes it is more likely that the two would enter into a partnership that would enhance the scope of the former while providing a way for the latter to monetize their content. On the heels of an announcement that Facebook developed analytics to track moods and emotions online, she states:

> Maybe this is a better way of monetizing Facebook is to go out, speak and share your research with research organizations. I'm sure Gallup would enter into an interesting partnership and Gallup sure figured out how to get money out of research.

Listening as a strategy allows businesses to easily collect information about their clients through their online presence. User visibility in this context is often motivated by reasons external to the business, such as peer-to-peer sociality. Yet businesses gain from interpersonal scrutiny by accessing that information. While users may be aware this is going on, they do not know exactly what information is collected. Jared suggests users benefit from this visibility, in that they can voice their opinions; yet he also acknowledges his radio station exploits this visibility to develop their market. While users may enjoy intrinsic rewards through social media, businesses enjoy extrinsic, monetary rewards. Even in the absence of a multitude of ways to monetize Facebook, the costs associated are so low that listening on social media seems like a low-risk strategy. In terms of configuring an audience, listening takes advantage of users on social media platforms, content generated by these users on such platforms—and increasingly beyond these platforms—and sophisticated search and analytic tools in order to target a specific audience. These efforts produce a kind of "optimized" flow (c.f. Smythe 1977, 6) between users, third-party mediators and advertisers, and this listening allows the social life presented through social media to be increasingly visible and measurable.

Conversations with Users

While the previous section examined how businesses use Facebook to take advantage of the visibility of users and produce them as "consumers" of/for particular products, a conversational approach dovetails a brand's own self-production and enactment using social media strategically. The focus on conversation is in line with relationship marketing, which has become the default metaphor (and practice) within marketing (see Pridmore 2013); however the integration of social media has been suggested as yet another paradigm shift in terms of market relations. Internet business guru Don Tapscott suggests to businesses: "Don't focus on your customers—engage them. Turn them into prosumers of your goods and services. Young people want to co-innovate with you. Let them customize your value" (2009, 217). Both marketing literature and the respondents in this research emphasize two-way communication and engagement as opposed to broadcasting. This resonates with Andrejevic's (2007) description of mass-customization, where user input is presented as empowering consumers, but clearly benefits telecoms and other businesses.

The kinds of engagements that occur through social media are diverse and diffuse. Though this can bring a democratization of visibility, it also allows businesses to exploit social media. Popular marketing literature describes this as empowering for individuals in their relation to businesses. Shirky (2008) claims these tools allow users to voice their complaints, make suggestions, and be less removed from production process. These features are not inherently problematic, but they can be configured by businesses to maintain an exploitative relation vis-à-vis their user base. In particular, connecting back to Smythe once again (1981), this process can be seen to enable unpaid labor by further extracting value from users.

Cultivating a conversation-based engagement to produce a particular consumer audience presupposes these users want to directly engage with businesses on Facebook. However, sometimes users welcome a corporate presence on Facebook, and sometimes their presence is seen as a privacy violation. Joana, who is developing a Facebook presence for a brand she represents, is concerned with whether or not she should announce herself as officially tied to that brand. At times, some businesses may choose to remain anonymous or hidden in their Facebook efforts. Yet Ben, a founding partner at a search engine optimization company, contends that users value businesses that are visible and willing to converse. This strategy is deemed effective when coping with negative feedback about a brand or product:

> I think people just want to be heard and if you are prepared to listen to people and if again […] [m]ost people will stop at that point. At that point, people will be like "You know what, okay. At least they are trying."

Activity on social media often has multiple benefits. Joana describes the returns she was experiencing when building a significant presence on Facebook:

> We're getting really marketplace information, like we're getting consumer feedback pretty much first hand if we can see those comments and also it becomes really word of mouth marketing at some point when people are reading each others' comments and commenting on each other.

An optimized social media engagement combines the collection of market information with viral promotions and advertising. It combines the strategic advantages of radical transparency and listening for businesses.

Respondents claim Facebook is optimized for maintaining connections with users, which means exercising caution when promoting business content. Corey, president of a new media marketing agency, favours a conversation-based approach to Facebook, noting those who bombard the audience they have sought to develop with branded content are likely to lose that audience. Instead, he suggests businesses need to learn about their market by watching and strategically engaging with it online. Marc echoes the idea that Facebook can foster long-term relations with clients, as users find that mutual visibility to be less intrusive, thus leading to a richer engagement on their part.

The conversational approach resonates with the way application developers build their products. Developers make perpetual revisions to their products based on in-game and external feedback generated by their users. They watch over and listen to users, and respond to this information by improving their product. Likewise, businesses revise corporate strategies based on information garnered from users. In terms of how Facebook shapes the production process, respondents claimed it helps integrate the productive participation (read: labor) of users. By targeting key populations and maintaining a lasting engagement with them, Facebook allows businesses to channel user input at an early stage and frequently return to these users. Damien, who runs a software development company, takes advantage of user activity in public discussions to add value when developing and revising products. This approach harnesses user input to the benefit of the company. Martin, who develops Facebook applications, echoes this approach:

> We often launch an application before it's completely perfect, and then fix it as it goes. We have an "always in beta" kind of mindset. Again, spending two years developing an application and then releasing it, it's better to just get something out there, watch your customers, talk to them about how to make the game more fun, and then bring those changes in as it goes. It's pretty interesting and fun that way. Sort of a ready-fire-aim approach.

Here, a conversational approach fits with a development cycle modelled after Facebook's own way of operating, as ongoing revisions are based on user visibility and response.

Based on technical features detailed below, both Facebook and developers are able to closely watch user behaviour on the site, as well as complaints and recommendations they may indicate. As such, Facebook and businesses are able to modify their services based on what they know about users. Respondents suggest the optimal way to use social media for some businesses is to selectively target a group or population, gather information they broadcast, and eventually converse with them in a strategic manner. Businesses on Facebook are able to make use of user participation by collecting relevant feedback from them. This suggests that Facebook is more like an enclosure than a conventional database as it is able to gather a wider range of input, including possibilities to make products and services more valuable. This is all made possible by allowing and encouraging the increased participation of users and the presentations and performance of daily life that is occurring within these digital "enclosures" including its relevance as a developed "audience" for particular businesses. Most marketers favour increased visibility of users, indicating that being able to identify them leads to a more disciplined user, and more valuable data. For Damien this is "the flip side of getting rid of the anonymity: the quality of conversation goes up."

A conversational approach to market surveillance is informed by users' familiarity with interpersonal visibility on social media. Users are more comfortable making their lives public to businesses if they feel a sense of reciprocation. As such, corporations manage their presence on Facebook through employees who use peer-to-peer sociality to produce an audience—that is, a targeted group of Facebook users that willingly participate in marketing conversations through Facebook. This approach can be seen to resemble Dallas Smythe's understanding of audience labor (1981), part of a shift from an attention economy to an engagement economy (McGonigal 2008). Audience labor involves an active construction of goods and services by unpaid laborers, and in this case this labor is cultivated and promoted by marketers eager to capitalize on the knowledge produced through these ongoing conversations. It is a personal information economy that goes beyond one-way surveillance as it seeks to take advantage of this social medium to enroll users into the production process.

The focus in this section is on marketers and consultants who develop an audience through social media, revealing the strategies employed to promote audiences-as-users that knowingly or unknowingly provide rich marketing knowledge. Audience labor in this case cannot be understood outside the efforts/labors of these marketers or employees, to say nothing of the other factors required to enact such an audience. That many of these respondents have launched marketing initiatives that focus on social media platforms, instead of responding to top-down directives from their superiors, is noteworthy. Audiences are enacted on social media as a result of users' own intentions to participate in conversations of interest to them. This enactment is also the product of social media platforms that promote and distribute user content, and entry-to-mid-level workers in the information sector who combine knowledge of users and platforms with the broader mandate of their employer. Of course this also requires significant effort from social media companies to provide appropriate platforms and tools for social interactions for the production of audience labor, something which is continually undergoing revision and subject to negotiation. By relying on particular tools and technologies, social media companies allow organizations, particularly marketers, to extend their engagement with the brand audience.

Technologies of Collaboration—Audience Building and Clicks of Visibility

While enabling market research through watching and engaging in conversations with consumers via Facebook is an effective strategy to build a brand audience, Facebook and other social media have furthered the potential for increasingly more ordered and proactive processes for marketing. The use of various technical practices extends information gathering and dissemination for businesses by enrolling consumers in an ongoing conversation. This allows consumers who participate to keep up with the latest products and services offered by a number of brands through "technical" practices including the "like" button on Facebook, the "following" on Twitter, using the "pin-it" feature for Pinterest, and more generally the ability to share content with others. This is all dependent on the way in which social media have set up the interface between users and these organizations. For most, after an initial agreement, the technical aspect of this participation quickly disappears. However, once activated, the agreement to "connect" or "log in," and other practices that rely on Application Programming Interfaces (API), permits consumer access to specific branded content, features, and services that are available both on social media and proprietary websites. In many cases, it also permits these organizations significant access to the users social media profile. Application Programming Interfaces at their basis are the means by software components are enabled to communicate with each other (for further information see "What is an API? Your Guide to the Internet (R)evolution" and "API versus Protocol"). In the case of social media, APIs can connect relevant data held by these systems with external databases and systems that can make use of this information. For instance, the geographic location of a tweet or a Facebook post may be relevant for an automated advertisement in that area or the associations with fellow recreational sports enthusiasts may be relevant for the market focus of a sporting goods manufacturer.

What is perhaps most interesting here is that these tools do not seem to be technical at all. They extend the marketing potentials for organizations as convenient mechanisms for consumers to engage with businesses in a simple format as part of people's everyday use of social media. Again, though not primarily marketing platforms, social media have made users more visible to marketers, and complicated protocols and processes of APIs or other forms of sharing/display are hidden behind simple one-click processes. Marketers are increasing leveraging these tools even as consumers increasingly rely on their convenience, and, it might be said, use them to facilitate their own labor in producing relevant profiles and gathering information about the brands and businesses in which they are interested.

These simple devices and practices transition consumer awareness of brands and services into ongoing and "sharable" conversations. The like button on particular Facebook pages (rather than just for particular status posts) for instance, allows product promotions, discussions and developments to appear on users' newsfeeds. Following a brand on Twitter or pinning a particular products (including things like crafts, recipes, fashions, etc.) to your Pinterest page can be seen to indicate a form of conversation that marketers can engage in an updated version of relationship marketing. Through this, consumers can both respond to brand messages and share and/or augment them. These "technical" practices make the conversations discussed above easier. Yet the argument here is that these do more than just allow for better conversations. Social media APIs are increasingly a crucial part of the production of brand audiences and, more specifically, an ally for marketers and a tool for profit for social media. With API interactions (calls) in the billions per day for larger social media companies (such as Facebook and Twitter) and others moving in this direction (Google, Microsoft), these stabilize conversations with consumers and serve to strengthen the durability of the audience as a commodity.

API Enrollment

Despite the ability to perpetuate a conversation, there are no clear directions for marketing through social media, although as suggested above, there are a number of social media gurus eager and willing to guide businesses in their practices. The indiscriminate aspects of social media—that (almost) anyone can comment or follow or augment shared messages—means that businesses have to pay more careful attention to what is happening in this context, lest their brand reputation be significantly slighted.

However, as is clear, unlike the development of previous audiences through advertising practices (see Pridmore and Zwick 2011; Miller and Rose 1997), social media is more difficult to control. In addition, the conversational aspects of social media—the sharing or responding to messages—are not easily quantifiable, at least in terms of relevant content. However, with further access to user information, businesses can begin to see and develop more measurable and focused interactions with consumers.

To do this, a number of brands rely on different platform or application interfaces that connect user actions with the brand, allow users to use social media accounts to "sign in" to branded websites or use other applications to access particular services or branded content. Through these interfaces, brands are able to gather significant data on consumers by having access granted to their personal information. At the very least, these businesses have access to consumers' basic information, once permission is granted to the application or as a form of identity verification (sign-in). This gives access (permission) to the following pieces of basic data: "id name", "first_name", "last_name", "link", "username", "gender", and "locale." More information can be requested of the user through specific permissions. Figure 1, 2 and 3 indicate the additional information that businesses can request from users through Facebook.

Select Permissions

User Data Permissions | Friends Data Permissions | Extended Permissions

user_about_me	user_activities	user_birthday
user_checkins	user_education_history	user_events
user_games_activity	user_groups	user_hometown
user_interests	user_likes	user_location
user_notes	user_online_presence	user_photo_video_tags
user_photos	user_questions	user_relationship_details
user_relationships	user_religion_politics	user_status
user_subscriptions	user_videos	user_website
user_work_history		

Basic Permissions already included by default

Get Access Token | Cancel

Figure 1

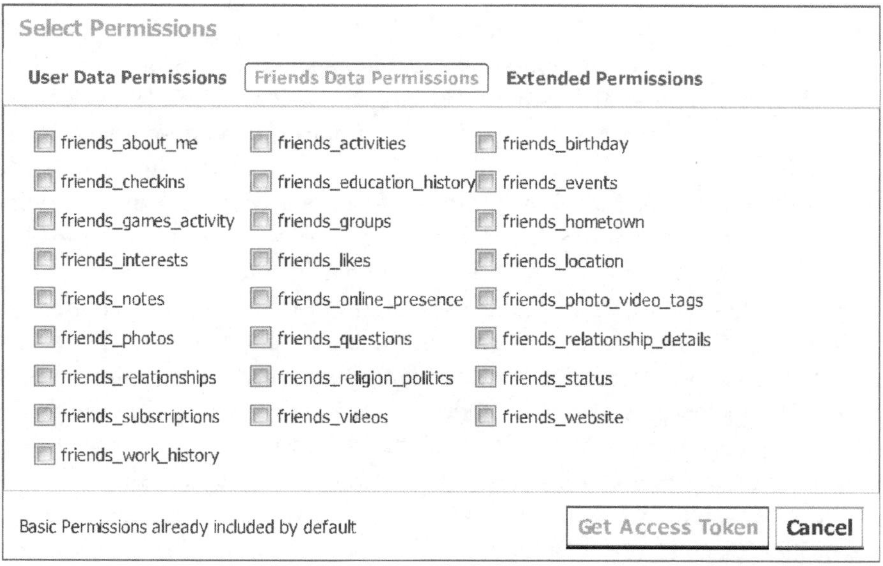

Select Permissions

User Data Permissions | Friends Data Permissions | Extended Permissions

friends_about_me	friends_activities	friends_birthday
friends_checkins	friends_education_history	friends_events
friends_games_activity	friends_groups	friends_hometown
friends_interests	friends_likes	friends_location
friends_notes	friends_online_presence	friends_photo_video_tags
friends_photos	friends_questions	friends_relationship_details
friends_relationships	friends_religion_politics	friends_status
friends_subscriptions	friends_videos	friends_website
friends_work_history		

Basic Permissions already included by default

Get Access Token | Cancel

Figure 2

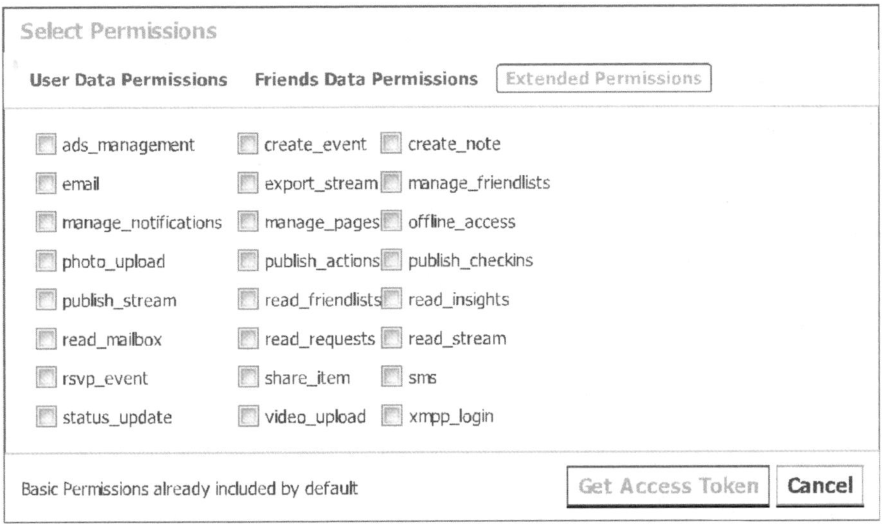

Select Permissions

User Data Permissions | Friends Data Permissions | Extended Permissions

ads_management	create_event	create_note
email	export_stream	manage_friendlists
manage_notifications	manage_pages	offline_access
photo_upload	publish_actions	publish_checkins
publish_stream	read_friendlists	read_insights
read_mailbox	read_requests	read_stream
rsvp_event	share_item	sms
status_update	video_upload	xmpp_login

Basic Permissions already included by default | Get Access Token | Cancel

Figure 3

These additional permissions encompass over 70 additional data fields beyond a user's basic information. Businesses can build upon the information that is available about these users to effectively segment and market to consumers and their "friends" who have already indicated an affiliation with the brand. Currently, most businesses are not requesting most of these permissions beyond email address and birthdate. Figures 4 and 5 indicate the process by which businesses use the Facebook platform to obtain these permissions from users. Figure 4 indicates a minimal request for data, while Figure 5 shows a substantial request from the application.

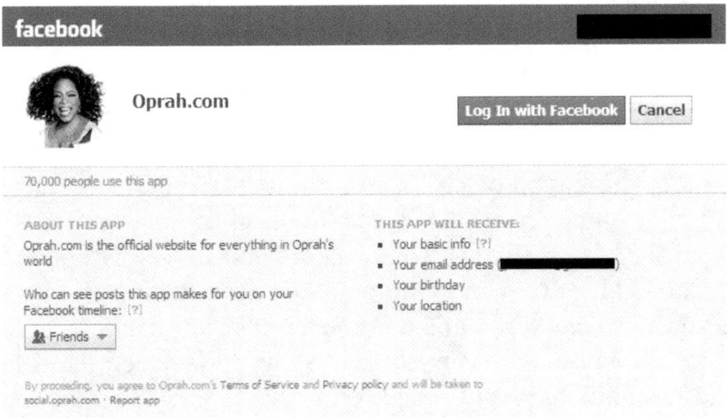

facebook

Oprah.com

Log In with Facebook | Cancel

70,000 people use this app

ABOUT THIS APP

Oprah.com is the official website for everything in Oprah's world

Who can see posts this app makes for you on your Facebook timeline: [?]

Friends

THIS APP WILL RECEIVE:

- Your basic info [?]
- Your email address ( )
- Your birthday
- Your location

By proceeding, you agree to Oprah.com's Terms of Service and Privacy policy and will be taken to social.oprah.com · Report app

Figure 4

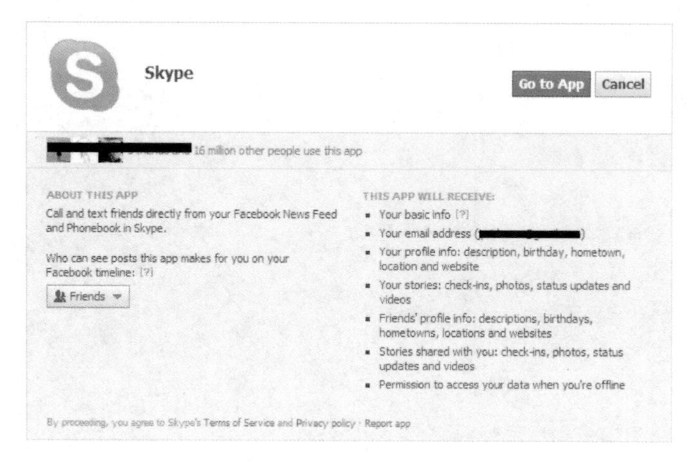

Figure 5

The use of these applications allows for initial and then subsequent access to personal and personalized information, data, and features for businesses. The screens in figures 4 and 5 are only seen once, and then the permissions granted to these users remain until either the terms, conditions, or permissions requested change, or until the user changes his or her password. A similar process happens in the use of applications for Twitter. Figure 6 indicates the initial request made of Twitter users to give permission to a particular site so that the user may continue to use the site. Beyond accessing certain information on the site, businesses in this case are able to update consumer profiles, learn whom the user follows and is followed by, and even post information (tweets) for that user from their site (for example, products purchased or articles recommended).

While Facebook and Twitter might be the most prolific social media offering APIs for users, these services are also provided by other social media and services like LinkedIn, Yahoo, Google and Microsoft Windows Live. Often these APIs are a form of user identity authentication and verification that replace or augment traditional user names and passwords. However, these APIs offer significantly more profile data about consumers and, in part because they are seen as less subject to forms of dissimulation—that is, users most often use their real names and identities (see Gross and Acquisiti 2005), and as such are seen as more reliable than traditional forms of identity authentication and verification. By using this information, from user statistics to profile building, business have begun to leverage the data from social media for an increasingly social form of Customer Relationship Management (CRM).

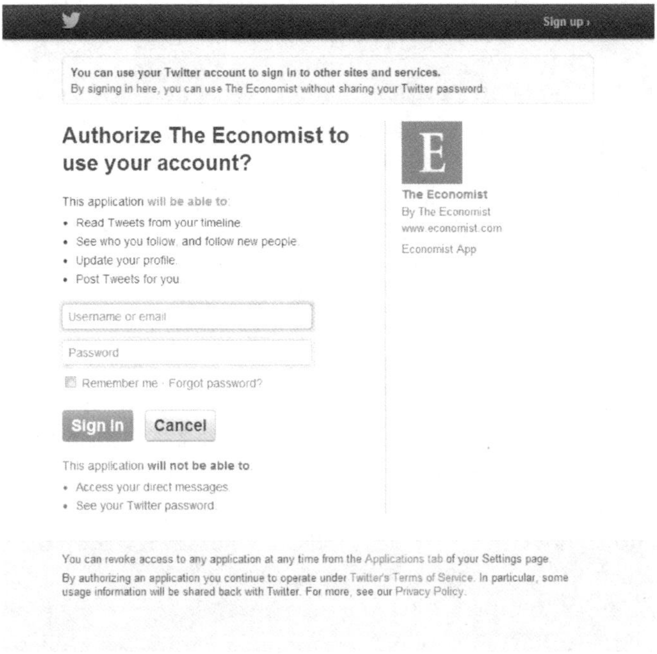

Figure 6

The significance in the use of APIs in their various forms is that these are distinct means through which marketing observation and/or marketing conversations are embedded into technologies. This should not be underestimated. APIs allow for connections to be made between users and businesses that may be seen as benefiting both the business and the users. Relationships are established and the visibility of relationships and brand associations is increased. APIs enable users to become promotional partners, extending the means of producing or enacting a particular audience. In the process, the desire for "word of mouth" marketing become more automated, more prolific and, in a sense, more tangible. Responses to particular marketing campaigns and/or product promotions or ideas can be measured. This moves beyond the production of an audience through advertising or even sustaining/reinforcing current conversations with that audience through social media. The technical apparatuses of social media allow engagements with the audience to be more repetitive and predictable. As such APIs serve to stabilize consumer practices in ways that makes the businesses become less of a secondary aspect within social media, and translates these into palatable "relationships" that mimic those with friends, family and colleagues.

Materialities and Audience Labor

What this work suggests is an interesting shift in how consumer surveillance is practiced. Though marketing has always been a means of, in our terms, producing particular audiences, the use of social media has seen some of this work delegated to pieces of software that extend the lifespan and value of produced audiences. Interestingly, the use of social media also shifts the reliance away from retailers or specific physical locales as the primary source of connection for most brands with consumers. However, in this case, this form of mediation has largely been transferred to social media platforms. This is, for us, a crucial point. While observing consumers and engaging them in conversations is important, and doing this through social media has become a default and expected way to do this, extending the production of an audience through the means of APIs allows marketers the ability to continue observations and conversations beyond specific time- or place-sensitive interactions (i.e. traditional advertising, retail contact, customer service calls).

In addition, practices of information gathering crucial to the personal information economy are simplified: businesses are no longer reliant on surveys, warranty cards, loyalty cards, or information passed on by retailers. Rather, they have instant access to significant consumer data that is arguably more reliable and representative of these consumers. There are of course significant issues that brand managers and marketers have to cope with in this process, however, including the opacity and volatility associated with social media platforms. The terms of service that structure relations between users and the platform—but also businesses and platform and users/audiences and businesses—are frequently revised. Yet it is increasingly understood that social media allows consumers to perceive that they have a "voice" in marketing practices, from changes in strategy, to complaints, to "prosumption." In this process, social media produces an audience ripe for marketing.

Our suggestion here is that audiences are not simply produced when individuals face a television set, computer screen or mobile device. They are in fact "performed" and "enacted" through a configuration of practices, including engagement on social media and the application programming interfaces that extend this consumer engagement. What is clear from our discussion above is that the continual enactment of brand audiences does not rely solely on the labor of these audiences. Social-media marketers are actively participating in this process and they are reliant upon the tools and developments of social media to extend and reinforce the production of these audiences. For these marketers, what is perhaps most interesting is that while many have backgrounds in marketing and brand management, most also have IT and consulting backgrounds. Often their enthusiasm for social media platforms means that they perform additional, unpaid labor for their employers (Trottier 2012). While audiences are visible to marketers through a series of labor practices that are occasionally deliberate, employees are charged with the task of facilitating a continual engagement with audiences on volatile social media platforms.

The enactment of social media audiences can be made meaningful by revisiting Smythe's understanding of the audience as a commodity, where engaging with advertisements is a vital component in the marketing process, and one that is typically not considered as a form of labor. This kind of mass-media labor is still present on social media, but it is accompanied by a further engagement with so-called audiences. In addition to providing direct feedback in response to marketing campaigns, audiences routinely supply personal information about themselves and their peers who often-unknowingly contribute to consumer relations. As a commodity though, these audiences are even more valuable in an era of social media, where they are produced and stabilized through new techniques and technologies, diffusing and expanding the work of watching as it is fused with the work of being watched.

Acknowledgments

The research leading to these results has received funding from the European Research Council under the European Union's Seventh Framework Programme (FP7/2007-2013) / ERC Grant no. 201853

CHAPTER TWELVE

Capital's New Commons: Consumer Communities, Marketing and the Work of the Audience in Communicative Capitalism

Detlev Zwick
York University, Toronto

Alan Bradshaw
Royal Holloway, University of London

Perhaps Dallas W. Smythe's most central insight from his analysis of the role of mass media in monopoly capitalism relates to how the superstructure (the Consciousness Industry) has become engaged directly in production. This production is two-fold: first, mass media produces the audience commodity, and second, mass media "beckons" into action the production of commodities through the production of the audience as commodity. By discovering the role of mass media in the development of an audience market, Dallas identified a key contradiction in the logic of this mode of commodity (re)production: audience commodification in turn commodifies people in audience markets who are consciously seeking non-commodified group relations. This contradiction between the needs of capital for managing demand and the desire of people to remain outside of capital relations, in the sphere of communication has become markedly more acute and visible in the age of participatory media. Resolving this contradiction has become the great challenge for contemporary marketing managers and, in their search for innovative ways to commodify the audience without antagonizing it, marketers turn increased attention towards the idea of consumer communities. The language marketers produce around such communities is particularly noteworthy as it appears to borrow more from Gerald Winstanley and Karl Marx than Milton Friedman and Michael Porter. So the question arises: are marketers becoming the new communists? Towards developing some initial answers, this paper considers the recent rise in marketers' interest in consumer communities and argues for the need to theorize this marketing innovation within the context of communicative capitalism's unique challenges for customer government, market control, and value creation.

The Real of Virtual Consumer Community

The last decade has witnessed the appropriation of the Internet as a global marketing communication network and thereby amplifies what was already recognized as the age of dotcom marketing madness (Jacobson and Mazur 1995). As a result, we witness renewed interest in the study of marketing practice in general and advertising practice in particular, with scholars trying to disentangle the implications of an increasingly global, networked, and decentralized mode of marketing communication (see e.g., Araujo, Finch, and Kjellberg 2010; Zwick and Cayla 2011). Some Foucaultians have sought to understand the world of ubiquitous marketing as a form of government (Beckett 2012; Skalen, Fellesson, and Fougere 2006; Zwick, Bonsu, and Darmody 2008), while others focus on marketing as a material social practice in the business of qualifying commodities and creating markets (e.g., Callon, Meadel, and Rabeharosoa 2002; Cochoy 2007; Slater 2011). Especially with the emergence of the Internet as a participatory medium, attention has turned to new ways in which companies attempt to incorporate the consumer directly into the firm's marketing processes (see e.g., Cova, Dalli, and Zwick 2011; Ritzer and Jurgenson 2010). Arvidsson (2006) discusses how consumer labor in online environments is first encouraged and then appropriated as economic value by the firm, and Andrejevic (2011), Pridmore and Lyon (2011), Manzerolle and Smeltzer (2011) and others analyze how marketing adopts the Internet as a technology of surveillance, manipulation and exploitation.

In this vein, our paper looks at the current popularity of virtual consumer communities in marketing practice. We focus less on the mechanics of such communities, i.e., how to set them up and how to manage and maintain them. Instead we seek to understand, from a critical theory perspective, what the *idea* of an online consumer community allows marketers to do and why we encounter this desire to think of consumers as community. In other words, we are interested in the ideology of the virtual community and what we call commonist marketing. We situate our discussion of consumer communities and commonist marketing within a larger critique of contemporary marketing's attempts to capture its users in intensive and extensive networks of entertainment, production, consumption and surveillance. Jody Dean (2010) calls this economic-ideological form communicative capitalism, a formation—simply put—that relies on the exploitation of communication.

Smythe (1981) in his most central work on the audience commodity also singled out the exploitation of communication via mass media as the central driver of what he called monopoly capitalism. Mass media, Smythe argued, turn the superstructure (i.e. the Consciousness Industry) into a key site for capitalist surplus production. This mode of production has two distinct dimensions. The first dimension is the production of the audience as a commodity for sale through mass media. A second dimension emerges when the media-produced (and thus now "real-existing") audience commodity "beckons" into action the production of consumer commodities.

This double dimension of media's productive role remains valid under communicative capitalism, and so does a key contradiction in the logic of this mode of commodity (re)production: audience commodification commodifies people in audience markets who are consciously seeking noncommodified group relations. Indeed, this contradiction between the needs of capital for managing demand and the desire of people, singularly or collectively, to remain outside of capital relations in the sphere of communication has become markedly more acute and visible in the age of participatory media. Resolving this contradiction has become the great challenge of contemporary marketing managers and, in their search for innovative ways to commodify the audience without antagonizing it, marketers have turned to the consumer community.

The idea of the online consumer community is not an entirely new idea. As McWilliam (2000, 43) pointed out quite some time ago: "*The popularity of communities on the Internet has captured the attention of marketing professionals. Indeed, the word "community" seems poised to overtake "relationship" as that new marketing buzz-word.*" Interestingly, though, the buzz about these communities subsided relatively quickly among marketing professionals, partly because the collapse of the 2000 technology stock market bubble threw into question the economic viability of the Internet as a driver of marketing and commerce. Only with the emergence of what we now call Web 2.0 around 2005/2006—the birth of the social web, as it has come to be known—did communities return to the forefront of marketing attention. And, in recent years we have seen a deluge of blogs, popular consulting books, and business conferences that deal generally with the concept of virtual consumer communities. Authors like Li and Bernoff (*Groundswell*), Tamar Weinberg (*The New Community Rules*), Brian Solis (*Engage!* With an exuberant, quasi-revolutionary Ashton Kutcher—co-founder of online marketing consulting outfit Katalyst—writing the foreword) to name but a few, captured large audiences interested in how marketers can find value amongst networked, communal consumers. It is hardly overstating the matter to say that many consultants and academics see a community of consumers everywhere they look.

Such a resurgence of the idea of community is interesting for a number of reasons rooted in the ideology of marketing. And we will discuss them in detail below. But one reason should be mentioned up front, namely that the actual existence of online communities is significantly overstated (see Arvidsson 2013). In other words, consumer communities are more of an idea, indeed a fantasy, than a reality. McWilliam was certainly correct in predicting the success of the consumer community as a buzzword. Not least due to his own seminal essay, the idea of community quickly gained significant purchase with marketers struggling to make sense of the way consumers acted in this emerging networked world of the Internet. From a sociological perspective, however, as Arvidsson also points out, user aggregations such as the "now defunct Geocities web space with its 'more than three million members' are not to be understood as communities, at least not in anything that resembles the significance that that term has originally held in social theory (not to speak of Facebook or YouTube that are most definitely *not* communities)."

Despite all the talk about communities—especially marketing scholars' and professionals' favorite version of such a community, the brand community—they remain difficult to locate as real existing social formations characterized by dense webs of interpersonal interaction and a durable attachment to a shared territory or identity (for the classic reference see Tönnies 1973). This is particularly true for social formations with any kind of commercial purpose, whether user-managed or company-managed. Indeed, online communities, such as they are, arguably conspire against meaningful community building because communication online is at the same time radically massified (and often anonymized) and highly fleeting (Dean 2010) leading to a decline of meaningful communication and lasting social engagement. Social media in particular have increased dramatically the number of people that participate in collective communication and, more importantly, *how* they participate. Short-lived and uncommitted forms of communication such as (re-)tweeting, clicking "like" on Facebook, or posting random pictures from one's recent get-away to New York City with few if any descriptive words have become dominant forms of "participation" online. Indeed in this non-enduring character of communal engagement social communication perhaps finds its parallel in the transitory and energetic but ultimately meaningless forms of "pure participation" such as the flash mob. As Arvidsson (2013) argues, social media accommodate and constitute many hit-and-run forms of engagement, especially when it pertains to brands and products: posting once or twice on a blog, looking up an online forum on motherhood to ask a question about a product and then never coming back again, and so on.

To be sure, online communities, when defined as a social formation of sustained and meaningful interpersonal interaction, are hard to find. Especially brand or product-related communities which, we submit exist more in the minds of marketers and marketing scholars than in actual reality. Rather, there is accumulating evidence, according to Arvidsson, "that when such direct forms of interaction do occur, 'membership' is highly transitory." Thus, just as buzz flows quickly in and out of networks of influencers and influenced, individual affiliations flow in and out of groups because affiliations are not based on strong webs of interpersonal interaction but on weak forms of mediated association (like re-tweeting a message).

Empirical evidence against the existence of virtual consumer communities is not surprising because promoting "hit and run" style communication is exactly what the social online network is all about. Such a mode of communication, however, forecloses the possibility of collective meaning creation and understanding. Instructive in this context are Slavoj Žižek's early essays on communication and meaning in cyberspace (1996, 1997), where he rejects the dominant view of the Internet as the coming of universal meaning and total understanding. Instead, for Žižek computer-mediated interaction ushers in the decline of symbolic efficiency as our ability to "transmit significance" from one individual to another and also from one setting to another diminishes. Ultimately, the decline in symbolic efficiency leads to a failure of transmission. In the final analysis, then, the concept refers to a type of communication inherent to cyberspace that cannot but produce radical uncertainty of what it is that is being said and on what grounds it is said at all. Therefore, virtual communication does not provide the conditions of possibility for collectively shared and co-produced meaning. What computer-mediated interaction does allow for and fosters is a universe of radically individualized utterances (without committed reference to previous statements by others) occurring in a space that is a collective space only in the most formal sense. In addition, Jody Dean (2010) points out that communication in cyberspace also reconfigures the relationship between communication and the communicating subject because of the ease with which the communicating subject can withdraw from his or her utterance. As Dean (2010, 7) puts it, "[S]ince exit is an option with nearly no costs, subjects lose the incentive for their word to be their bond."

Perhaps the form of community at stake is best exemplified in a typical Žižekian example—the Masturbathon (Žižek 2008); an event in England where hundreds of people masturbated in a mass phenomenon to raise awareness. According to Žižek, inasmuch as the event generates a collective out of individuals whose very narcissistic isolation forms the basis of their group immersion, a contradiction emerges. Precisely what is excluded from such a collective of individuals is their "intersubjectivity proper" and an "encounter with an Other." As Žižek (2008, 25) puts it:

> Is the typical World Wide Web surfer today, sitting alone in front of a PC screen, not more and more a monad with no direct windows onto reality, encountering only virtual simulacra, and yet immersed more than ever into the global network, synchronously communicating with the entire globe? "Masturbathon," which builds a collective out of individuals who are ready to share the solipsism of their own stupid enjoyment, is the form of sexuality which fits perfectly the cyberspace coordinates.

From a critical theory perspective, it is not surprising that the breakdown of the tissue of social relationships and experiences of intersubjectivity are re-stated as its opposite—as a growth of community. More specifically, Ben Fine (2010) has directed critical attention towards the rise of a related concept, social capital. As advanced in such texts as Putnam's influential *Bowling Alone*, the idea is that the more relations that people develop, the better off they will be. However, Fine critiques that the idea of social capital is a displacement of more traditional forms of community and civic society and amounts to a conceptual reconfiguration of the idea of human relations according to an entirely neoliberal agenda.

From this analysis, we can see that the nature and structure of (even what we call "social") communication conspire against the possibility of community (at least in any meaningful sense of the word). That does not mean online communities do not exist. We argue, however, that the ever increasing speed and volume of communication make communities less likely to persist, if not to form in the first place, and less and less relevant for marketing practitioners. However, in proceeding with this argument, it is important to problematize the idea of a disappearing authentic community used as a critical counter-foil to marketing mediated communities. An example of such problematizing is presented with reference to a further idea that exists in parallel with marketing communities—relationship marketing; a marketing practice predicated on the idea of fulfilling relationships emerging between corporations and their consumers. As Østergaard and Fitchett (2012) argue, the idea of exchange transactions taking the form of a relationship is very much an idea that belongs within the symbolic realm and so they analyze the discourse from a Baudrillardian perspective and accordingly conclude that what is needed is a third level of understanding that "moves beyond" either considering relationship marketing as a serious and authentic practice or as a shallow, superficial rhetorical strategy. Instead they argue that the nostalgic yearning for an authentic relationship or form of human intersubjectivity as much belongs to the inauthentic symbolic realm as the obstinately symbolic realm of relationship marketing. Hence Østergaard and Fitchett (2012) conceive of relationships as a construct that results from the complex interplay between signs, code and program which, in this case, become manifest as the market, marketing institutions and marketing technologies.

The Ideological Function of Consumer Online Communities

And yet, marketers hail the power of the virtual consumer community. Public relations and brand guru Brian Solis writes that online community cultivation is now at the essence of any business success, and Larry Weber (2009) identifies ways of how consumer communities can build one's business (to just name a few of myriad writers on the topic). In the absence of any convincing real evidence for such claims, we ask: why is the industry for consulting books on consumer communities booming? We propose the answer that the online consumer community fulfills an important ideological function in communicative capitalism. Recall Smythe's insight into the contradictions caused by monopoly capitalism's attempt to commodify audiences. He was very aware of the tension that would arise from mass media's attempt to produce and sell to the highest bidder an entity, a collectivity of viewers, that does not consider itself to be for sale. Hence the need of monopoly capitalists for an audience commodity culminates in the making into a commodity people who are consciously seeking group relations without and outside capitalist commodity relations. In the age of participatory media, this contradiction—to manage demand through the commodification of collectivities of people—has become more pronounced. Resolving this contradiction has become the great challenge of contemporary marketing managers and the consumer community has come to function as an ideology of marketing where the commodification of the audience can proceed without openly antagonizing it.

Below, we explore three specific ideological functions of the community mobilized by contemporary marketers in order to overcome a key contradiction of communicative capitalism: the commodification, or as Arvidsson and others (see e.g., Arvidsson 2005; Foster 2011; Zwick et al. 2008) put the matter, the valorization and subsumption under capital of the social media audience while appearing to do the opposite.

1) The function to dispel the belief that marketers actually do the marketing.

The idea of the online consumer community was associated immediately with the claim that these communities spell the end of marketing as we know it. In fact, in the popular marketing and consulting literature, the term un-marketing has come to describe how marketing changes in the age of social communication and the community from a company-controlled technique of persuasion and selling to a people-driven activity of friendly advice-giving and idea-sharing about brands and products (e.g., Stratten, 2010). Such a formulation of what virtual consumer communities mean for marketing is instructive because it points to the fascinating contradiction of appropriating for marketing practice a particular socio-cultural figure—the community—that appears to be against such appropriation and is in fact recognized as such by the marketing professional. Each consulting book on the topic spends a significant part of the text getting the reader into the "right marketing mindset" required to deal with this contradiction. Advice sometimes takes a more straightforward and sober tone as in Weber (2009) and Qualman's (2009) texts, which goes something like this: "communities represent the best and the worst of the social web. If you tread lightly and engage on the community's terms, you will be richly rewarded. If you hope to control the community by simply replicating marketing as usual, the community will turn on you." Sometimes, the tone turns more blustery and theatrical such as the pleas of Solis and Stratten. In the words of social media guru Ashton Kutcher (in Solis 2011, xiii): "New media is creating a new generation of influencers and it is resetting the hierarchy of authority, while completely freaking out those who once held power without objection [...] In the end, everything starts with engagement. This is our time. This is your time. Engage."

In the typical Silicon Valley style of *Wired* magazine, Kutcher, Solis, Kelly and others disseminate the news about the consumer insurgency we are witnessing (ix): "[T]he roles are reversing and individuals and brands have the ability to reach and rouse powerful and dedicated communities without ever having to pay for advertising. I'm just part of a bigger movement of empowering the people who care enough to change the world. Social media is socializing causes and purpose and inciting nothing short of a revolution." As Kutcher implores his clients, "[M]arketers, don't control us, support us; don't talk to us, listen!" And Kevin Kelly (2009), former editor of *Wired* magazine, proclaims that communitarian projects such as Wikipedia, Flickr, and Twitter "aren't just revolutions in online social media [...] They're the vanguard of a cultural movement" that will ultimately de-throne traditional marketer-to-customer marketing.

Brands, for Kutcher, need to "be there," but instead of running the show, they need to surf on the energy and passion of all those who "believe in themselves and their ability to push things forward." Marketing, for Kutcher, faces a huge opportunity by latching onto potentially legitimate movements of emancipation and anti-capitalist collaboration and he, apparently, sees no conflict here. In the world of virtual communitarianism, marketing and political struggle are heralded as coming together in harmonious matrimony.

At first blush, the case of Ashton Kutcher and his countercultural comrades might seem innocuous enough: another proselytizing proclamation in the recent tradition of cyber-entrepreneurial boosterism extending from the New Communalists (Steward Brand's *Whole Earth Catalog* [1986] and later his Global Business Network) to the cyber-libertarian writings of Howard Rheingold (1993) and John Perry Barlow (many essays in *Wired* magazine, the *New York Times*, and others) during the 1990s to, more recently, the neolibertarian thoughts of Kevin Kelly (e.g., 1998, 2010) and Christopher Kelty (2008): a discourse on individual empowerment, information sharing, and networked collaboration extolling the virtues of technology and total individual self-reliance as a personal and, in extension, collective ethic. In his 2009 *Wired* essay entitled "The New Socialism: Global Collectivist Society Is Coming Online," Kevin Kelly defends the communism of open source by reassuring us that "[I]t is not class warfare. It is not anti-American; indeed, digital socialism may be the newest American innovation." For Kelly and his cyber-neolibertarian audience, this new digital socialism is palatable because digital socialism is socialism without the state, operating in the realm of culture and commerce, rather than government.

Similarly, the anti-state, communitarian ideal of the new cyber-marketers presents itself as a countercultural approach to the militarist, bureaucratic, and corporate forms of marketing. Consumer communities reverse to its opposite—the oppressive, top-down communicative violence characteristic of the pre-digital age. Marketing is now cast in terms of collaboration, flexibility, and the ability to generate and share information which requires treating consumers as agentic and empowered equals who have the will to determine their own preferences, opinions and communication and definitely are not manipulated dupes. In the age of cyber-communitarianism, as per the discourse of its proponents, marketers must give up their aspirations to absorb consumers into technocratic projects of corporate control. Instead, marketing must become *communist* by enabling local practices, supporting practices that free consumers from commercial-corporate attempts of persuasion, and helping to cultivate a variety of ways of being. Understanding how communicative capitalism ensnares speech and action, commonist marketing is produced in networks that favor decentralization over centralization, the multitude over the unified, bottom-up power over top-down control, and experimentation over the planned.

This, then, is what is meant by the countercultural aspiration of un-marketing: *to do marketing without the marketer*, even as an army of un-marketing experts consult for companies on how to market their products effectively. Hence, what is really at stake in the push for this kind of "un-marketing marketing" is not an end to marketing, but a desire of marketers to make themselves disappear as *the other* of the consumer. Indeed, we claim that the ideology of the virtual consumer community aspires to make marketing out of, and through, the other.

2) The function to establish the belief that marketers no longer control consumers.

Espousing an especially exuberant kind of techno-utopianism, cyber-community consultants such as Solis, Weinberg, Kutcher, and Stratten (but also cultural anthropologist Christopher Kelty, albeit somewhat more nuanced) reject any aspiration by marketers to pursue what they see as outdated marketing projects aimed at manipulation and control of demand. In communicative capitalism, according to this discourse, control of consumers is not only *not* possible but *unethical*; it runs up against the ethos of the network—an ethos of self-organization, individual freedom, collaboration, and experimentation. Hence, from the perspective of the cyber-commonist marketer, the cybernetic definition of consumer control is the antithesis of centralized and corporate marketing, because the latter presupposes—or at least aspires to—a complete knowledge of each individual component of the overall system, which is a notion that must be rejected as impossible to achieve in these open, self-organizing and decentralized consumer community systems (Terranova 2004).[1] Put differently, from the perspective of online community, marketers' consumer control means control of a community, which can only be done from within because the community is respected as a self-organizing, self-determining and always emergent organism.

In commonist marketing the consumer subject is addressed as a moral individual with bonds of obligation and responsibilities for conduct that is assembled in a new way: a consumer in his or her community is both self-responsible and subject to certain emotional bonds of affinity to a circumscribed "network" of other individuals—unified by some kind of shared passion or interest, perhaps by a moral commitment to environmental protection, or to child safety, to exploring teenage angst or animal welfare, or personal hygiene. Communities thus constructed, whether by the marketer himself or by a collective of individuals, must be understood as a particular moral community—or as Miller and Rose (2008, 91) put it "a network of allegiance with which one identifies existentially, traditionally, emotionally or spontaneously, seemingly beyond and above any calculated assessment of self-interest and monetary expectations." Simple-minded forms of self-interested communication with thinly masked commercial interests must fail in this environment. Neither the medium nor the message can any longer be controlled, and thus neither can the recipient of communication.

Hence, we arrive at another contradiction: there is a tension between the right to free self-determination of the community and its members and the generally unquestioned right of the marketer to be a part of this determination (recall Kutcher's claims above). The management of this tension becomes the craft of the commonist marketer. It requires a new technique of management, or put differently, marketing becomes governmental at this particular moment when communities of consumers become at the same time a technical problem and a new technique of marketing.

Initially, community in general, and the virtual consumer community in particular, is invoked by marketers as a possible antidote to the psychological simplification and sociological isolation of the individual generated by traditional marketing. This idea of community as an authentic expression of consumer desire and interest is now deployed in the marketing field as part of the language of critique and opposition, directed against remote and bureaucratic marketing. In contrast, community-touting marketers utilize the language of community to comprehend the problems they encounter in dealing not just with populations partial to marketing messages but, more importantly, with difficult zones—"the anti-Starbucks community," the anti-sweatshop community, and so on (Miller and Rose 2008). Consumer communities here become a point of penetration of a kind of ethnographic sociology—now called netnography by marketing and consumer researchers—into the vocabularies and classifications of these communities. As commonist marketers intensify their investigations of collective life and its moral order, what began as a language of resistance to and critique of traditional marketing was quickly transformed into an expert discourse and a professional vocation in it its own right—community is now something to be programmed by chief community officers, developed by community development officers, policed by community monitors, and rendered knowable by netnographers pursuing "consumer community studies." Even as consumer communities are touted as essentially uncontrollable and unmanageable (see Gabriel and Lang 1995), commonist marketing experts made them zones to be investigated, mapped, classified, documented, interpreted, their vectors explained, to enlightened marketing managers-to-be (or enlist new doctoral students) (Miller and Rose 2008).

This discussion, of course, borrows from what Foucault (1991) termed government. Unlike the more top-down approach of disciplinary power, aimed at shaping the actions of individuals through the imposition of orders, rules, and norms, government works from the bottom up and represents a form of power that "acts through practices that 'make up subjects' as free persons" (Rose 1999, 95). It is, hence, a political form of power that aims at generating particular forms of life (Rose 2001, 1999), which in the context of new strategies of customer management, means "the provision of particular ambiences that frame and partially anticipates the agency of consumers" (Arvidsson 2006, 74).

Thus, commonist marketing's central idea is that control over consumers and markets can best be achieved by providing managed and dynamic platforms for consumer practice (cf. Lury 2004) which, on the one hand, free the creativity and know-how of consumers and, on the other, channel these consumer activities in ways that are desired by the marketers. Customer management, then, as the exertion of political power to produce particular forms of life, clearly does not mean domination because marketers presuppose and, in fact, expect, the consumer subject to act, innovate, tinker and run free. Therefore the marketing challenge posed by the consumer community model rests with establishing ambiences that program consumer freedom to evolve in ways that permit the harnessing of consumers' newly liberated, productive capabilities. In short, marketing with and through virtual consumer communities is a technique of the conduct of consumer conduct.

3) The function to dispel with the belief that marketers create value.

It is the task of marketing to make sense of and shape the market, to understand the motivations of consumers, and to find ways and means for connecting a product with a buyer. In other words, marketers of products and services must find a path to ensure that consumers perceive the offer presented to them to be of sufficient value. Marketing, then, is a particular mode, or strategy, of valorization.

When in the early 2000s management scholars Prahalad and Ramaswamy (2000, 2002, 2004a, 2004b) began to write a series of essays suggesting that the locus of economic value creation is shifting from the firm's research and development department to the interaction between the firm and the consumer, traditional conceptions of marketing as valorization strategy became challenged in a fundamental way. Accordingly, the writings of Prahalad and Ramaswamy caused great excitement among scholars of markets and marketing, giving birth to an area of research that is now, a decade later, commonly referred to as value co-creation. The term denotes that the production of value that takes place increasingly via the interaction between firm and consumer is the outcome of both collaborating in manufacturing products, services and, increasingly, communication. As anthropologist Robert Foster (2007, 715) points out, "this engagement has been identified as a trend, dubbed 'Customer-Made' and defined as 'the phenomenon of corporations creating goods, services and experiences in close cooperation with experienced and creative consumers, tapping into their intellectual capital, and in exchange giving them a direct say (and rewarding them for) what actually gets produced, manufactured, developed, designed, serviced, or processed."

Co-creation, as Ritzer (2009) reminds us, is neither historically new nor specific to twenty-first century communicative capitalism. Rather, by recognizing that production and consumption are two sides of the same coin, co-creation is intrinsic to all forms of capitalist and noncapitalist economies. Nevertheless, recent social transformations such as the emergence of the internet and, in particular, its user-generated version commonly called Web 2.0 with its more social-communitarian character, have moved practices of co-creation to the center of a firm's economic value creation (Ritzer 2009; Ritzer and Jurgenson 2010). Accordingly, recent marketing and management discourses of co-creation aim to reconfigure the production of use and exchange value—previously considered internal to, as well as the sole purvey and "competitive advantage" of, the firm—as increasingly dependent upon the active participation of formerly passive(ied) consumers (see e.g., Donaton 2006; Lagace 2004).

In addition, marketing practitioners like to allude not only to the inevitability of rising consumer power but also to the strategic imperative of voluntarily handing over control to consumers in order to ensure future value creation (Zwick et al. 2008). Behind this "surrender" is marketers' widespread belief that consumer masses provide a stock of almost unimaginable creative and innovative talent that awaits exploration, development and exploitation by smart companies (Prahalad and Ramaswamy 2000; Thomke and von Hippel 2002; von Hippel 2005). The growing role of customer co-creation in producing economic value and driving innovation in communicative capitalism has seen many commentators, including some from the critical camp, wonder whether we are witnessing a shift towards a new type of capitalism—a kind of "co-creative capitalism" where firms become enablers and resource providers for customers to create economic value (Arvidsson 2006; Lusch and Vargo 2006). The newly emerging consumer subject has many names: prosumer, produser, protagonist, post-consumer, consum-actor, etc. (see e.g., Cova and Dalli 2009; Ritzer and Jurgenson 2010). What these terms have in common is that they refer to a vision of customers as active and productive creators of economic value rather than passive users of goods and services, and that such production of value substitutes, if not replaces, traditional forms of value creation by the marketer.

The role of commonist marketers, then, is to find new ways to valorize communities, or more generally, the activities of all new commons, increasingly to be found online (see e.g., Bonsu and Darmody 2008). Herein arises a difficulty because at the same time that the collaborative ethos of communities becomes a viable alternative to forms of collective value production and collective ownership, commonist marketers must find ways to align these productive potentialities with the need of capital to capture and privatize the economic value produced in communitarian associations—associations that may even consider themselves as operating outside the logic of capitalism (Dyer-Witheford 2009). In short, the challenge for the commonist marketer is not only to understand that he or she is no longer in charge of producing all of use and exchange value but to find ways to nevertheless appropriate all the economic value from the cultural, technological, social, and affective labor of the consumer masses (see e.g., Zwick and Ozalp 2011).

Conclusion

Marketing managers love the idea of the virtual consumer community replete with dedicated brand ambassadors, idea generators, and problem solvers that revel in the pleasure of self-determined productivity and fun. The reality, as described in the many case studies found in the books of Solis, Weinberger, Stratten and others, is often a lot more sobering and boring. Conversations among consumers are more likely to be purpose-driven and pragmatic (e.g., getting advice on how to fix a software problem or how to get a discount for a specific product), customer participation is often transitory, and relationships between community members are weak and socially insignificant. And yet, virtual consumer communities represent a strong fantasy for marketers operating in the context of communicative capitalism, not least because of a veritable avalanche of practitioner-oriented consulting books and articles declaring their strategic importance and marketing benefits. Undoubtedly, there are some virtual customer and brand communities that attract a sizeable crowd, although when measured against the total number of customers of the brand, the number is likely to be trivial. In addition, empirical evidence seems to suggest that the notion of community is simply not appropriate to describe the often transitory and generally uncommitted flocking behavior of consumers online. Finally, drawing on Žižek and Dean we have suggested a theoretical basis from which to evaluate the (im)probability of virtual consumer communities to form and persist.

Yet, even if virtual communities represent an unlikely entity, we claim that its plays an important ideological role in the transition and legitimization of a new form or marketing that is based on notions of bottom-up peer marketing, consumer empowerment, and democratized value creation. By its proponents, consumer communities are elevated to the status of a subversive countercultural force that challenges the oppressive "command and control" structure of traditional corporate marketing approaches. The community comes to represent the new reality of communicative, rather than corporate, capitalism: a capitalism that is based on individual empowerment, self-realization, respect for the will of customers, free information sharing, and networked collaboration afforded to us all by the virtues of new communication technologies. As Ashton Kutcher knows well, marketers must adapt. For the cyber-libertarians—on whose ideological coattails today's commonist marketers ride—communicative capitalism promises freedom from both a regulatory (freedom-denying) government and a monopolistic (information exchange-denying) corporate class, while at the same time providing the conditions for unencumbered spaces of collectivity, solidarity and self-determination (see e.g., Kelly 2009). "Open source" becomes the new paradigm for everything, from writing software code to building a new society. Not surprisingly, then, Kelly's vision of the coming of global virtual communism includes the idea of a non-state/non-corporate community as at the same time constitutive of, and constituted by, this virtual communism.

Marketers must, nevertheless, find ways to reach customers, appropriate value and sell things; but they must now do so within the discursive frame of communicative capitalism. In other words, marketing must find ways to persuade and control consumers, push messages and products, and create economic surplus without appearing to do any of these things. It is for this reason that we refer to the community as an ideological mediator because, to put it in Lacanian terms, the transition from one form of marketing to another is a transition on the level of the symbolic. Note how in their attempts to overcome today's three main antagonisms of marketing—Who does the marketing? Who has control? Who creates value?—commonist marketers never challenge the existence nor the purpose of marketing itself. In the final analysis, marketing cannot change its objectives. To survive, it must, however, always re-create the symbolic structure through which we come to understand what marketing is. Marketing thus remains a technique of consumer commodification, just as Dallas Smythe observed long ago. What has changed since his ground-breaking work is the ideological sophistication with which theorists and practitioners of marketing pursue the commodification of the audience.

Note

1 It would be fair to criticize this argument for setting up a straw man because such knowledge is impossible to achieve in any business function. In addition, theories and practices of traditional marketing have never been based on complete knowledge of each individual consumer: marketing strategies as such have always been virtual, exercised as a potentiality—that is, as a desire, aspiration, or threat.

PART THREE

THE POLITICAL ECONOMY OF MEDIA TECHNOLOGIES: THE INTERNET, MOBILE DEVICES, AND INSTITUTIONS IN INFORMATIONAL CAPITALISM

CHAPTER THIRTEEN

From Googol to Guge: The Political Economy of a Search Engine

Micky Lee
Suffolk University

Googol, or 10^{100}, was the name that Sergey Brin and Larry Page chose for the search engine when they were graduate students at Stanford University. Google—now synonymous with "search engine"—is a misspelling of googol. "Guge" is how Google is pronounced in Mandarin Chinese. The Chinese characters of Guge roughly mean "valley song," a name that neither reflects technological prowess nor resonates with Chinese speakers.

The anecdote of Google being a misspelling can be seen as a "cute" story showing the founders' naïveté about the business world. Another often recited cute story is that Google was not incorporated when Brin and Page received a check of $100,000 USD from the venture capitalist Andy Bechtolsheim. Both stories seem to reinforce that Brin and Page, like many American inventors, deserve their great fortune because they have something useful and unique to offer. In popular thinking, because of the raw talent and grand vision of these inventors, their companies should not be criticized. Instead, their inventions should be viewed as technology that positively impacts society, and as knowledge that leads humankind one step away from darkness. The interface of the Google homepage, along with the Apple logo and Edison's light bulb, are seen to inspire awe and wonder, not to receive criticism and analysis.

Guge, however, tells a different story. When Google entered the Chinese market in 2006 and complied with the censorship rules of the Chinese Government, some U.S. politicians called the company, along with Microsoft, Yahoo, and Cisco the high-tech Gang of Four. Google was criticized by human rights advocates for its assistance in helping the Chinese Government limit citizens' freedom of speech and access to information. In December 2008, the server of Google.cn was hacked into. The Chinese government was believed to have masterminded the scheme in an attempt to read the e-mails of human rights activists. Because of this, Google decided to withdraw from China in March 2010. It subsequently moved the Chinese

Headquarters to Hong Kong, where Internet regulations are outside the jurisdiction of the People's Republic of China. When visiting the uncensored search engine, users in China saw a blank page and experienced slow connection speed. Since its withdrawal, Google has lost 20% of Internet search market shares to Baidu, the most popular search engine in China. By September 2012, it only had 15% market shares (Tejada 2012). The Guge episode is an unhappy one in the Google story: not only did Google compromise its vision of "organizing the world's information" and abandon its motto "Don't be evil," but it has also failed to capture one of the fastest-growing Internet markets.

In the time since I have offered a political economic critique of Google (Lee 2010a, 2010b, 2010c, 2011), there has been an exponential growth of critical studies of search engines, as evident from the number of published articles. To build on existing literature, this book chapter aims to: first, update a political economic critique of Google by summarizing and commenting on critical studies published since 2009; and second, to relate Dallas Smythe's observation of technological development in China to Baidu. For the first aim, I argue that to understand the political economy of Google necessitates an understanding of Google's advertising system, central to which is the concept of the audience. For the second aim, I argue that technological development should be examined in relation to corporate structure, research and development (R&D) investment, and the collaboration between corporations and educational institutions.

Google's AdWords

The political economic approach to communication conceptualizes technology as a commodity produced by corporations. The goal of corporations is to maximize profit. Political economists study the process of capital accumulation and examine the evolution and renewal of capitalism. Applying this approach, my article "Google Ads and the Blindspot Debate" (Lee 2011) examines the Google AdWords system. Ninety-six percent of Google's profit comes from advertising (Google 2012). Although a user's search query generates both search results and advertisers' links, Google prides itself on offering objective search results that are not influenced by the money that advertisers offer to Google. The Google founders were initially skeptical of advertising because ad content is not related to search results. The AdWords system solves this problem by generating search-related advertisements. For example, the keywords "corgi t-shirt" generated two ads from vendors that sell dog breed merchandise. Both vendors bid for the keywords "corgi t-shirt" on AdWords. If I click on one of the two advertising links, that vendor will pay Google the amount that it bids for. This is called the "cost-per-click" rate. The quality score of the keywords "corgi t-shirt" for that vendor will go up. A high quality score means that the keywords "corgi t-shirt" will have a lower bidding price for that vendor in the future; also, that vendor's ad will occupy a more favorable position next time a user searches for "corgi t-shirt." However, a user may actually want to search for t-shirts of the Corgi brand of toy cars. In this case, ads that sell dog breed merchandise are unlikely to be clicked on. The quality score of the keywords "corgi t-shirt" of that vendor will go down. By rewarding advertisers who can successfully predict the keywords that users type in and by punishing those who cannot, Google hopes that AdWords can be relevant and useful to online search.

Nevertheless, AdWords is only profitable because it is a vertically integrated system in which Google controls every step of the process by providing search results to users, by selling "keywords" to advertisers, and by providing statistics to marketers. Television stations, advertising companies, and ratings firms have a close relationship (Meehan 1993a), which Google takes one step further, integrating the three types of business into one. Because of this, advertising executives complain that Google wants to alter the rules of the advertising industry (Delaney 2006). Google defended its practices by suggesting that AdWords promotes objectivity because advertisers with limited funds can compete with those with a deeper pocket (Mangalindan 2003). Google also suggested that AdWords promotes objectivity because the quality score is calculated by a number of factors, including whether users click on the ad links and if the landing page looks "clean" (i.e., without flashing buttons and blinking banners).

Despite Google's claims, a vertically integrated advertising system may not be that objective because it is difficult to determine what the exchange value of keywords is. In broadcast and print media, advertising rates depend on both the scarcity of space and time, and the audience size. A full-page advertisement in the *New York Times* costs more than a classified ad in a regional newspaper. Similarly, an advertising slot during a national event such as the Super Bowl costs more than that of a less watched program aired at an unpopular time. In contrast, Internet search creates unlimited time because millions of searches take place at any given moment even though only a small number of users see the same ads. In addition, keywords are not exclusive; a number of advertisers can bid for the same keywords. Because of the non-exclusivity and non-scarcity of keywords, their exchange value should be very low, if not close to zero. This is clearly not the case for Google ads: although there is no scarcity of keywords and although there are a small number of viewers for each ad, the cost-per-click rate for the keywords "car insurance" is $20 USD, and that for "debt consolidation" is $10 USD. In comparison, the cost-per-click rate for "Karl Marx" and "Adam Smith" is cheaper, at only one U.S. dollar each.

The concept of the audience commodity is crucial to understanding why Google is able to charge advertisers for keywords even though they are neither exhaustive nor exclusive. Maxwell (1991) has effectively argued that the exchange value of the audience commodity is imaginary because it is an abstract concept, not an aggregate of audience. The Google advertisers do not care what flesh-and-blood users do when they search for keywords: do they search during work hours? At leisure time? Are they happy when they search? Are they anxious? What advertisers care about is statistics such as: how many times users search with certain keywords; how many times users click on the ads next to the search results; and where the ads appear. To illustrate this, the audience commodity of the keywords "car insurance" has a higher exchange value than that of "Karl Marx" not only because the keywords "car insurance" are searched for ten times more than "Karl Marx," but also because users who search for "car insurance" are 12,300 times more likely to click on a related ad than those of "Karl Marx." (The statistics are available at Google AdWords.) On average, an ad returned with the keywords "Karl Marx" is only clicked on once every 25 days. The statistics do not show the *actual* number of users who search for certain keywords (presumably some users search for the same keywords multiple times) and users' interest level (casual search or purposeful search). Even if the *same* user searches for "car insurance" and "Karl Marx," the exchange value of the two audience commodities is *different*. This reinforces Maxwell's claim that the exchange value of the audience commodity is imaginary.

Critical Studies of Google Since 2009

Since 2009, a number of critical studies of Google have been published. The focus tends to be more on prosumers and on the free labor performed by Google users rather than on the advertising system and the audience commodity. Here I argue that the key to the political economy of Google as a search engine is not the free labor performed by the prosumers, but how the advertising system works. Another troubling trend in critical studies of Google is the de-emphasis on (or even negligence of) the Marxist dialectical, historical materialist approach. A dialectical approach recognizes that the concept of labor would have to be understood *in relation to* those of commodity and value. In addition, each of the three concepts has a dialectical pair: concrete labor vs. abstract labor; resources vs. commodity; use value vs.

exchange value. An historical materialist approach recognizes that capitalism has a history, and that it constantly evolves and renews itself. What is deemed "new," especially in the study of technology, should be understood as a crystallised moment of the circulation of capital. This will be further elaborated after reviewing recent critical studies on Google.

The Googlization of Everything (Vaidhyanathan 2011), *Search Engine Society* (Halavais 2009), and *Deep Search* (Becker and Stalder 2009) are three scholarly books that look at the social and cultural impacts of Google. Some specific discussed impacts are: PageRank is claimed to be objective but its algorithm prioritizes sites that are already popular (Halavais 2009; Lobet-Maris 2009; Vaidhyanathan 2011). The search results provide good over best answers to queries (Halavais 2009). The personalization function of Google "fractures a sense of common knowledge or common priorities" (Vaidhyanathan 2011, 139). Google compromises the privacy of users by showing their faces and houses in Google Maps (Vaidhyanathan 2011), by using unexpired cookies, and by making individual data searchable (Halavais 2009). Further, Vaidhyanathan (2011) has pointed out that Google is a company, not a force to do good for society. He coined the term "googlization" to show how Google has permeated culture and has affected how users view themselves, the world, and human knowledge. Lastly, Vaidhyanathan has cautioned that Google should not be seen as a social and cultural steward because companies rarely last for more than a century.

The aforementioned volumes have hinted at Google's commercial interests but have not delved deeply into the political economy of the company. Because of this lack, Fuchs (2011a) (see also Fuchs 2012b; Fuchs and Winseck 2011) has thus made a significant contribution to the understanding of Google's capital accumulation process. Built on Marx's M-C-M' (money-commodity-more money) model, Fuchs (2012b) presents Google's capital accumulation model, as denoted by M-C…P_1-P_2…C'-M'. The process begins with Google investing money (M) to buy capital (C)—in the forms of labor power and technology. Google employees provide services (P_1) for users. Because most services are provided free of charge, they are not commodities. The popularity of Google's free services leads to a large number of users performing unpaid labor (P_2) by both generating data through searches and by producing web content. The Google users (C') are then the double objects of commodification because "(1) they and their data are Internet prosumer commodities themselves; (2) through this commodification their consciousness becomes, while online, permanently exposed to commodity logic in the form of advertisements" (Fuchs 2011a, para. 21). Finally, M' is the money that Google makes from advertising. Fuchs has laid out a more complete picture of how Google accumulates capital than the scholars referenced above. However, as I will explain further, the notion of prosumers providing free labor is problematic.

Other studies that also discuss the political economy of Google include Kang and McAllister (2011) and Pasquinelli (2009b). Kang and McAllister (2011) also suggest that the audience and its consciousness are the commodities, and that users provide free labor for Google. Pasquinelli (2009b) suggests that the political economy of Google is the political economy of PageRank. Value accumulation comes from the economies of attention—which depends on the attention capital of the whole network—and of cognition—which depends on Google being a "rentier" of the Internet. To Caraway (2011), "the media owner rents the use of the medium to the industrial capitalist who is interested in gaining access to an audience" (701). Pasquinelli (2009b) has overlooked that technology only increases productivity, but it does not create surplus value. Fuchs (2012b) has effectively shown that it is erroneous to assume that PageRank produces profits. In addition, Pasquinelli (2009b) has ignored that PageRank preceded the Google corporation and Google's advertising system. In other words, the political economy of the non-profit search engine google.stanford.edu is different from that of Google.com not because of PageRank, but because of advertising.

Some other articles critically examine online free labor and user-generated content (thereafter UGC). These studies do not look at Google specifically, but their assumptions of free labor and UGC have implications on an understanding of the political economy of Google. The notion of the audience providing free labor for both the media and advertisers in exchange for free content has been advanced by Dallas Smythe (1977). Smythe contended that the audiences sell free labor to broadcasters when they watch television. He concluded that watching is a form of work. Smythe's notion of free labor has received much attention in critical studies of new media. Online users now not only passively watch and read content from the Internet, but they also actively search for information and produce content that can be indexed by search engine and searched by other users. Napoli (2010) believes that "the notion of the work of the audience becomes much more concrete in an environment in which the creative work of the audience is an increasingly important source of economic value for media organizations" (17). He further suggested that users "willingly engage in the work of the marketers" (19) and work for both advertisers and media organizations.

The assumption that users are active producers of content has probed critical scholars to re-focus on the centrality of labor in capitalism. Marx stated that commodified labor is necessarily abstract, objective, and alienated. Can these assumptions be made of the free labor provided by Internet users? Petersen (2008) said no because the new mode of production has a kind of human and subjective capital that is variable and uncontrollable. Workers create new forms of subjectivity and knowledge after being freed from the assembly line. Andrejevic (2009a) believes that UGC is produced with immaterial and affective labor.

Immaterial labor consists of activities that produce the cultural content of the commodity; these activities are usually not considered work. Informational labor is a form of immaterial labor because it involves activities that are not considered "work," such as those that "define and fix cultural and artistic standards, fashions, tastes, consumer norms, and public opinion" (Lazzarato 2006, 142). Immaterial labor thus blurs the boundary between leisure and work, and it produces social relations rather than goods.

Similar to immaterial labor, affective labor produces and manipulates human affects, contact, and proximity. Hardt (1999) believes that affective labor constitutes collective communities and subjectivity. It produces social networks, forms of community, and biopower. To him, affective labor can serve as a crucial ground for "anticapitalist projects" (p. 89). However, Andrejevic (2009a) cautions that the unequal power relations between corporations and users structure what constitutes "free" choice. Further, users' free participation is capitalized as a form of productive labor.

To conclude, the notion that the audience works in both traditional and new media environments has received criticism. Artz (2008) suggests that the audience does not voluntarily consent to sell their labor to television stations. Caraway (2011) adds that there is no commercial transaction between the audience (as the owners of their labor power), the advertiser, and the media organization.

Political Economy as a Dialectical, Historical Materialist Approach

In this and the next sections I first critique the concepts of prosumers and free labor by arguing that labor is treated as a static object, not a component in relation to value, commodity, and time in the circulation of capital. I then proceed to argue that existing critical studies on Google have focused on the "newness" of technology and have neglected to situate the political economy of Google in the continuity of capitalism.

A dialectical approach seeks to understand and represent processes of motion, change and transformation (Harvey 2010). Marx emphasized the *relations* between concepts, one of which is that between labor, value, and commodity. As explained by Harvey (2010), the *commodity* form is a universal presence within a capitalist mode of production. Commodities are all bearers of the human *labor* embodied in their production, and are material bearers of exchange *value*. Exchange value is the sum of surplus value and the price at which labor is bought by capitalists, which in turn is the cost that is "necessary to maintain and reproduce the life of the worker and the subsistence of labor, in general" (Heydebrand 2003, 151).

In addition to the relation between labor, value, and commodity, Marx has also pointed out the relation between concrete labor and abstract labor, and that between use value and exchange value. Praxis is primary to human existence; therefore, human labor is applied to transform natural resources into goods of use value. The questions then are: how does concrete labor become abstract labor? How does use value become exchange value? How do concrete labor and use value that are qualitative become abstract labor and exchange value that are quantitative?

Scholars who wrote about prosumers rarely define what free labor is and hardly problematise what it means in the new media environment. It appears that the "free" in free labor is free as in "free" beer rather than free as in "free"dom. The notion of free labor implies that there also exists labor with a cost (that is, commodified labor) and that this free labor is quantitatively similar to commodified labor.

These scholars neglect that free labor is an historical concept. Children do not provide labor because their labor power cannot be legally sold in most countries. The seriously ill and the dying may also have no free labor to sell because of the lack of health, hence the lack of labor power. Nowadays there is more awareness that mothers provide free labor to the household and to the community. Free labor is also a class concept. Individuals with independent means do not sell labor power for a living; their labor power is by default non-existent. The Queen of England may work, but she does not labor. Lastly, free labor implies workers possess certain levels of skills offered with consciousness and intention (Kang and McAllister 2011)—a baby banging fists on a computer keyboard does not constitute free labor.

Likewise, free labor in the new media environment is assumed to be performed by individuals who possess labor power and who have a certain level of competencies. At its most basic, free labor is disposed through conducting a search on Google and viewing online advertisements. This requires access to a computer and some level of literacy. At the advanced level, power users—those who have a registered account with Google (Vaidhyanathan 2011)—send e-mails, save files in Google Cloud, and upload videos on YouTube.

If the above are the assumptions of free labor in the new media environment, then there are at least four problems. First, online labor does not materially transform natural resources into goods (Harvey 1982). As I have argued (Lee 2011), watching television cannot be considered work because an advertisement watched once or a million times is still the same ad; there is no material transformation during watching. Similarly, a Google ad clicked once or a million times remains the same one; it is neither exhausted nor consumed. However, if a user digitalizes content (for example, scanning books for Google Books), then a material transformation occurs, through which surplus value is created.

Second, there is no distinguishing between concrete labor from abstract labor. To Marx, the former is exercised with a definite aim, is heterogeneous and qualitative while the latter is used to produce surplus value, is homogeneous and quantitative. To give an example, an individual who bakes a loaf of bread for self-consumption exercises concrete labor. A baker who makes a loaf of bread for sale exercises abstract labor. The concept of "free labor" seems to imply a type of *abstract* labor that is provided at no cost. That is, if the abstract labor is spent elsewhere other than using Google or producing web content, the laborer will be paid. However, in actuality, UGC is produced with concrete labor. Communicating on a social network, uploading a video online, and writing on blogs are leisure activities; they are more like baking one's own bread and tending one's own vegetable garden. In these cases, the individuals dispose of labor at their own will and at their own freedom. I concur with Artz (2008) and Caraway (2011) that the laborers are not forced to exercise their own labor. The only difference between online activities and offline activities is that users' data and web content are later capitalized by companies while homemade goods are usually consumed, not capitalized. In this case, UGC is more like fixed capital—traditionally defined as buildings, fixtures, and machinery that are necessary for production. Fixed capital such as machinery is used to "bring the power of past 'dead' labor to bear over living labor in the work process" (Harvey 1982, 204). If there is nothing online to search for, then search engines do not exist. A factory cannot just have a building without machinery. UGC can also be seen as non-exclusive, non-exhaustive resources that can be capitalized. It is like spring water being bottled and sold as a commodity. Lastly, although users may be aware of their data being packaged for sale and of UGC being made searchable, they do not *work* for one single company. A blog can be searched using Google or any number of search engines. UGC is not exclusively owned by a company, the intellectual property is owned by the content creators (at least at the time of writing).

The third problem of the notion of free labor is that scholars do not pay attention to the relation between value, labor, and time. Roughly speaking, there are four circumstances under which online content creating and web searching are conducted. The first kind is professionally produced content by workers who are paid to do it. For example, *New York Times* bloggers and Amazon employees are paid to provide content. In this case, the labor is not free because it has been bought by the owners of the means of production. The second kind is technology-assisted paid work. For example, teachers searching for online information for class preparation. Workers do not produce online content, but technology assists the tasks. In this case, the labor is not free either. The third kind is leisure activities performed during work hours. Employers have complained that workers spend working hours sending personal e-mails, shopping online, and browsing the Internet. In this case, the labor is not free because someone else has borne the cost. The last case is UGC uploaded and searches done at users' leisure time (that is, unpaid time). In this case, as already argued, the "free time" is more like concrete time, not abstract time that can be sold otherwise.

The fourth problem with the notion of free labor is that the exchange value of the labor power producing UGC is assumed to be of the same exchange value as that producing professional content. In other words, it is assumed that the surplus value that is created by both amateurs and professionals is the same. This is clearly not the case. There is no need to mention that anyone—amateur or professional—can give financial, legal, and medical advice online, but free information has not driven down the cost charged by professional financial advisors, lawyers, and physicians. The exchange value of professionals' time is still high. Similarly, the labor cost of producing professional videos is significantly higher than that of amateur videos. Advertisers complain that there are not enough high quality and professional videos on YouTube ("Google to Help Broker Video Ads" 2009) because the online video provider is clogged with amateur videos. To advertisers, the exchange value of a YouTube ad spot is then related to the exchange value of the labor power that produces the content. Therefore, not all content and labor power have the same value.

Are the New Media Really that "New"?

Scholars are often so blinded by the "newness" of technology-enabled media that even critical scholars have lost sight in situating the apparently new phenomenon in the continuity and transformation of capitalism. A Marxist approach is necessarily historical materialist. Therefore political economists aim to "[examine] the dynamic forces in capitalism responsible for its growth and change. The object is to identify both cyclical patterns of short-term expansion and contraction as well as long-term transformation patterns that signal fundamental change in the system" (Mosco 2009, 26). Political economists ought to examine the inherent contradictions in capitalism and how capitalism evolves and renews even though—or especially because—it is not a sustainable political economic system. It is thus imperative to examine how corporations seek new capital once they encounter an over-accumulation of capital.

The discussion of UGC seems to imply that content production by users is new. However, users have always produced content with analogue technology prior to the so-called digital age: from writing in a diary on paper with pigment to photocopying fanzines; from recording one's singing on a cassette tape to taking pictures with a roll of film. The "newness" in UGC lies in the digital distribution and consumption of content (Napoli 2010), not the production of content.

If the focus is shifted to the distribution and consumption of UGC, then there are three implications for understanding Google. First, while Google can function with one-tenth or even one-hundredth of current web content, it cannot survive with a smaller number of users. Most of the users do not scroll past the first page of search results, hence less content will not significantly affect the search quality. In contrast, a smaller number of users may drive Google to bankruptcy because it fully depends on advertising revenue. Fewer users mean fewer visits, fewer visits mean fewer clicks.

Second, the market for content distribution is very concentrated with one company dominating one single market: Google dominating the search engine market, Internet Explorer the browser market, Facebook social networking, and YouTube video. Google's competitors are other large corporations such as Microsoft and Yahoo. This highly concentrated ownership can also be seen in traditional media such as television, radio, and newspapers. Therefore, there is nothing new about media concentration in the new media environment.

Third, because the distribution of web content is concentrated, these companies constrain, if not dictate, how users understand content. Previous studies (Halavais 2009; Vaidhyanathan 2011) have rightfully suggested that PageRank rewards popular websites and punishes unknown ones. Companies with plentiful resources such as Amazon.com, Apple.com, and eBay.com have already learned to maximise their benefits at the AdWords system and have understood how to be prominently listed in search results.

If the focus of UGC is shifted to distribution and consumption, then the concept of the audience commodity is as germane to the political economy of commercial media as ever. Smythe's (1977) article has posed one of the most enduring questions to political economists, which is: *what is the political economy of commercial media?* Smythe's mention of commercial broadcast television is better seen as a particular example of commercial media rather than *the* universal and defining example. Smythe wrote the article during the "golden age" of broadcast television when there was a limited number of channels; when few families owned more than one television set; when technologies such as digital video recorders were not in place for viewers to "skip" ads; and when television programmes could only be watched at the time of broadcast. The "golden age" of television has also created the ideal American family with the head of the household, presumably the breadwinner, deciding what the family watches (Meehan 1993b). It is unknown if this ideal mode of watching was the most prevalent during the "golden age." It is also highly doubtful that this ideal mode is at present widely practiced in North America and the rest of the world. By accepting Smythe's example as a particular mode of broadcast television, political economists may better attend to the understanding of the movement of capital via studying the political economy of media—be it old media like television or new media like search engine. Because the object of study is capital, the heart of the question is still the same in both old and new media environments: how does money search for more money through the process of commodification? (Harvey 2010).

To Google, advertising is core to its revenue stream. Since 2009, Google has been rapidly expanding its advertising business. In addition to text ads, it also offers display ads on YouTube and mobile ads on phones with the Android platform (Efrati 2010). Google has also changed its initial stance from bypassing advertising agencies to eagerly working with them by assigning designated employees to advertising accounts (Efrati 2012b). The political economy of display ads is not entirely similar to that of text ads because the ads displayed on YouTube have little to do with the search content. The political economy of mobile ads is also different because by working with phone companies, Google provides a free-of-charge Android platform to users. For the case of the mobile phone, the vertically-integrated system of mobile ads is further strengthened by locking the hardware with the software. How Google advertising works may be theorised again, but the analysis may add little to the understanding of the nature of capital.

On the other hand, what is new is the relation between productive and finance capital. The financial performance of Google (or any public company) is closely watched by institutional investors. Google, once a darling of the stock market, is now seen as an underperforming company. In the fourth quarter of 2011, Google had *only* a 7% increase in profit from the third quarter; and its revenue rose by *only* 25% when compared to a 33% increase in the third quarter (Efrati 2012a). A *Wall Street Journal* article openly expressed disappointment at Google's performance (Letzing 2012).

Because Google charges less for mobile ads, investors worry that the increased use of mobile phones for searches will drive down Google's revenue. Investors clearly do not care how much advertisers pay. The *WSJ* article concludes that "[Google] must continue to make its economics work for shareholders" (Letzing 2012, B2). This quote illustrates whom Google should please—it is not the users, not the content producers, not even the advertisers, but the investors. Institutional investors compare public companies by their financial performance, not by the quality of the goods that they produce, let alone by the positive social impact of the goods. It is hardly far-fetched to suggest that institutional investors are oblivious to how well the average citizen and the national government fare. The average U.S. household income has hardly risen from 2011 to 2012; the U.S. government also does not have a 25% increase in national revenue. The excessive power that investment banks have over public companies may be something new to the trajectory of capitalism, something that occurred since the 1990s in the centuries-long history of capitalism. Marx did not write extensively on the relation between productive and finance capital and political economists in the communication discipline have only paid more attention after the 2008 financial crisis. This gap of knowledge should be a "new" area of attention.

After Guge, What?

In the last section, I turn to Baidu, the most popular search engine in China, to see if a "home-grown" technology may pose a threat to Google. I begin by summarizing Smythe's article "After Bicycles, What?" (1973/1994) and then relating his observations to the quick rise of Baidu, which had already been a dominant player in the Internet search market before google.cn was launched. After google.cn withdrew from China, Baidu dominated the market with a 70% market share. Robin Li, Baidu CEO, said the company has no problem with complying with the censorship rules of the Chinese Communist Party because Chinese entrepreneurs are used to state regulations (Fletcher 2010).

Existing evidence may lead to a premature conclusion that one day Baidu will be the world's number one search engine in terms of popularity and revenue: in the fourth quarter of 2011, the earnings of Baidu increased by 77% from the third quarter. However, in 2010, only 32% of Chinese were online. To globalization enthusiasts such as Thomas Friedman (2007), technology has flattened the world because it has enabled a firm from China to compete with Google (and by association, the U.S.).

In 1971–2 Dallas Smythe visited China to study ideology and technology (Zhao 2011). He pointed out to the Chinese Communist Party officials that they could consider making a two-way television system which enables the receivers to send responses to the broadcasters who in turn replay the responses to the public. The officials replied by saying that technology is neutral. Smythe rejected this suggestion by arguing that: (1) technology is culturally and politically determined because "what constitutes science at any given time and place reflects the world view and the political structure of society of the particular culture at that time and place" (Smythe 1973/1994, 235); (2) resource allocation through R&D is a political decision; and (3) technology facilitates a certain kind of "technique as employed in practice after the innovation" (1973/1994, 235). A comparison of Baidu and Google hence requires more than a comparison of the companies' services, sizes and revenues. It requires a comparison of the broader political, economic, and social conditions under which technology is invented.

Elsewhere, I wrote that "Baidu.cn is not a search engine in China, but a global search engine in the Chinese language" (Lee 2010, 21). In addition to the Chinese language, the company has also launched search engines in Japanese, Thai, and Arabic (Chao 2011b). Non-alphabetic languages are Baidu's specialty. Now Baidu hopes to enter the fast-growing Internet market in Brazil (Chao 2013). Even though Google is not the most popular search engine in China, its influence on the aesthetics and technology of Baidu is obvious. In fact, it is questionable if any search engine launched after Google can avoid copying, referencing, and being compared to the Google search engine. Critics said that Baidu is unoriginal, it copies the interface from Google, Wikipedia, and Amazon (Fletcher 2010). Baidu defended itself by saying that unlike Google, Baidu does not innovate for innovation's sake (Chao and Back 2010). In addition, Baidu users also do not seem to care that it is not original (Chao 2009). Developing technology by the locals for the locals remains a challenge. Similar to Google, the homepage of Baidu is minimal. Underneath the logo is the query box, which is longer and wider than the Google one (Chao 2009). Listed above the query box are: news, search, discussion board, knowledge, MP3, image, video, and map. All these services are offered by Google, except the MP3 service to download MP3 files for free (Chao 2009). This service may only be original in Baidu because of relaxed Chinese copyright laws and because the Apple iTunes Store has no Chinese interface.

The majority of Chinese Internet users experience Google through Baidu. More so than Google, Baidu's and google.cn's various services promote consumption. For example, Google.cn lists goods for online shopping underneath the search boxes, and the Baidu shopping site is like Amazon.com (Lee 2010c). While critics like to idealize the Internet as a tool of liberation and they are concerned about the filtered content in Baidu, Baidu users are not particularly interested in censored topics such as Taiwan, Tibet independence, and human rights; instead they are interested in entertainment and leisure news (Fletcher 2010). In the same vein, it is unrealistic to assume that the average Google user in the U.S. searches for topics such as democracy, the U.S. Constitution, and the various Amendments.

Baidu's similarity to Google goes beyond the interface and the services. Baidu favours a U.S.-styled corporate structure with an international Board of Directors over a more traditional family-based business. Its management team consists of overseas-educated Chinese who have worked for Fortune 500 companies. Baidu management can be seen as a collection of neoliberal technocratic elites who assist the party to implement a top-down, state-led "information revolution" to fulfill the Four Modernizations (development in agriculture; industry; national defence; and science and technology) (Zhao 2007, 2011). Similar to Google, Baidu received initial funding from a Silicon Valley venture capital firm. The Chinese Communist Party prohibits foreign ownership in the media, but it allows foreign investment in technology companies because they do not produce content. Yet sensitive topics on Baidu discussion boards have alerted the party officials, who have ordered Baidu and other technology companies to receive political training (Chao 2011a).

Baidu and Google are both public companies. Taking advantage of Chinese laws allowing local companies to list overseas, Baidu became a NASDAQ company in 2005 and is the biggest U.S.-listed Chinese company in terms of market value (McMahon and Fletcher 2011). As previously stated, the corporate structure of Baidu does not adopt a traditional Chinese business model where familial relationship and filial piety (that is, sons submit to fathers) dictate power relations in companies. Yet filial piety may limit the power of non-state-owned corporations in China. Baidu has found it difficult to list in Chinese stock exchanges because of the domination of state firms (McMahon and Fletcher 2011). This illustrates that neoliberals have failed at a wholesale privatization of the information society in China (Zhao 2007, 2011).

In addition to state regulation, some infrastructural problems in China also prevent the country from nurturing "home-grown" technology companies. Even with its fast-growing economy and huge population, China may find it hard to invent "home-grown" technology that serves local needs. The invention of technology like Google is an outcome of the U.S. highly developed academic-military-industrial complex, in which federal funding is channeled to the university, the military, and the industry for R&D. Brin and Page's alma mater Stanford University exemplifies how the engine of the academic-military-industrial complex runs. At Stanford, faculty are expected to apply for external funding for tenure and promotion. Working as consultants for the government and corporations is encouraged. Technology transfer policies such as the Bayh-Dole Act of 1980, and the Federal Technology Transfer Act of 1986 stipulate that publicly-funded projects be commercialized. According to the OECD, the total R&D funding in the U.S. was $401,576 million USD while that in China was $92,066 million USD in 2008.[1] To capture the fruits of innovation, Stanford University has the Office of Technology Licensing to assist faculty with applying for technology transfer, thus capitalizing on invention. The university also has a network of alumni and venture capitalists in Silicon Valley. Eric Schmidt, an alumnus of Stanford and the ex-CEO of Google, serves on the university's Board of Trustees.

Furthermore, U.S. immigration policy welcomes highly educated technical workers to stay and work. Companies such as Microsoft have lobbied the Congress to increase the quota of work visas for highly educated and skilled immigrants. Sergey Brin of Google migrated to the U.S. with his parents, a college professor and a scientist, from the former USSR. In comparison, China lacks any strategic immigration policy to attract foreign talents. It is hard to imagine that China would attract foreign nationals to permanently stay and work.

Lastly, China does not have a world-class educational system. According to CIA World Fact, the average Chinese has twelve years of formal schooling whereas the average American has 16 years. The literacy rate of the Chinese population is 92.2% while the percentage is 99% in the U.S. China's higher education institutions are also not competitive with U.S. schools. Top universities like Peking University and Tsinghua University, although dubbed as the Oxford and Cambridge of China, cannot compete with top notch U.S. universities in terms of resources, international influence, reputation, and faculty quality. If the *Times Higher Education* ranking of the best universities in the world is an indicator, Stanford University is the second best whereas Peking University and Tsinghua University are No. 46 and 52 respectively.[2] However, the number of Chinese people studying overseas has been sharply increasing, and this may have a long-term impact on "home-grown" technological development in China.

Recall the "cute" story of Larry Page and Sergey Brin being handed a $100,000 USD check by a Silicon Valley venture capitalist (who is a German native emigrated to the U.S. after Stanford). This story would not happen just anywhere in the world. This story represents an outcome of the academic-military-industrial complex, which reinforces U.S. domination and political economic advantage. Smythe (1973/1994, 238) said that "there is no *socialist* road to western capitalist technological development." The case of Baidu illustrates this statement well because technology has already been imagined and dreamed by inventors in advanced economies. These dreams and imaginations also require an abundance of resources to come true. Without the resources, dreams remain dreams.

Notes

1 OECD. StatExtracts. http://stats.oecd.org/Index.aspx?DataSetCode=GERD_FUNDS. Last accessed: May 7, 2012.

2 The World University Rankings. *Times Higher Education*. http://www.timeshigher education.co.uk/world-university-rankings/. Last accessed: February 14, 2013.

CHAPTER FOURTEEN

"Free Lunch" in the Digital Era: Organization Is the New Content

Mark Andrejevic
Centre for Critical and Cultural Studies
University of Queensland

The spectre of Dallas Smythe haunts the recent popularization of the aphorism, "When something online is free, you're not the customer, you're the product"—uttered perhaps most famously by Jonathan Zittrain (who admitted he did not coin it) but appearing in a variety of reflections, critical and otherwise, on the emerging online economy (Zittrain 2012). It is an aphorism that should be read alongside a second set of claims about personal data as "the new oil" (Rotella 2012) and a new "asset class" (World Economic Forum 2011). The common sense wisdom of the online economy, in other words, has neatly aligned itself with a logic that Smythe (1977) identified several decades before its emergence. This fact does not necessarily vindicate the entirety of the substance of Smythe's analysis for contemporary digital media, but it does suggest that his formulation has something to tell us about the character of the changing media landscape.

There is little doubt these days that data about audiences is a commodity. This is all but indisputable in an era in which entire business models have grown up around the collection and sale of personal data about consumers, audiences, and citizens. There is also little debate about the fact that audiences participate in significant ways in generating this data—after all it is about them and their activities. To the extent that there remains fruitful room for debate and discussion, I think that it can be found in claims that the generation of this data is a form of value-generating labor and that such labor might in turn be framed in terms of exploitation (see, for example, Fuchs 2010b). For this, after all, was what was at stake in Smythe's analysis: that audiences are engaged in forms of productive activity that generate surplus value for the media organizations that produce audience commodities.

One of the significant developments of digital interactive media has been to redouble the audience, which becomes not simply a target to be sold to advertisers, but also the source of an increasing range of information commodities about audience preferences and behavior. Exposure to audiences is not simply sold to advertisers, but data about audiences is also sold to those who use it to customize goods, services, and marketing appeals. This redoubling is not entirely novel: selling audiences means measuring them, and these measurements (in the forms of ratings) thus become a commodity unto themselves. Audiences for commercial fare have always been tracked and monitored, and the information about them has therefore served as a source of value. That is to say, from the inception of commercial broadcasting, the work of being watched developed alongside that of watching. Compared to the vast data frontiers opened up by digital developments, however, the scope of ratings commodities in the mass media era is relatively crude and task specific (while nonetheless remaining the result of highly sophisticated formulas for probability sampling). In the era of the digital data trove, new forms of data generation and collection take on a wide range of roles, decoupling themselves from strict service to the commodification of advertising time and space—although these remain an important part of their remit.

Recent work updating the notion of the audience commodity and of audience labor for the digital era has focused on the changing character of what is sold to advertisers: no longer time or space, but user activity or key words or searches (see, for example Lee 2011 and Bermejo 2009). These analyses are useful insofar as they highlight the shifting logics of advertising in an era in which "flow" (Williams 1974) is displaced by search, and space is virtually unlimited. Clearly traces of earlier logics of advertising remain: inserting ads in front of YouTube videos recapitulates the logic of the sale of time (15 seconds before a video starts, for example), and ad size and placement still has a role to play in many forms of online advertising. However, the digital medium is a much more malleable and thus customizable medium—which means that spaces and times can be more narrowly framed and targeted (that is, a particular space can show one ad to one user and a different ad to another user on the same page). Moreover, online ads generate their own feedback—creating additional data that can be folded back into the process of customization. The upshot is that emerging advertising regimes are becoming ever more data-intensive, a fact which helps explain the demand for increasingly comprehensive information about audiences. At the same time, new strategies for audience management draw on these new forms of data to predict consumer behavior and response, based upon ongoing forms of interactive, controlled experimentation.

If the ratings game was largely an anonymous one, insofar as small samples were used to generalize about large groups of unknown viewers, new forms of audience measurement are increasingly personalized and the resulting advertising commodity forms are correspondingly diverse and flexible. Online advertisers can buy highly specialized and customized sets of potential viewers, but only if the data exists to sort through available users and select the appropriate audience. In this regard, the production of audience commodities parallels that of other types of commodities, including for example, customized products that come with personalized prices, and marketing strategies that disaggregate markets in order to more effectively capture forms of so-called consumer surplus (the "extra" that some consumers are willing to pay for a commodity with a fixed mass market price).

As Smythe (1977) pointed out, to speak of the audience as a commodity is also to invoke the notion of audience labor: the fact that audiences must do something of value for advertisers in order to be worth their asking price. This is the critical substance of Smythe's formulation: an interest in the work that gets done by audiences—and therefore in the ways in which the realm of media consumption can be approached as a site of production. The work that audiences do, according to Smythe, "is to learn to buy particular 'brands' of consumer goods, and to spend their income accordingly. In short, they work to create demand for advertised goods" (1977, 6). The fact that not all viewers see the ads or respond in anticipated fashion is taken into consideration by Smythe's argument, which considers the overall transformations associated with the rise of consumer society at the aggregate rather than the individual level. Whether or not a particular user responds in a particular way is largely immaterial with respect to the substance of his claims—not least because this diversity is also factored into marketing calculations. What matters is that the rise of a consumer society would have been impossible without a pervasive and powerful advertising industry. As Smythe's analysis in *Dependency Road* suggests, viewers of advertising "work" at becoming trained consumers—at embracing a consumer-oriented lifestyle, the values that go along with it, and the vocabulary of images and associations upon which it builds. The media industries are not the sole participants in the creation of the audience commodity and its productivity from the perspective of capitalism; they are assisted in this endeavour by the range of social institutions that produce and reproduce consumption-driven lifestyles, including the school system, family, and peer groups.

The thrust of Smythe's argument in this regard is not simply critical—in the sense of challenging the values of consumerism and the forms of dependency it reproduces—but also, and crucially, from the perspective of this chapter, analytical. The significance for Smythe of treating the audience as comprising not simply a commodity, but one that takes on the attributes of commodified labor, is to identify forms of de-differentiation between the realms of consumption and production proper (the infrastructure—superstructure distinction he repeatedly challenges). But he is also crucially interested in critiquing a symptomatic tendency in media analysis (both popular and academic) to buy into the distinction between media content and advertising, overlooking the close relationship between the two. Smythe describes this distinction in terms of the convenient fiction of the "free lunch" (the content which attracts viewers, free of charge) that masks the continuity between the structure and the function of the two realms: "The fiction that the advertising supports or makes possible the news, entertainment, or 'educational' content has been a public relations mainstay of the commercial mass media" (1981, 37).

The result of foregrounding the "free lunch"—the reward that attracts audiences to advertising-supported outlets—is that its relationship to the surrounding or associated advertising content is obscured. One of the insights of the analysis of the audience commodity is the reversal of the relationship between the "free lunch"—understood as the primary object of interest and analysis—and the advertising content: "the characteristics of the free lunch must always be subordinated to those of the formal advertisements, because the purpose of the mass media is to produce audiences to sell to the advertisers" (Smythe 1981, 38). The constitutive fiction of commercial media content, in other words, is that of the conceptual and ideological firewall between the "free lunch" and the "work"—the news coverage and entertainment content on the one hand, and the advertising that supports it on the other. This observation has played a productive role in the critical analysis of media content prior to the advent of digital media, and it has become a mainstay of critical political economic analysis (see, for example Nichols and McChesney 2000; Thomas 1995). Unsurprisingly, the more literary-theoretical inflected strands of cultural studies—particularly in the American context—have tended to be less effective in marking this continuity (despite the early invocation by Williams (1974) of the importance of attending to "flow" in the study of television), perhaps because of the tendency to conceive of (particularly televisual) texts as self-contained entities. Thus the proliferation of content-based and textual analyses of various commercial television shows (including news programming) has tended to foreground the "free lunch" as the focal point of analysis, treating the "work" and the "coverage" as separate from the disposable content in which it is embedded. This tendency is further exacerbated by the increasingly flexible and targeted character of advertising; a certain sense of textual unity is produced by the fact that programming is disarticulated with the surrounding advertising content as it travels through time and space. Over the years, a particular episode of a television show like *MASH* or *Friends* has been embedded in a wide array of ads associated with different time periods and locales, making it easier to treat the text as ideologically disconnected from its commercial environs and origins.

The critique of ideological analysis has also played a role in the tendency to isolate and fetishize the "free lunch." The notion that ideology critique is misguided, simplistic, or reductionist has licensed an implicit dismissal of the significance of advertising content in favour of polysemic texts (almost exclusively the "free lunch" content) and the freedom of "active" audiences to make their own meanings. The dismissal has run the danger of becoming as reductionist and overly simplistic as its target. The production and reproduction of consumer values and a consumption-oriented lifestyle is a flexible and complex process: there is room to accommodate a wide range of "interpretations" that nonetheless accord with a portrayal of the normalcy and normative character of consumer capitalism.

The net result in the theoretical realm has all too often been a recapitulation of the message that marketers and advertisers themselves promote: the (active) audience/consumer as king, whose whims are merely catered to by the marketplace. As David Morley has observed, the backlash against various forms of ideological critique (primarily in British and American cultural studies) has taken the form of the assertion, "that the majority of audience members routinely modify or deflect any dominant ideology reflected in media content [...] and the concept of a preferred reading, or of a structured polysemy, drops entirely from view" (Morley 1993, 13). Perhaps not surprisingly, such theories reflect a sensibility that is, as Morley notes, "readily subsumable within a conservative ideology of sovereign consumer pluralism" (Morley 1993, 14). The champion of the active audience is, in this respect, aligned with the consumer who dismisses the notion that advertising might have any effect on him or her. The dramatic rise of consumer society as a historically distinct formation, and the huge resource mobilization that contributed to it fades into the background as a seemingly natural expression of human desire—as if a world in which shopping is entertainment and entertainment shopping is the natural outcome of human liberation. If advertising and its associated regimes of both material and media production are deprived of their formative power, consumer capitalism can be framed either implicitly and perhaps inadvertently (as in the case of some theoretical positions that dismiss the critique of ideology) or explicitly (as in the case of the marketing industry) as the spontaneous (and inevitable?) result of individual choice. People want to consume, and the market gives them what they want. Too often the distinction is elided between a simplistic notion of direct effects (easily critiqued and debunked) and a more nuanced understanding of the relationship between media portrayals and the ways of life in which they are embedded (impossible to dismiss or set aside without undermining serious analysis). This elision is a result of what Morley describes as the unfounded, "assumption that reception is, somehow, the only stage of the communications process that matters in the end" (1993, 15).

The power of Smythe's analysis relies on the insight that an exploration of the site of consumption reveals the productive character of media audiences. In the era of monopoly capitalism, he argues, "the prime purpose of the mass media complex, [in addition to making money] is to produce people in audiences who work at learning the theory and practice of consumership for civilian goods and who support (with taxes and votes) the military demand management system" (1977, 18). This is not to say that the "complex" always succeeds, or that it achieves the purpose outlined by Smythe self-consciously or single-handedly, but that commercials are not simply an external add-on to the forms of media content they support.

A savvy attitude toward the advertising economy and the audience commodity in which it trades is the hallmark of what Slavoj Žižek (1998) describes as the operation of fetishism in contemporary, ideologically reflexive contexts. It is a fetishism that is paralleled by that of the "free lunch"—the content isolated from its commercial context. The ability to abstract this content is underwritten by the seeming irrelevance of advertising, an appearance which is in turn underwritten by the combination of implicit faith (in the working of an economy in which advertising is valuable to producers) and explicit disavowal (of the notion that advertising and marketing might have an important role to play in shaping social relations). As Žižek puts it, "the fetishist inversion lies not in what people think they are doing, but in their social activity itself." Interpassive subjects are reflexively savvy subjects who allow the system to do the believing for them. If "typical bourgeois subjects" (including academics and mainstream pundits) are "in terms of their conscious attitudes, utilitarian nominalists—it is in their social activity, in exchange on the market" that their faith in the power of the commodity is confirmed (Žižek 1998; unpaginated). In other words, if the diverse players in the media economy acted in accordance with their expressed savviness, the entire system would break down. No one would purchase advertising, and the apparatuses for supporting access to information and communication resources would collapse. To paraphrase Žižek, then, the faith in this system is externalized in the form of the actions that support it. Objectively, the system believes for us (in our place) in the efficacy of the audience commodity and thus in the labor performed by audiences.

One obvious component of Smythe's argument that needs updating for the digital era is the focus upon the "free lunch." In the interactive, user-generated-content era, the content is not necessarily directly assembled by media producers, but is often created by the audiences themselves. In this regard, their work is redoubled (or trebled): not simply in the form of watching (ads) (or of generating data about their behaviour and preferences), but in the form of the content creation on sites like Facebook and Twitter that attracts audiences/participants. Google, for example, makes its living by organizing information other people have created and drawing upon the meta-information that they have also generated (the links that famously inform its algorithm). It only recently entered the direct content creation business by commissioning "made for YouTube" programming. In the online economy, in other words, the notion of audience labor becomes somewhat more palpable and concrete than in Smythe's formulation. Audiences are visibly, measurably, expending effort that results in marketable commodities: not just the content they create, but the information they generate about themselves in the process. This work of generating user content and performing online services (such as chat-room moderation, for example) has been dubbed "free labor" by Terranova (2000) and that of generating data about what one does in the process, I have elsewhere described as "the work of being watched" (Andrejevic 2002).

If new media providers do not attract viewers by generating content themselves, they nonetheless provide a platform that enables viewers to generate the content that will entertain or inform them. Somewhat more precisely, this chapter argues that an important element of the "free lunch" of the digital era—that which lures users to a site and over which producers maintain control—is often not the content proper, but the organization of information. The content entices users, but some of the most powerful commercial digital business models spare producers the work of creating it. By developing a platform, application, or service, the digital lure for users does double duty: enticing them to create the very content that draws them to the site. Consider the example of some of the big players in the digital media space: Amazon.com, eBay, Facebook, Twitter, and Google. These companies are not primarily content providers, rather they provide services based on the content provided by others. The role they play is in helping to organize the information landscape, telling us where to find information, entertainment, products, and services, and allowing users to share information with one another. Content, of course, continues to play an important role in the digital media economy, but its proliferation and the widespread availability of more information than it is possible for any individual user to keep track of means that information organization has an increasingly important role to play.

Organization, search, and retrieval are, in a sense, the new "content"—the new "free lunch" provided by big digital media companies.

The question then posed for a critical analysis of the political economy of new media is somewhat different than that outlined by Smythe. If the new "free lunch" is information organization, search, and retrieval (in addition to content), how are these influenced by the commercial imperatives that structure for-profit media industries? If the "free lunch" is always subordinated to the characteristics of the formal advertisements, what happens when the character of this free lunch shifts from content to information organization? The outlines of such a contemporary critique would include an interrogation of the ways in which the organizational schemes themselves reflect the overwhelming tendency of the "free lunch" to "reaffirm the status quo and retard change" (Smythe 1981, 39). This line of critique strikes me as a crucial one—the extension of the concerns of the critical political economy of the mass media into the digital realm. The role of organization is, in a sense, to impose a new form of scarcity upon the information glut of the Internet age. If, once upon a time, mass mediation imposed scarcity through the limitations of content and distribution, in the digital era, it imposes scarcity through the activity of organizing access to information—that is, determining which content will be prioritized for which users. We might analogize this distinction to that between modeling and carving in sculpture: modeling creates forms by taking limited amounts of clay and building them up, whereas carving unearths forms from solid blocks of raw material. If we describe companies like Twitter, Facebook, or Google as organizational intermediaries, their role is to carve increasingly specialized information landscapes out of the overwhelming amount of data that populates the online information sphere.

If the critique of media content sought to excavate the imperatives that shaped news coverage and entertainment portrayals, the critique of what might be described as organizational intermediaries focuses upon the often opaque logics that structure the increasingly specialized and targeted informational landscapes we inhabit. As in the case of allegedly objective journalism on the one hand or market-driven entertainment content on the other, organizational schemes embrace the promise of invisibility (in the form of a natural result for a search) or transparency (giving users exactly what they want) as means of masking their own built-in imperatives. The promise of Google, for example, is that of a reference resource: to provide the best results for a particular query like a dictionary provides the most common definition for a word. Many users would likely be surprised to know that this particular "dictionary" provides different sets of results for different users based on the information Google has about them. Moreover, as Pariser

(2011) has compellingly argued, the results are not generated solely by the imperative of providing the "best" result to individual users. Indeed, the very question of what counts as "best" is a vexed one. For users, the term "best" might refer to the result that most closely captures what they imagine themselves to have been looking for. As Pariser notes, however, for organization-brokers, it may well be something different: results that are more likely to get users to click on a particular link or perhaps to click on an advertisement. As Pariser puts it, "it's not necessarily what you need to know; it's what you want to know, what you're most likely to click" (Goodman 2011). The parallel with commercial media content in this distinction is clear: the difference between what, for example, an informed citizenry might need or want to know and the entertainment fare that is most likely to maximize ratings and advertising compatibility. There is no one "best" result for a particular search. There are different definitions of "best" that reflect different imperatives, and those who develop the algorithms and have access to the databases control which imperatives prevail. Pariser describes the response of one Facebook engineer to his question about the possibility of, for example, building civic imperatives into the site's operation by filtering the results to facilitate exposure to a broader range of opinion and information. According to Pariser, the engineer said,

> what we love doing is sitting around and coming up with new clever ways of getting people to spend more minutes on Facebook, and we're very good at that. And this is a much more complicated thing that you're asking us to do, where you're asking us to think about sort of our social responsibility and our civic responsibility, what kind of information is important. This is a much more complicated problem. We just want to do the easy stuff. (Goodman 2011)

The easy stuff, in this context, means that which reflects commercial imperatives. Thinking beyond those imperatives, by contrast, poses a serious challenge. The default "neutral" background, then, is that provided by the market, whereas anything that intervenes in and alters those outcomes is implicitly figured as "political"—not just biased, but difficult. When it comes to the goal of the commercial algorithm, Disney CEO Michael Eisner's observation about the role of the media retains its relevance: "We have no obligation to make history. We have no obligation to make art. We have no obligation to make a statement. To make money is our only obligation" (Lewerenz and Nicolosi 2005, 83).

Commercial web sites, utilities, and platforms similarly have no obligation to do anything other than succeed economically, which is why the ready embrace of applications like Gmail, utilities like Twitter, and sites like Facebook by educational and public sector entities is alarming. Universities and public school systems are embracing Google applications like Gmail, online calendars, document storage and document sharing as cheap alternatives to licensed software systems, and government agencies, universities and schools, amongst others, are using Facebook and Twitter to support their public communication and outreach initiatives. In short, commercial entities with commercial imperatives are being treated as if they are public services. To give a relatively small but telling example, not long ago, when I was sitting on a university committee considering the issue of how best to deal with library books that no one had checked out in the past 15 years, the committee was told that all the books would be pulped. I asked a representative of the university library whether it might make sense to scan the books first, to preserve a cheap and easily storable record. The library had considered the option, I was told, but had decided it would be cheaper to wait until Google did the digitizing for them (presumably using still extant copies of the books stored elsewhere). The answer took me by surprise, not least for its implicitly self-defeating assumption that eventually the tasks of a public university library would be more efficiently and effectively undertaken by an overseas, commercial organization. Perhaps Google might end up scanning some of the volume (but perhaps not—the university was not partnering with Google on its library project at the time); even if it did, there was no way of knowing in advance what the terms of access would be, how long these would last, and whether indeed, the archive would be protected from the vicissitudes of Google's own economic fortunes (which at the time were admittedly looking quite secure).

It was intriguing and a bit scary to see a representative of the public sector and, presumably, the public interest, so willingly confer the public trust to a private, for-profit, overseas entity. In "old" media terms, it would be like deciding to turn the university library over to Barnes & Noble. Once upon a time this would have been unthinkable—after all, the whole point of public libraries was to address the shortcomings of a market-driven system. It is perhaps a sign of the times that it has become so easy, in some quarters, to imagine that the library function ought to be turned over to a company like Google, which, as Siva Vaidhyanathan has put it, "is an example of a stunningly successful firm behaving as much like a university as it can afford to" (2011, 187).

In a world in which universities are facing tight budget constraints and in which Google has become the de facto information organizer for the online world—the platform onto which much of our knowledge and culture continues to migrate—the temptation to offload aspects of the stewardship of public knowledge onto the private sector is strong and seemingly unavoidable. The university library's decision to hope that Google might one day digitize its unused books—a hope that helped absolve it of the responsibility of preserving some part of an archive, no matter how disused—represents a relatively small-scale example of what Vaidhyanathan calls "public failure," which occurs when, "instruments of the state cannot satisfy public needs and deliver services effectively" (2011, 6). This is not to blame the university, which had to make the best use of its scarce resources, but rather to point out the structural pressures that make such a decision seem optimal. It is also to anticipate the hazards of the failure to consider the relationship between the "free lunch" offered by Google (in terms of both organization and content—albeit content provided by someone else) and the commercial imperatives of a company that is, for all intents and purposes, an advertising company. As schools fold their communicative functions into Google, Twitter, and Facebook, they turn access to tremendous amounts of data over to the private sector in exchange for the discounts and convenience provided by these platforms. At the same time, however, they run the risk of relying upon organizational and communication schemes that have their own commercial imperatives "baked in" to the resources we use to inform ourselves about the world, to communicate with one another, and to educate our children.

It is precisely because these companies are not content providers themselves that they can be portrayed as post-ideological. Smythe's critique remains germane because it enjoins us to consider the production regimes that structure the relations between the affordances of the platform or application and the process whereby free access is valorized. The importance of such an approach is that it further invites us to think through the relations between the reliance on advertising, the collection of personal information, and the customization of the information environment: to see these as dialectically linked in an economically productive fashion. Google does not produce goods or services for sale in a conventional sense (it even offloads the creation of the ads it serves onto ad purchasers)—the only product it produces for sale is the audience for its various advertising products. If we accept the proposition that exposure to advertising is valued by producers—enough to be a multi-billion dollar industry—the value of tracking, profiling and customization follows logically. As Jhally and Livant (1986) have argued, one way to make exposure to advertising more efficient for advertisers is to custom-tailor the process: to ensure that consumers are exposed to those ads that are most likely to be effective (see also Gandy 1995). Thus, the rationalization of advertising exposure relies upon increasingly detailed data about consumers. Customization and flexibility makes it possible to generate data about consumer behavior and response. That is, once it is possible to sort and target consumers, it becomes possible to create controlled experiments that determine which ads in which circumstances are most effective at generating a desired response. In this regard, the data generation, collection, and mining processes are self-reinforcing. Customized appeals generate higher resolution detail about consumer response, which in turn enables more sophisticated forms of customization and new forms of experimentation (that generate even more data).

If the political economic critique of the mass media raised concerns about the way in which commodification narrowed down the scope of available information and opinions—thereby potentially constraining public awareness, understanding, and deliberation—the critique of organizational intermediaries raises a related set of concerns about the ways in which customization of the information landscape undermines the civic role of mediated communication in democratic societies. For example, the "affirmative" culture identified by Bunz (2013) reproduces itself by promoting the forms of positive reinforcement associated with having one's preconceptions and interests confirmed rather than challenged or expanded. Thinkers like Cass Sunstein (2009) raise concerns about the forms of polarization associated with enclave deliberation and the displacement of civic ideals by consumer models for information distribution and use. Significantly, as Fuchs (2010b) has pointed out, Smythe's approach highlights the continuing significance of class against the background of the democratizing promise of the Internet. In its updated form, Smythe's approach invites us to consider the ways in which the apparently "free" service provided by organizational intermediaries is an integral component of the production logics that embed the priorities of commerce into the algorithms that shape our information landscape. Everyone with Internet access can use Facebook, but Facebook alone can structure users' information environment (deciding, for example, to privilege the newsfeeds of those with similar beliefs and attitudes in order to exploit the positive reinforcement of shared affirmation that keeps users on the site as long as possible) and mine the information generated by users (for specific examples, see Pariser 2011). As Fuchs puts it, "All humans produce, reproduce, and consume the commons, but only the capitalist class exploits the commons economically" (2010b, 193).

The fetishization of content underwrites the fantasy that it is somehow detachable from the infrastructure that supports it. The thrust of Smythe's argument is to dismantle this fantasy, a task which remains a pressing one on the digital era in which, we are told, everyone (or at least a lot more people than before) can create their own content—but not, significantly the structures for organizing, sorting, and retrieving it. We can make our own Web page, but not our own Google; we can craft our own Tweets, but not our own Twitter (at least not without a fair amount of expertise and venture capital).

By focusing on practices seemingly associated with the superstructural realm of cultural consumption, Smythe nonetheless reminds us of the importance of control and ownership over productive resources, including the means of information sharing, organization, and retrieval. Even in the digital era, matter still matters—especially the expensive kind, such as network infrastructure, data storage facilities and processing power.

Finally, of course, Smythe's analysis holds open important avenues for rethinking the relationship between production and consumption—in particular for considering the productive aspect of activities that take place beyond the time and space of the workplace proper. His (1977) observation that increasingly all non-sleeping time under capitalism is productive time anticipates the Autonomist Marxists' description of a "social factory," just as his formulation of the audience commodity anticipates the notion of "free labor" (Terranova 2010). More than that, his formulation requires a concrete consideration of the ways in which non-work activity reproduces and shapes productive time proper. This is a particularly fruitful line of inquiry in an era in which leisure time become an increasingly important form of networking, self-promotion, and training in response to the flexibilization and casualization of labor. If, for Smythe, leisure time helped reproduce labor power in part by inculcating practices and attitudes that accorded with the imperatives of consumer capitalism, in the digital era, non-work activities like blogging, tweeting, and social networking, are becoming, at least in some realms of professional activity, an important form of public relations, outreach, self-branding, and self-promotion (Hearn 2008). What once seemed almost metaphorical (although this certainly was not the thrust of Smythe's analysis), that watching TV could be a form of labor, now seems all too literal: bloggers do for free what other people get paid for, and often they do it in order to enhance their professional profiles, to build audiences for their paid labor (as in the case of professional writers or academics who blog about their publications, talks, and so on). If watching TV is not directly recognizable as labor, sitting down to craft an essay, doing research to make a status update or a Tweet more interesting or attention-grabbing is more clearly recognizable as a form of unpaid work. In this regard, Smythe anticipated a tendency that has developed dramatically in the digital era: the enfolding of a growing range of non-workplace activities into the realm of production proper. Furthermore, his analysis of commodification provides a unique perspective on what might be described as the re-introduction of scarcity in what has seemingly become a context of surfeit. If, once upon a time in the mass media era, control over information relied upon barriers to production and distribution of content, the digital erosion of these barriers has coincided with what might be described as new barriers to organization, search, storage, and retrieval. Borrowing John Durham Peters' (2009) formulation, we might describe the success of new media companies like Twitter, Google, and Amazon.com in terms of the rise of "logistical media." In historical terms, such media include those seemingly content-free media that organize time and space (the title of Peters' essay on the topic is "Calendar, Clock, Tower"). Against the background of the proliferation of data and information, the organizational function becomes increasingly challenging, resource-intensive, and indispensable. As Peters puts it, Google's "power owes precisely to its ability to colonize our desktops, indexes, calendars, maps, correspondence, attention, and habits" (2009, 8). In the digital era, the power of data mining lies further in the ability of the algorithm to organize decision making processes based on information provided by others. The scarcity lies not in the information itself, but in the ability to put it to use in new and powerful ways. There is a spiraling productivity to the economy of logistical media insofar as each attempt by users to avail themselves of organizational tools (by, say, using Google to find information or Twitter to share it) generates more information that, in turn, needs to be organized if it is to be put to use. Smythe's analysis usefully reminds us of the logic that underlies this spiralling productive process, the role that users play in it, and the challenges that it poses for the role of information and deliberation in democratic societies.

CHAPTER FIFTEEN

Technologies of Immediacy / Economies of Attention: Notes on the Commercial Development of Mobile Media and Wireless Connectivity

Vincent Manzerolle[1]
University of Western Ontario

The Era of Ubiquitous Connectivity

This chapter contextualizes and expands upon Smythe's contributions to the critique of capitalist media within an environment increasingly defined by the rapid global development and adoption of mobile devices and ubiquitous wireless connectivity (UC). Specifically it theorizes the evolutionary trajectory of mobile media and wireless connectivity within the context of Smythe's analytic focus on the audience commodity as: a) the organizing principle of commercial media; and b) a central component in the development of "consumption relations" including those "that motivate the population to buy consumer goods" (Smythe 1973/1994, 239–240) necessary to informational capitalism. By *informational capitalism*, I mean a version of capitalism whose dialectic between forces and relations of production and consumption revolves around technologies specifically designed (and marketed) to enhance, capture, transmit, and store human capacities such as creativity, communication, co-operation, and cognition (see Fuchs 2009; Manzerolle and Kjøsen 2014). Under the condition of UC, these consumption relations are increasingly shaped by a contradictory milieu where the seeming abundance of information is countered by a growing scarcity (and prospective degradation) of attention itself.

As Smythe has noted (1981, 7), the competition for attention is an essential aspect of the demand management strategies that underpin the organization and development of commercial media. This competition engenders an emphasis on "technologies of immediacy" which tend toward real-time consumer engagement, targeting, and purchasing opportunities realized in the "twinkling of an eye" (Harvey 1990, 106). In the popular press, the development of sophisticated consumer devices, for example Internet-enabled mobile devices (IMD) like smartphones and tablets,[2] are cast as unproblematic forms of empowerment and liberation. The current popularity and profitability of IMDs bears the imprint of this competition for attention, as wireless connectivity has commercially developed beyond simple tools for voice and text communication. Indeed, they now represent a potentially lucrative site (or "platform") for expanding billable data, real and virtual purchases, and ultimately reconstituting the audience commodity as a collection of discrete individuals produced by an explosion of contextual data. As such, the recent commercial and technical development of IMDs and related services has demonstrated a shifting emphasis from the "use value" of communication to the "exchange value" of mobile data.

The implications of this shift are all the more important because IMDs are increasingly treated as staples of everyday life by growing numbers of consumers. IMDs have become ubiquitous mediators of personal communication and the production and consumption of information, culminating in their development into "remote controls for everyday life" (Chen 2013).[3] In the United States, annual household spending on mobile devices and services increased from $1100 USD in 2007 to $1226 USD in 2011 (Troianovski 2012).[4] It is important to note that this increase occurred despite a deep and sustained economic downturn where consumer discretionary spending generally decreased. The *Wall Street Journal* reported that, "Americans spent $116 more a year on telephone services in 2011 than they did in 2007, according to the Labor Department, even as total household expenditures increased by just $67. Meanwhile, spending on food away from home fell by $48, apparel spending declined by $141, and entertainment spending dropped by $126" (Troianovski 2012).[5]

For these reasons, IMDs, devices designed and marketed to be "always on" and "always connected" (see Manzerolle 2013), offer a vital analytic opportunity to not only re-assess (and potentially expand) Smythe's original critique of capitalist media, but to link it to forms of mediation that express the prevailing acceleratory logic of capital's circulation and reproduction (Manzerolle and Kjøsen 2012).

Materiality, Mediation, and the Infrastructure of Being

In beginning with the material thing—IMD; the technical object—I draw some inspiration from Marx's (1976) opening chapters of *Capital Volume 1* which strategically begins with an analysis of the commodity in order to set the stage for a more systemic imminent critique of bourgeois political economy. Social relations are (re)produced as lived experience, but artifacts offer the material trace of these experiences and their specific political economic pretexts (e.g., wage labor), although fetishization, Marx explains, conceals these pretexts (e.g., prosumption). As Marx wrote, "The hand-mill gives you society with the feudal lord; the steam mill, society with the industrial capitalist" (Marx 1984, 102). One should not take Marx's observation to be espousing a deterministic, causal relationship between social and technological change. Rather, as Barney (2000) suggests,

> What Marx appears to be saying in this aphorism is that certain technologies are indicative of, or significant to, particular productive relations. He may be going so far as to posit that these technologies facilitate particular relations, but, unlike the determinist reading, this is well within what is suggested by "giving." (35)

Similarly, I argue that informational capitalism gives us the IMD. This is not to imply a deterministic and causal relationship, but rather to demonstrate how human capacities are organized and articulated by the prevailing mode of production and its specific technological apparatuses and related forms of mediation. Thus capital, in its informational form, compels a quixotic search for a mode of stabilization partly dependent on mediation by ubiquitous connectivity. For example, this condition is an essential component in mobilizing the intellective capacities of both workers and consumers towards social relations conducive to informational capitalism (a process that regularly encounters resistance, friction, and failure). The specific articulation of these capacities, and the extent to which they are mobilized in the service of capital, partly depends on the technical composition of the available media.

Smythe's critique of capitalist media reinforces the fundamental inseparability of political economic and ontological levels of analysis. In ontological terms, mediation can be thought of as articulating the relationship between different modalities of human experience (e.g., introspection, sociality, and citizenship). The essence of modern technology, Heidegger writes, is not only a "mere means" to an end, but also a "way of revealing" and "enframing" human potential (Heidegger 1977, 13–29). Building on Heidegger's concern with the *essence* of technology, Darin Barney reframes the "question concerning technology" to deal with mediation. Barney writes that,

> Heidegger understood the essence of technology to be located in its mediation between the ontic and the ontological—between the practices of existing beings and a thoughtful engagement with the Being of those beings. Technological practices, like all existential activities, are ontologically significant to the extent they express something at issue in terms of Being. (Barney 2000, 204–205)

Insofar as Being is increasingly mediated by complex, capital-intensive technological apparatuses, media, or what Marx terms "general intellect"—as the "infrastructure of Being"—act as tethers to the dialectic of forces and relations of production that underpin historically contingent political economic structures.[6] As I will discuss below, this mediation offers insights into the limits and barriers associated with the articulation of human capacities, specifically centering on the competitive *channeling* and *tuning* of attention itself.

As McGuigan notes in the introduction to this volume (4–5), Smythe's and Innis' research emphasize the material constraints that shape attention (i.e. media). For Innis answering this question involved comparative historical research guided by a new heuristic and conceptual framework emphasizing the materiality of media in socio-historical contexts. Smythe employs a similar type of historical and materialist analysis, though one specifically directed at capitalist media and what he calls the "Consciousness Industry." In this sense, the audience commodity can be understood as a "real abstraction"[7] that materially governs the organization of commercial media systems influencing crucially "the things to which we attend" or the *things paid attention to* or *thought with* (Carey 2009) in order to accelerate the circulation/turnover of capital by attempting to mobilize consumers with greater intensity towards the final and essential moment of capital's reproduction: the moment of exchange.

The era of ubiquitous connectivity defined by personalized devices like smartphones is an expression of this logic. Indeed, packet-switched wireless data connectivity creates the potential to maximize possibilities for exchange (for example, in the development of mobile payment and location based services) as well as the real-time logistical data about user behavior and location. In both cases the capabilities inhering in the device are essential to the functioning of the vast and highly complex technological system tethering individuals wirelessly to commercial networks. In relation to Smythe's overarching critique of the capitalist organization of commercial media systems, what distinguishes media adapted to a condition of UC can be best explained with reference to their *ubiquity, immediacy,* and *personalization.*

Ubiquity here refers to both the perceived and actual colonization of digital media devices and, in this case, the technical capability to remain connected at all times through devices designed to be "always on" and "always on you."

Immediacy refers to a perceived instantaneity (or simultaneity) enabled by the devices and infrastructure of UC, tending toward real-time, networked communication, and a collapsing of spatial distance. Connectivity (comprised primarily of both the transmission and reception of digital data) is relatively unencumbered by spatial and temporal constraints, effectively tied to the specific location of individuals. In spatial terms, immediacy refers to a perceived direct relation or connection—a proximal experience of "nearness" (Tomlinson 2007, 74). In temporal terms, immediacy refers to something current or instant occurring without seeming delay or lapse in time (74). More generally, immediacy highlights the tendency of contemporary media to accelerate the circulation of information. It reflects the general condition of speed-up that is experienced phenomenologically at the individual level as equal parts euphoria and anxiety (or as an experience of the technological sublime, as Leo Marx [1964] might characterize it). At the same time, it can also be expressed at the level of a political economic compulsion, as in David Harvey's (1990) conception of space-time compression. John Tomlinson has referred to this pervasive technological milieu as an expression of the "condition of immediacy" (Tomlinson 2007, 72–93)—as a relatively "new" narrative that encompasses culture, economy, and everyday life.

Personalization refers to the tendency of contemporary media to materially incorporate the identity, information, and relationships of a particular user. The identity of the user is deeply embedded both in the commercial development of digital media as well as in its technical composition (e.g., SIM cards, NFC chips, unique device identifiers). Indeed, personalization of digital media is implicit in concepts like "the filter bubble" (Pariser 2011), "the daily you" (Turow 2011), or "monadic communication clusters" (Gergen 2008). Each of these terms attempts to capture how contemporary media customizes our content and services, for example, through the embedding of algorithms that learn the habits of particular users (Mager 2012). The personalization inherent in IMDs suggests: (1) an intensified transformation of public space into private space (Fortunati 2002); (2) an expansion from connected places to connected people to connected everything. Personalization through UC thereby privileges possessive individualism (Macpherson 1964) as well as consumer-centric market mechanisms to deliver access to connected technologies and services (e.g., through the use of spectrum auctions).[8]

Mobile Media, Personal Data, and Digital Prosumption

Mobile devices represent a now ubiquitous, yet personal, consumer technology perfectly suited to the construction of scalable mobile audiences. As nodal points in a vast feedback loop, mobile and ubiquitous technologies like IMDs are really personalized communication devices hooked into the user's specific social networks and tuned to the user's consumption-mediated or consumption-defined interests, needs, and behaviors. As tools of digital prosumption, these devices contribute to a central area of contemporary capitalist accumulation: personal data (Elmer 2004; Lace 2005; Manzerolle and Smeltzer 2011; Tucker 2013).[9] A report from the World Economic Forum (2012) entitled "Rethinking Personal Data: Strengthening Trust" suggests that personal data is the key economic resource of the twenty-first century. The report states that:

> The explosive growth in the quantity and quality of personal data has created a significant opportunity to generate new forms of economic and social value. Just as tradable assets like water and oil must flow to create value, so too must data. Instead of closing the taps or capping the wells, all actors can ensure that data flows in a measured way. (5)
>
> Historically, the strength of a major economy is tightly linked to its ability to move physical goods. The Silk Route, the Roman roads and the English fleet all served as the economic backbones connecting vast geographies. Even though it is a virtual good, data is no different. Data needs to move to create value. Data alone on a server is like money hidden under a mattress. It is safe and secure, but largely stagnant and underutilized. (7)

This important sub-industry of the information economy shapes the development and deployment of consumer ICTs as they help accelerate the consumption and production of data in order to capture and sell the attention of users.[10] Specifically, the personal data economy, as a site for capital investment and accumulation, amplifies myths about the emancipatory and/or empowering nature of digital prosumption (e.g., Google, Facebook, and Apple).

The economic necessity of personal data to contemporary capitalism has contributed to the renewed popularity of a post-industrial archetype—the *prosumer*—a figure that, since its popularization by Toffler (1980), embodies the convergence of production and consumption within the purview of an empowered and autonomous user-consumer of ICTs (see Comor 2011). As the capabilities for producing and consuming data ubiquitously (e.g., through IMDs) become more widely adopted, the prosumer becomes the ideal user embedded in the technologies and services available, as well as the target of marketing/advertising. The prosumer, however, is in fact the techno-utopian representation of the sovereign consumer championed by neoclassical economists (McGuigan 2000; Babe 2006a). In accordance with neoliberal theory, this figure provides a digitalized version of human rationality premised on self-interest. Thus it is not surprising that Web 2.0 reflects a neoliberal form of individualism that posits consumer sovereignty in the creation of user-generated content—a symbol of the empowerment of rational individuals over networks.[11]

Importantly, IMDs serve roles other than just communication. By associating mobile communication access with fashion and status through, for example, the branding and design of the iPhone or BlackBerry, such devices reflect possessive individualism—a form of agency central to capitalist hegemony (MacPherson 1964). Possessive individualism refers not only to the goods one possesses, but also to the capacity to sell one's labor; it provides a basis for a labor market in which individuals sell their productive capacities as commodities. In so doing it creates a homology between the commodities one consumes and the labor one sells. By creating channels for personally identifiable data flows, IMDs are part of a commodification process that cuts across traditional distinctions between work and leisure. Thus the popularity of the prosumer and prosumption as terms celebrating the collapse of media production and consumption provides cover for the exploitation of free or unpaid labor by commercial interests (Comor 2011; Scholz 2013). The growing ubiquity of IMDs, particularly those that exist at the convergence of computing and mobile telephony, are paradigmatic technologies illustrating this point.

Fundamentally then, prosumption supports the sale of devices and services, while also enabling the creation of a secondary market of personal data.[12] Because the Internet does not have an "identity layer" (meaning personal data is scattered and fragmented), Cavoukian (2012) estimates that a given user "releases over 700 items of personal data per day" (3). The bulk of all digital data produced globally carry some "fingerprints" that identify the person (or persons) of origin (Ungerleider 2013), for which IMD are particularly well suited given the nature of their technical functioning.[13] Wireless devices and services offer the possibility of real-time, highly precise and contextual data about users which is now becoming a new revenue source for the commercial entities that control this data (e.g., wireless carriers, Facebook, Google) (see Leber 2013). The major challenge for telecommunications and media conglomerates is in properly channeling the user's prosumption—whether in the form of text messages, email, file sharing, video uploads, blogs, or photojournals—into the expansion of the personal data economy in order to maximize return on investment (ROI) particularly in light of costly infrastructure, excess capacity, and expensive R&D projects (World Economic Forum 2011).

The personalization of consumer technologies, including IMDs, creates scalable audiences with varying degrees of heterogeneity and segmentation. Because UC underpins the logic of prosumption I have just discussed, the drive to implement "mobile strategies" as key to future profitability on the part of many Web 2.0 companies (specifically Facebook; Pepitone 2013)[14] signals how UC is now a dominant paradigm in the development of commercial digital media in the near term.

Audience, Abstraction, Capacity

In the case of Smythe's provocative (and controversial) concept of the audience commodity, the "work" of the audience is materially embedded in, and articulated through, the capitalist development of ICTs directed primarily at "demand management." The audience commodity emerges as a logistical necessity in the sphere of circulation, where surplus value is *realized*, as opposed to the sphere of production, where surplus value is *created* (Lebowitz 1986).

The audience commodity is not a material thing, but an abstraction that gains a reality in the commercial organization of media systems. It is an abstraction produced by the logic of acceleration inherent in capitalism's sphere of circulation. Following economic historian Karl Polanyi (2001), the audience commodity might be considered an "essential element of industry" and a central "organizing principle" of communication media (76). Just as land, labor, and money are "obviously not commodities" in an "empirical sense" (76), the audience commodity is a fictitious commodity that serves a logistical and acceleratory function in reproducing capital both generally and specifically. The extent to which these fictions become real—that is, *treated* as real—depends on historical context. Specifically, it depends on the social relations that govern both the spheres of production (e.g., wage-labor) and circulation (e.g. prosumption), as well as, and this is the point that Smythe alludes to, the specific organization of communication itself, including the systems and technologies that articulate and mediate communicative capacities. Since these capacities are themselves limited, media—as "attentional forms" (Stiegler 2010)—are a means of tuning and channeling attention, and as such, directed towards mobilizing these finite human capacities. Mirroring the sale of labor as "labor power" in the sphere of production, the abstraction of the audience commodity allows the sale of "audience power" in the sphere of circulation.

Although Smythe was highly dismissive of his work, Innis' (1964) concept of bias conceptualized as *capacity* here provides a tool for analyzing the relationship between dominant media and the specific articulation of intellective capacities, insofar as the former influence the articulation of the latter through time and space. In this sense, the concept of capacity refers to an "index of potential" (Parker 1985, 76). Capacity maps a crucial intersection between ontological and political economic considerations as it entails, "analyses of the limitations and opportunities faced by people in their day-to-day lives and the factors that may influence them in any given place and at any particular time," implying that "physical and intellectual limitations and opportunities are both influential and dialectically related" (Comor 1994, 111).

The specific articulations of intellective capacity not only reflect the social settings and various media that allow the social subject to act, but actually orient the individual to the world; that is, they open up a set of potentialities—actions, thoughts, concepts, and values—that reflect pre-existing ways of living, relating, and thinking by active agents. Thus while the myth of UC (Manzerolle 2013) suggests a new era of limitless or infinite social connectivity, by foregrounding the technical mediation of intellective capacities we highlight the *limits* or *constraints* shaped by a specific political economic milieu (which includes the habits of thought and action that are continuously produced and reproduced; Parker 1985, 88).

In a commercial/capitalist system, this mobilization is subsumed by the needs of demand management, and the overall logistics of circulation that culminate in the determining moment of "exchange"—the key reproductive moment for capital both specifically and generally.

Thus commercial media are organized to mobilize consumers to go out and help produce the moment of exchange. The unpaid "work" done by consumers in the sphere of circulation is increasingly necessary since this participation helps conserve and realize surplus value; as the commodity form spreads through culture, consumers play a crucial role in facilitating competition by redistributing wages within the market. Thus the audience commodity appears as a necessary abstraction in the sphere of circulation. Its reality is given by consumption relations (e.g., prosumption), technical capabilities, and by a specific economy of attention.

The broadcasting model that defined the rise of the audience commodity (see Jhally in this volume), and the more contemporary forms of fragmentation that mark Internet users, are successive evolutionary steps in the ever-expanding circuit of capital comprising the integration of both spheres of production *and* circulation. Through Smythe's emphasis on the capitalist application of ICTs, the sphere of circulation can be seen as productive in two senses: (1) it literally facilitates the expanded/accelerated circulation of commodities and thus the realization and accumulation of surplus value; and (2) it facilitates the subjective reproduction of the wage-laborers themselves as subjects of capital. In so doing it enables the reproduction of the wage-relation *in general* by compelling consumers back to work so as to consume an expanding bundle of goods through the willing, and sometimes involuntary, acceptance of new and novel needs.

Technologies of Immediacy / Economies of Attention

The growing dependence on this unwaged labor, absorbed in the "production of circulation," the colonization of personalized devices in free leisure time has spurred-on the monetization of user-generated content (UGC). Consequently, the consumption relations that inhere in the prosumption activities associated with IMDs help maximize the productive use of leisure or unwaged time. With this capability, economic and cultural pressures re-shape the consumption relations that inhere in, and are enabled by, ubiquitous connectivity:

> Mobile communication anytime, anywhere, increases social accountability. The revival of 'dead' moments not only gives us extra time, but also makes us open to real-time monitoring and control. Mobile communication etiquette seems to involve the norms of 'being always available' and 'reciprocating messages/calls you get.' (Arminen 2009, 97)

This engenders, Arminen continues, "normative pressure for availability [while it] also allows [for] an increase in accountability, a continuing monitoring of communicative parties" (97). Similarly, as Fortunati (2002) writes, mobile phones enable users to progressively "single out the pauses in their actions, the pores, the cracks in time, so as to get hold of and to make communicative use of them" (517).

In this sense "free" time helps translate the unused capacity associated with the fixed cost investments in infrastructure into profitable services (and devices) but also creates the means to generate potentially valuable personal data. This data serves a dual purpose as it is used both to commodify personal information and to enhance, rationalize, and personalize marketing and advertising in exchange for user's attention. Like the abstract nature of the audience, the monetization of attention requires new techniques of measurement through "attentional assemblages" (Terranova 2012) of digital media.

The productive capacity of the prosumer also extends beyond this largely passive and logistical role of providing ever more detailed commercial data.

In contrast to traditional mass media audiences, in the Web 2.0 era "users are also content producers: there is user-generated content, the users engage in permanent creative activity, communication, community building and content production" (Fuchs 2009, 82). Web 2.0 and related myths offer up a fetishistic valorization of UGC, which conceals the more expansive "commodification of human creativity" (82). Because these creative capacities are now unleashed both technologically and symbolically, the explosion of UGC mirrors the equally rapid expansion of a flexible, precarious, and contract-based workforce, particularly in media industries (Neilson and Rossiter 2005; Gill and Pratt 2008). In addition to the perception of empowered users across a variety of technologically mediated settings, Web 2.0 reflects a new web-based marketing approach that strategically employs UGC in the production and targeting of commercial messages. Mobile media are evolving into the penultimate expression of Smythe's original premise regarding the capitalist development of ICTs and the audience commodity.

The concise definition of "mobile marketing" outlined by the Mobile Marketing Association seems to reinforce Smythe's premise: "Mobile Marketing is a set of practices that enables organizations to communicate and engage with their audience in an interactive and relevant manner through any mobile device or network."[15] The words "interactive" (i.e. digitally networked) and "relevant" (i.e. personalized, context aware) are most telling here, particularly as mobile marketing develops in and through the interactive (re)production of the digital, socially networked, and commodified self. The resulting commodification is two fold: on the one hand, the commodification of self and sociality through the consumption of digital devices, networks, and devices; and on the other, the commodification of the prosumer as a saleable and ultimately productive audience for potential advertisers and marketers.

Mobile devices and wireless connectivity have therefore developed from basic communication technologies into platforms for the articulation of the audience commodity with four primary purposes:

- To expand the range and quantity of virtual consumption (games, entertainment content, software, information services).
- To increase the volume of payable/metered data increasing the average revenue per user (ARPU) for telecommunications providers.
- To create a channel for targeted and context specific commercial (or political) messages.
- To enable and expand the production of UGC, thereby supplying companies developing web 2.0-centric business models with free content.

The construction of a mobile audience commodity emerged amidst the explosion of IMDs and the widespread Web 2.0 euphoria beginning in the mid-2000s. AdMob, incorporated in 2006 and acquired by Google in 2009 for $750 million USD,[16] is highly valued because of its prospective ability to monetize data traffic to and from personal devices. In so doing, it produces and sells mobile scalable audience commodities through the generation of detailed user information across a number of different metrics and includes the collection of data about application and website use. Promotional material for AdMob proclaims that it will offer "brand advertisers the ability to reach the addressable mobile audiences."[17]

It goes on to note, "(m)obile advertising provides you with targeted access to mobile users, and is easy to buy and measure."[18] More recently, Google has re-configured and optimized its Ad Sense service to exploit the growing use of mobile web browsers (Rowinski 2011). Not to be outdone, Apple acquired mobile advertising company Quattro Wireless (founded 2006) for $275 million USD, in order to release its own mobile advertising platform in April of 2010—iAd—which provides similar access via its iPhone handset users. While Google built an advertising empire based on search engine optimization and keywords from the ground up, Apple's relatively late entry into mobile advertising demonstrates the perceived profitability of this area because, until then, it focused primarily on revenue from hardware and software sales. The iAd platform was Apple's first concerted foray into advertising.[19] iAd has particular relevance for commercial brands, as the official website explains:

> iAd reaches millions of iPhone, iPad and iPod touch users around the world in their favorite apps. With the iAd Network, you can reach the Apple audience, the world's most engaged, influential and loyal consumers. Each ad is shown only to the audience you want to reach, in the apps they love and use the most. Our highly-effective targeting leverages unique interest and preference data that taps into user passions that are relevant for your brand. Whether they are reading the news, playing a game or checking the local weather, your ad will make an impact.[20]

Another important company seeking to construct a mobile audience commodity, Millenial Media, founded in 2006, is the largest independent mobile advertising platform. Partners and advertisers include AOL, *New York Times*, Zynga, Bank of America, McDonald's, Disney, Pepsi, UPS, IKEA, and MasterCard. Millenial Media provides an assortment of targeted, rich media advertisements using various forms of graphic banners, interactive, full-page, and video ads. Because of its reach and influence in mobile advertising—with roughly 91 million U.S. mobile users—it was sought after by Research in Motion (RIM) as a means of competing with both Apple and

Google in the mobile advertising space. In the end RIM was unable to acquire Millenial Media in part because of the high valuation assigned to AdMob and Quattro Wireless. As a consequence, Millenial Media raised their asking price beyond what RIM was willing to pay. By the time RIM released its own advertising platform in September of 2010, simply named BlackBerry Advertising service (much like iAd it was a platform for application developers to monetizing in-app advertising for which RIM would take a percentage), mobile marketing and advertising was already worth an estimated $3.5 billion USD in 2010, with projected mobile ad spending reaching $24 billion USD in 2015 (Middleton 2007). The massive explosion of "apps" has led some industry analysts to speculate that in-app ad buying could rival traditional Internet advertising (Rowlands 2013). By the end of 2010 the three most important IMD companies—Apple, Google, RIM—operated and were generating revenue from their own proprietary mobile advertising platforms.

Digital and ubiquitous media have given rise to another high-tech iteration of the audience commodity. Companies like Nexage and Rubicon provide real-time bidding for mobile users' attention by inserting video and rich media advertisements into mobile applications and websites. Here Smythe's concept audience commodity reaches its apotheosis. As one critic of the process explains, "Real-time bidding creates the possibility for companies to tag you wherever you are going, without you knowing or having the ability to influence it" (Singer 2012). In effect, what Nexage, and similar companies like Tapjoy or JumpTap, do is partner with publishers or applications developers looking to monetize the attention of their users and then create an exchange (or auction) for potential advertisers or marketers to bid for access to a specific user. Based on a given user's profile, potential advertisers use sophisticated algorithms to bid in real-time for a chance to have their respective messages displayed on a mobile device. As the *New York Times* explains: "The odds are that access to you — or at least the online you — is being bought and sold in less than the blink of an eye. On the Web, powerful algorithms are sizing you up, based on myriad data points: what you Google, the sites you visit, the ads you click. Then, in real time, the chance to show you an ad is auctioned to the highest bidder" (Singer 2012).

Similarly, location-based services are poised to take full advantage of the contextual nature of mobile data usage. For example, location-based mobile app provider Waze's CEO Noam Bardin explains that "not only are customers being offered something that is relevant to them because they may be close to a Taco Bell, but the advertiser is also getting very specific information, which it can use to tailor future offers" (Reardon 2013). He continues, "The real value is in seeing which people arrive at different locations based on various offers. It's powerful. We can influence where people go" (Reardon 2013). The real value is derived from tracking and targeting users with "proximity information, like Taco Bell promotions, "because, as the CEO goes on to explain, "If you can't attribute and track the value of the advertising, you can't get the money for it" (Reardon 2013).

The industry term used to describe the quantification of attention in this way is "impression." Thus companies like Nexage, Rubicon, or Milllenial Media can offer prospective clients a rate on a given number of impressions. Although Google and Apple are the dominant players in mobile advertising and marketing, the explosion of both mobile users and mobile content (applications, websites) has created a similar explosion in the means whereby the attention of users can be monetized (Rowinski 2011).

Thus the logic of monetization, and the high valuation assigned to these mobile media platforms, fundamentally hinges on user attention as the primary commodity produced and delivered to advertisers or data merchants.

This logic reflects the overall scarcity, and resulting quantification, of attention; what some theorists, economists, and marketers refer to as the "attention economy" (Davenport and Beck 2001; Lanham 2006; Falkinger 2005). Michael Goldhaber (2006) describes the attention economy as "a system that revolves primarily around paying, receiving and seeking what is most intrinsically limited and not replaceable by anything else, namely the attention of other human beings." On this point, Bauman (2007) writes, "In the cut-throat competition for the scarcest of scarce resources—the attention of would-be consumers—the suppliers of would-be consumer goods, including purveyors of information, desperately search for the scrap of the consumers' time still lying fallow, for the tiniest gaps between moments of consumption which could still be stuffed with more information" (40). To effectively tap into the attention economy marketers need to create interactive, participatory, or emotional connections with potential consumers; and for many, mobile is viewed as the penultimate medium for engaging with consumers in these ways.

Assessing the iPhone's success offers an important example of how consumption relations, communicative capacities, and the competitive search for attention are drawn together in a given technical object. Though experienced as a specific and highly personal consumer technology, the iPhone is better understood more broadly as a platform for both monetizing attention as well as expanding the range of virtual consumption and the production of valuable personal data. The iPhone was fully integrated into iTunes, which provided an instant and straightforward way of selling iPhone-specific software or apps, among other digital content like videos and songs. Through iTunes, Apple created an app ecosystem that allowed software developers a direct channel to monetize their software. This generated a virtuous cycle for the iPhone platform because it offered a clear monetary incentive to develop software. Importantly, iTunes was already familiar with many users (introduced through the widely popular iPod MP3 player) who entrusted Apple with their credit card information for the easy purchase of applications. In so doing, iTunes helped rapidly expand the range of things the iPhone could do—from location-based services to video gaming—thereby increasing the appeal of the device and its ecosystem to consumers.

The app economy, seemingly overnight, fundamentally changed the relationship between handset manufacturers, software developers, telecommunication providers, and users. As a ubiquitous virtual storefront, iTunes offered a means of transforming mobile users into an active audience of potential consumers of devices, applications, and other virtual goods, while at the same time creating a highly personalized channel for generating marketing data and targeting advertising.

Personalization, Democracy, and "Present-Mindedness"

The audience commodity is not only another abstraction crucial to the circulation and realization of surplus value, but one that sheds light on how specific communication systems also shape the prevailing habits of mind, including the capacities for thought and action conducive to democratic institutions. The commodified personalization that is a hallmark of the era of ubiquitous connectivity arguably contributes to a closed symbolic world; one in which the control and preferences of the user are embedded in the very software and algorithms themselves. In contrast to the embodied flesh and blood individual, the digital self becomes a self-propelling algorithm that, if left uncontrolled, will work to personalize the symbolic and communicative landscape. While our dominant technological milieu adapts to, and reinforces, the creation of small "monadic communication clusters" (Gergen 2008), individuals are tacitly encouraged (or enabled) to disengage from the human beings around them, as they are committed to their respective social networks, rather than civil society.

The prospective degradation of democratic institutions in an era of personalized media is mirrored at the physiological level. Nicholas Carr (2010a), and others (Stiegler 2012; Terranova 2012), have suggested that this media condition may be altering the structures of the brain, thereby foreclosing the capacity to think in particular ways (i.e., "deep attention"). Carr (2010b) writes,

> The Internet is an interruption system. It seizes our attention only to scramble it…The penalty is amplified by what brain scientists call switching costs. Every time we shift our attention, the brain has to reorient itself, further taxing our mental resources. Many studies have shown that switching between just two tasks can add substantially to our cognitive load, impeding our thinking and increasing the likelihood that we'll overlook or misinterpret important information.

As Terranova (2012) argues, in a media environment defined by personalization, information—conceptualized as the process of being *informed*—describes the various techniques and technologies for "consuming attention" (4). Paying attention to what others do on networked social platforms triggers potential processes of imitation by means of which network culture produces and reproduces itself; e.g., "reading and writing, watching and listening, copying and pasting, downloading and uploading, liking, sharing, following and bookmarking" (7–8).

Thus the perceived abundance of information—conceptualized as a non-scarce, non-depletable resource—is countered by a growing scarcity and fragmentation of attention itself. Terranova writes that, "[b]y consuming attention and making it scarce, the wealth of information creates poverty that in its turn produces conditions for a new market to emerge. This new market requires specific techniques of evaluation and units of measurement (algorithms, clicks, impressions, tags, etc.)" (2012, 4). On this note, consider Smythe's description of the changing role of "information" in media systems:

> The function of "information" transfer, which in the 18th century was the province of the press and the post office, is now diffused through this broad complex of institutions. And the flowering of computers and information processing has added a new level of meaning to the "informational" function of the "communications" complex—a function of serving as the means of production, exchange, and consumption of "information" in the sense of Norbert Wiener's definition, " a name for the content of what is exchange with the outer world as we adjust to it, and make our adjustment felt upon it. (Smythe 1994c, 248)

Similarly, Herbert Simon writes that, "What information consumes is rather obvious: It consumes the attention of its recipients. Hence a wealth of information creates a poverty of attention, and a need to allocate that attention efficiently among the overabundance of information sources that might consume it" (quoted in Terranova 2012).

As such, attention is made more scarce but is also "degraded" (4). The personalization of our media environment epitomized by IMDs enables the regular intervention of a ubiquitously enabled siren's song competing for smaller and smaller slices of our attention. In this sense, Google's massive market capitalization ($271 billion USD as of March 2013), indeed its entire business model, can be related to the various ways by which it monopolizes and monetizes attention (Lee 2011; Pasquinelli 2009a).

Similarly, the implications of personalization on politics and culture seems to reinforce a tendency towards fragmentation, the creation of parallel communicative universes defined by closed symbolic structures of circular affirmation and group polarization (see Turow 1997a). This is the unreflexive tendency Innis tried to warn us against, for it is in society's ability to self-reflect, self-critique, that it is able to self-correct. At the level of political economy, we might consider the processes of personalization as one of symbolic enclosures in which the structure of wealth and privilege are reproduced in separate social and financial networks in ways that exclude non-participants (creating the equivalent of online gated communities). Overall, personalization is merely a cover for privatization, which in a post-Fordist neoliberal era means a growing precarity of labor, increasingly made replaceable or disposable by the automation enabled by personalized media.

We can think of the growth of personalization in the era of ubiquitous connectivity as a feedback mechanism that flows through our personalized media. Historian of technology Otto Mayr (1971a, 1971b) wrote two articles about Adam Smith and the debatable influence of feedback technologies (the steam engine in particular) on the intellectual genesis of liberal economic theory. According to Mayr, the concept of a self-correcting, self-regulating system was the paradigm, the chief metaphor of the free market, in which the flows of goods, money and prices would create a self-correcting system that could maximize social welfare for the most number of people. We are now seeing that personalization of this sort falls closely in line with the beliefs and values of typical liberal market theories, using personalization and ubiquitous connectivity as a means of efficiently and instantaneously matching services and products with consumers (Manzerolle and Kjøsen 2014).

In this, capitalism's cybernetic imagination (Robins and Webster 1999), we can find buried Shannon's mathematical formula of communication, described as a noise-reducing feedback system (1949). This cybernetic imagination is preoccupied with the search for perfect information—the elimination of noise—that constitutes a mathematically perfect communication system, yet one subservient to the expanding algorithm of capital circulation and accumulation (Manzerolle and Kjøsen 2012). It is no surprise then that our means of communication and our means of exchange, of payment, are converging together. While personalization creates nearly perfect information about users, commodified or commodity-defined, in the context of technologically mediated "social networks," noise will increasingly constitute those voices, opinions, and messages that do not already conform to our personally cultivated algorithm. Such occlusions thereby reinforce a *present-*

mindedness (Innis 1964, 76) suitable to the impulses and work routines mediated by a state of ubiquitous connectivity.

Notes

1 Thanks are due to Atle Mikkola Kjøsen, Edward Comor, and Lee McGuigan for providing valuable feedback. Portions of this chapter were developed in Manzerolle (2013) and Manzerolle & Kjøsen (2012, 2014).

2 Citigroup estimated smartphone sales to increase 50% percent year-over-year in 2013, with further 61% increase in smartphone shipments in 2013. Expect 1.5 billion units by 2014 (Citi Research 2013, 11).

3 This dependency was acutely exposed during the service outages that followed in the wake of hurricane Sandy in 2012 (Wortham 2012). By contrast, a chronic dependency is evidenced in the growing percentage of users that sleep next to their phones (44 of all mobile users, 66% of smartphone users; Smith 2012), despite their tendency to disrupt sleeping patterns (Gaudin 2012). A national survey of Americans revealed that a third of respondents would rather give up sex for a week than their smartphones (Jackson 2011). This dependence has been associated with forms of obsession and/or addiction by some psychologists (Gibson 2011; Gaudin 2012). More profoundly, dependence on networked technologies like smartphones and Google have been associated with changes in the structure and function of the brain itself (Carr 2008)—changes revealed through the growing use of brain pattern imaging technologies (Davidow 2012). These figures suggest that the title of Smythe's penultimate tome, *Dependency Road*, might also include forms of social and psychic dependency that crystallize around specific communication systems/organizations, which themselves bear the imprint of broader political economic interests.

4 Households with multiple smartphones often spend far more for wireless services than for cable TV and home Internet (Troianovski 2012). Unsurprisingly, "The trend has been a boon for companies like Verizon Wireless and AT&T Inc. U.S. wireless carriers brought in $22 billion in revenue selling services such as mobile email and Web browsing in 2007, according to analysts at UBS AG. By 2011, data revenue had jumped to $59 billion. By 2017, UBS expects carriers to be pulling in an additional $50 billion a year" (Troianovski 2012).

5 Mobile video is one of the key drivers of mobile data revenue, bandwidth, traffic, and a significant area of growing advertising revenue. According to Cisco (2013), mobile video constitutes 51% of mobile data traffic. Cisco projects mobile video will account for 2/3 of all mobile data by 2017.

6 Following Harvey's (2006, 99) explication of Marx's concept, by productive forces I mean the power to transform nature through the development of new technologies (e.g., spectrum technologies); and by relations of production I mean the social organization and implications of the "what, how, and why of production" (e.g., wage labor) (99). Using Smythe's focus on demand management, we can also think of the forces and relations of consumption as increasingly articulated in and through IMDs.

7 First coined by sociologist Georg Simmel, but implicit in Marx's critique of economic categories, the concept of real abstraction describes how abstraction "precedes thought"

and "social activity" (Toscano 2008a, 70). As Toscano further explains, "abstraction is primarily thought of as the effect of a spatio-temporal action or process" (70). Thus an analysis of abstraction entails a focus on the specific media and forms of mediation that confer it a material reality through situated human interactions and institutions (i.e. social relations, media systems and technologies). In this sense the audience commodity, insofar as it is the "organizing principle" of commercial media, is a real abstraction. For a detailed discussion of the concept see Toscano (2008a, 2008b); Reichelt (2007).

8 This fact is partially evident in the re-allocation of spectrum from traditional broadcasters, among others, to telecommunications providers for use in highly profitable mobile broadband services/devices (Wyatt 2013). This is particularly true for the "digital dividend" (700MHz) freed up by the digital switchover of broadcast television (Wray 2009).

9 The personal data economy comprises companies that exploit consumer data for internal use, sale in a secondary market, or to provide specialized services and analysis. The World Economic Forum (2012) distinguishes three types of personal data that might be treated as an economic asset. Volunteered data, data offered voluntarily by users, such as photos, blog posts, video, and so on. Observed data is data captured, controlled and owned by an organization, often without the knowledge of the data-creating individual. Inferred data, "involves information computationally derived from all the data volunteered and observed" (19). The secondary market for personal data is estimated at $2 billion USD in 2012; however this is a measure only of companies collecting data from third-parties (e.g., Azigo, Mydex) (Cavoukian 2012).

10 This marketing orthodoxy is usefully summarized by the following quote: "There is one overriding, simple, but powerful message for all twenty-first-century marketing, media, and advertising executives: insight about consumers is the currency that trumps all others" (Vollmer and Precourt 2008, 29). As one response to the commercialization of personal data for marketing purposes, a recent proposal in France would tax Internet companies based on profits associated with data mining and the commercialization of user data, affecting companies like Google and Facebook (Pfanner 2013).

11 At a recent industry conference an IBM VP described the rise of the "empowered consumer era" enabled by IMDs and the personalization of commercial offers: "customers are quite willing to share information with businesses they trust if they believe they are going to get value in return…They want you to make offers to them—not blind offers" (quoted in King 2013). To do this companies need to engage in "social listening, seeking out customer-created content, creating a single view of a customer across multiple channels, and engaging consumers through personalized channels and empowering them to operate as advocates for a brand" (King 2013).

12 Personal data is seen as a particular area of growth for the telecommunications industry since they are privy to detailed data stemming from the usage of IMDs (World Economic Forum 2011). Identification and authentication services alone are projected to reach $52 billion USD by 2020 (World Economic Forum 2011).

13 Both Google and Apple have recently faced scrutiny about their collection of precise locational data about individual users (Cheng 2011); similar concerns have been directed at app makers (Bonnington 2012) and telecommunications providers (Eckersley 2011).

14 Indeed, the recent commercial interest in both "big data" and "cloud computing" by established technology companies like IBM, Microsoft, Oracle, and others, suggests the

widening appeal of UC as an all-encompassing commercial goal. Gartner research projects worldwide enterprise spending on cloud services to increase from $91 billion USD in 2011 to $109 billion USD in 2012, reaching $207 billion USD by 2016 (Gartner 2012). Though important, I will not address this broadening of the myth of UC. For further critical analysis see boyd and Crawford (2012); Franklin (2012).

15 http://www.mmaglobal.com/node/11102 Last accessed: August 26, 2013.

16 At the time this was Google's most costly mobile-related acquisition, won in a competitive bidding war with Apple.

17 http://www.google.co.in/adwords/watchthisspace/admob/ Last accessed: August 27, 2013.

18 http://advertising.apple.com/brands/ Last accessed: August 27, 2013.

19 Apple's press release explained this bold move into advertising: iAd, Apple's new mobile advertising platform, combines the emotion of TV ads with the interactivity of web ads. Today, when users click on mobile ads they are almost always taken out of their app to a web browser, which loads the advertiser's webpage. Users must then navigate back to their app, and it is often difficult or impossible to return to exactly where they left. iAd solves this problem by displaying full-screen video and interactive ad content without ever leaving the app, and letting users return to their app anytime they choose. iPhone OS 4 lets developers easily embed iAd opportunities within their apps, and the ads are dynamically and wirelessly delivered to the device. Apple will sell and serve the ads, and developers will receive an industry-standard 60 percent of iAd revenue. (http://www.apple.com/ca/pr/library/2010/04/08Apple-Previews-iPhone-OS-4.html)

20 http://advertising.apple.com/brands/ Last accessed: July 30, 2013.

PART FOUR

TOWARD A MATERIALIST THEORY OF COMMERCIAL MEDIA IN A DIGITAL AGE

CHAPTER SIXTEEN

Commodities and Commons

Graham Murdock
Loughborough University

Re-reading my interchange with Dallas Smythe with the full benefit of hindsight, I see more clearly than I did at the time how securely our differing approaches to the questions facing a critical political economy of communications were rooted in our contrasted cultural and political formations. The different directions taken were always, for me, a matter of "both/and" rather than "either/or." I want here to recover their roots and suggest why both remain central to critical analysis.

I was born in 1946, the year after armed conflict in Europe ended. In 1945, Britain's war leader, Winston Churchill called a general election. He lost, bringing in a reforming Labour government that set out to write a new social contract. This revised settlement between the people and the state was underpinned by an extended conception of citizenship. This guaranteed everyone the right to participate fully in social life and help shape its future forms but also placed them under an obligation to contribute to the quality of collective life. Translating this ideal into a tangible reality required a substantial increase in public facilities paid for out of taxation. Some of the resources required were material: a universal pension scheme, a public health service, subsidized housing, a welfare safety net for those who fell on hard times. Others were cultural: a universal secondary education system, a new body to promote the arts, a renewed public library system, and an extended public broadcasting system.

These facilities, funded collectively and openly accessible, constituted a cultural commons that operated alongside and outside the commercial system and the circuits of paid-for consumption. A number of recent commentators have followed Michael Hardt and Antonio Negri in rejecting or marginalizing the role of public institutions in creating shared resources, and identifying the "commons" with the knowledge, information and solidarities generated through active participation in everyday, self-organized, social activity (Hardt and Negri 2009, vii–ix). This is unhelpful. In our lives beyond consumption and commodification we engage with communal life as both citizens and commoners, tax payers and volunteers. The key issue is how these involvements, and the organizational forms and moral economies that underpin them, intersect and relate (Murdock 2011).

The effort to generalize the ideals and practices of citizenship fundamentally shaped the social and imaginative landscape in which I grew up. I was a voracious user of my local public library and gained the full benefit of the new school system. My father had been apprenticed as a printer at fourteen. I was the first in my family to have the chance to go to university. But the cultural resource that made the most difference to reshaping popular access to information and analysis and representing a changing social milieu stemmed from decisions over the future of broadcasting. The technology of the radio spectrum held out the promise of creating universal, immediate, and comprehensive cultural provision. The public service definition adopted in Britain promised a theatre, library, lecture room, and concert hall without walls, an unprecedented opportunity to provide everyone not only with entertainment, but with the information and knowledge required for participatory citizenship. This conception pivots around the idea of education in its original Latin sense of opening people to new knowledge and experiences. It aims to broaden horizons, introduce unfamiliar ideas and arguments, challenge comfortable stereotypes, and invite people to walk in other people's shoes and see the world from unfamiliar vantage points. It ideally animates every genre of programming, from current affairs and documentary to drama and comedy. It encourages listeners and viewers to develop their own capacities for self-realization and helps cultivate the recognition and respect for difference, empathy with social distress, and informed critique of entrenched power, essential to a public culture of tolerance, solidarity, equity and justice. It was precisely this educational impetus that Dallas saw as comprehensively blocked by the commercialism of the U.S. system.

The formal provisions of The Radio Act of 1927 in the United States awarded broadcast licences on the condition that they were used for purposes that were in "the public interest, convenience and necessity." But as Dallas noted, from the outset commercial operators in the U.S. had argued that "the public weal will be served if broadcasting, like grocery stores, uses the conventional business organisation, subject only to legal restraints on its profit-seeking activity" (Smythe 1952, 104). This formulation neatly conflated the public interest with whatever was most "necessary" and "convenient" for the pursuit of business interests. As Chief Economist of the Federal Communications Commission (FCC) between 1943 and 1948, Dallas confronted this contradiction on a daily basis. In 1946, in a concerted effort to reassert the educational principle that broadcasters have a "common responsibility...to enrich and broaden their listener's understanding and personalities" (Smythe 1950, 463), he helped draft the landmark FCC report on *Public Service Responsibility of the Broadcast Licencees*. This document, popularly known as the "Blue Book," required broadcasters to eliminate "advertising excess," provide a "well balanced program structure," and carry "programs devoted to the discussion of public issues." But as Dallas later recounted, it was immediately met with a barrage of criticism and complaint from the business lobby who "demanded [to] be released from any review of its service performance and [...] be treated exactly as are newspapers," which enjoyed Constitutional protection against government infringement on their operational freedoms (Smythe 1950, 468). Faced with this onslaught the "FCC gave ground" and retreated from enforcing the full range of the Book's "general policy statements" (ibid, 468).

Two years later, in 1948, technical problems that threatened to degrade signal reception prompted the FCC to impose a "freeze" on granting licences for the emerging medium of television. This opened a space for the renewal of arguments for the introduction of a system organised around public service ideals rather than commercial imperatives. By then, Dallas had moved to an academic position at the recently founded Institute of Communications Research at the University of Illinois. In a post-war climate of intensifying anti-communism his trenchant opposition to Franco during the Spanish Civil War labeled him as a "premature anti-fascist," a designation which, coupled with his commitment to labor and civil rights reforms, made his position as a public servant increasingly untenable (Melody 1993, 295). In the summer of 1949, the University hosted an invitation-only colloquium, funded by the Rockefeller Foundation, to discuss "the nature of public service media" and the "validity of mass education as an ideal," which Dallas attended (Balas 2011). But he also made a more public intervention. In preparation for FCC hearings on the future allocation of channels, he was asked by the newly formed Joint Committee on Educational Television to compile evidence to support arguments for reserving some channels for educational purposes. The key findings came from a content analysis of every program shown on the seven stations operating in New York in one week in January 1951 (Smythe 1951, 1954). The results revealed schedules dominated by entertainment where information, that could be seen as potentially "educational," was presented as a "hodge-podge of isolated, shallow bits" with "little effort to provide the systematic background information to aid the public in understanding and evaluating the material presented" (Smythe 1951, 16) and where every moment of screen time was surrounded by "advertising so closely intermixed with contests, news, stunts and so on that it is impracticable to determine where advertising stops and program begins" (ibid,

15).These findings reinforced his already firmly held view that "advertising support alone" would not "be able to build the kind of broadcasting system we have been led to—or have a right to—expect in the near future" (Smythe 1950, 471). As an alternative he advocated stations supported by Federal and State funds but operated under the control of boards "independent of both political and private economic pressures" (Smythe 1950, 474).

Returning to the argument in 1960, in an unpublished paper, he saw no point in continuing to press commercial operators to comply with public service requirements since "although broadcast channels are public property, they have been used in the past 10 years as though they were private property" (Smythe 1994b, 86). Faced with the FCC's lack of will he argued that the commercial operators be freed from any obligation to serve the public interest in return for paying rent on the radio frequencies they used, raising money that could fund a new non-profit public service corporation. A system partly supported by public monies was finally introduced as PBS, in the autumn of 1970, but there was no contribution from the commercial sector, leaving it compromised in its operations by the constant push by Republican politicians to cut its state and federal funding and by its overreliance on voluntary donations to make up the shortfall.

In contrast, public service principles were embedded at the heart of the British broadcasting system almost from the outset. After a brief period as a private monopoly established by a consortium of radio set manufacturers, the British Broadcasting Company became a public entity, the British Broadcasting Corporation, with exclusive rights to deliver radio services. Barred from accepting any form of advertising it was funded entirely from a compulsory, dedicated tax on the ownership of radio sets. Established in the aftermath of World War I, when the prospect of social unrest preoccupied the Establishment, it devoted itself to pursuing its core educational project, cementing the nation, and helping to manage the mass democracy that had arrived with the final extension of the vote to women in 1928. It promoted the official cultural canon and encouraged the sober exercise of political and social participation by disseminating expert opinion and rational recreation.

When television services resumed after World War II, the BBC's monopoly control over broadcasting was extended to the new medium. My family, like many others in Britain, acquired their first television set to watch the Coronation of Elizabeth II in the summer of 1953. The BBC's almost sacral coverage epitomized the Corporation's generally deferential stance towards the core centers of political and intellectual authority and its promotion of a particular, and highly ideological, construction of the nation as an imagined community and world power. But the political and cultural landscape was shifting. The period of post-War reconstruction was also a period of cultural recomposition. Deference was being edged out by scepticism and satire. New, assertive, working class voices were emerging in the novel, theatre and film, speaking from the old industrial heartlands and impoverished neighborhoods. Movements were gathering momentum among young people, women, and recently arrived migrants from the former colonial territories. Groups who felt that their experiences and aspirations were being marginalized or misrepresented were pressing for greater representation in core cultural institutions.

Tensions between popular taste and culture with a capital "C"—professional production and vernacular contribution, expert and lay knowledge, grounded experience and sustained analysis—had been present from the outset of public service broadcasting but they intensified by fractures of gender, generation and ethnicity that cross-cut established divisions of class. As a consequence, the promise of inclusion on the basis of recognition and respect, which lay at the heart of the social contract of citizenship, became the site of sustained struggles over the terms of cultural representation and popular participation. As the central cultural resource for citizenship, public service broadcasting found itself at the center of debates over revisions to the terms of visibility, voice, and participation. In working through these questions, however, it had to confront the challenges presented by competition from commercial television.

The resumption of television services, after their suspension during the War, breathed new life into long-standing pressures to end the BBC's historic monopoly on broadcasting. In 1950, the economist Ronald Coase published *British Broadcasting: A Study in Monopoly.* Carrying a foreword from the Director General of the BBC, John Reith, and offering a comprehensive survey of the arguments being aired, it appeared as an entirely disinterested intervention. Coase stressed that it was not his "intention to come to a conclusion" but simply "to review the arguments" as a contribution to reasoned discussion (Coase 1950, ix). In fact the book was informed by his own adamant support for commercial competition. This conviction was anchored by two formative experiences. Firstly, as he later admitted, working for the Central Statistical Office during the War had cemented his growing "prejudices" against central planning (Coase 2004, 198). Secondly, the year he spent in the U.S. in 1948 studying commercial broadcasting there convinced him that it was a system worth endorsing. Where Dallas, with his insider's knowledge, saw a system that was comprehensively failing to provide adequate resources for citizenship, Coase saw a dynamic market serving consumer demand. He carried this message back to the U.K. arguing that "it is reasonable to assume that the force of competition would operate as a stimulus to improvements of all kinds" (Coase 1950, 185). At the same time, he felt strongly that he was swimming against the tide of prevailing opinion that saw "A monopoly held by a public authority, as in the case of British Broadcasting" possessing unique virtues (ibid, 196). The following year, prompted by his "lack of faith in the future of socialist Britain" (Coase 2004, 199), he emigrated to the United States where his free market advocacy eventually won him a Nobel Prize in economics. The citation, summarizing his contribution to the discipline, specifically mentioned his seminal paper on "The Problem of Social Cost." This built on arguments he had outlined the year before in a paper critiquing the FCC. Where Dallas saw too little regulation, Coase saw too much.

In 1951, a young law student, Leo Herzel, intervened in the debate over how the FCC should handle the rival systems for transmitting color television signals offered by CBS and RCA, advocating that it should withdraw from making judgements and let the market decide by auctioning television licences (Herzel 1951). After the article was published, a friend of his in the economics department at Illinois suggested that he come and debate the issue with Dallas, which he did (Herzel 1998, 525). In his printed response, Dallas argued forcefully that it was precisely because "the educational and cultural rights and responsibilities of broadcasting are unique" (Smythe 1952, 104) that market assumptions about the optimum organisation of economic life could not be applied in the same way as to other goods. In 1959, however, Coase returned to Herzel's case for auctions in his own article on the FCC. The economic arguments he mustered in support of his central claim that "economists and policy makers generally have tended to overestimate the advantages that come from government regulation" (Coase 2004, 202) provided the deregulation lobby with seemingly powerful theoretical ammunition. It wasn't until the Balanced Budget Act of 1997 that his ideas were finally implemented and the FCC was authorized "to assign radio and television licences via competitive bidding" (Hazlett et al. 2011). But well before this his arguments had played a "central role in redefining the "public interest" (Moss and Fein 2003, 410), shifting it from the promotion of resources for citizenship to the promotion of competition and consumer choice.

Coase's earlier pessimism about the prevailing consensus in Britain in favour of public service monopoly proved to be unfounded, however. His departure for the United States coincided with the election of a Conservative administration bent on enlarging the scope of private enterprise. The BBC's monopoly was a prime target and in 1955 the government bowed to business lobbying and opened the way for commercial television services funded entirely by advertising revenues. The new service styled itself as Independent Television, ITV. The term "independent" carried a double meaning. Structurally, not being funded by government was promoted as a release from possible political pressures. At a personal level, it promised greater freedom for consumer choice.

Initiated in London the new service spread rapidly across the main population centres. Franchises were awarded for geographical regions rather than cities. Only one company was allowed to operate in each region giving them monopoly rights to television advertising revenues in their coverage area. Some public service requirements were imposed but as one successful franchise holder gleefully declared, the franchises were still "a licence to print money." Programming was markedly more populist from the outset. It mobilised both popular scepticism towards the Establishment and the desire to hear more vernacular voices. The first fueled innovations in current affairs documentary programming, the second nurtured the hugely successful soap opera, *Coronation Street,* set in a working class area on the edge of Manchester. Most of the core components of the schedules however set out to maximise audiences by working with already established tastes and tapping into the personal aspirations generated by rising real wages. An increasing number of households were beginning to glimpse the possibility of moving from maintaining basic living standards to building a life style. In a country that had only recently emerged from a long period of postwar austerity, and where food and clothing had remained rationed for some years, the displays of abundance and style in the advertisements and the surrounding programs pointed to life beyond drudgery and struggling to make end meet.

Viewers were addressed not as citizens or workers but as consumers. Appeals to the solidarities of labor and the reciprocities of citizenship, with their injunction to contribute towards the common good, were displaced by a discourse of individual redemption through purchases. This celebration of the market's promise of affluence and opportunity for all was perfectly caught in the Conservative Party slogan of the time: "You have never had it so good." The social contract that had elected a left-of-center government after the War was steadily dissolved in the warm ideological bath of consumerism. The figure of the worker, particularly the unionized worker, came increasingly to represent backwardness and disruption. The new heroes of advancement were heroes of consumption, with the rising celebrities of television playing an increasingly important role. Their on-screen personas, lavish houses, and glamorous leisure activities came to epitomise the rewards of consumption. In the stock exchange of styles, they acted as universal investment advisors. The American personalities, featured in imported commercial programming, operated as an advanced guard. Programs like *I Love Lucy* held out the promise of attainable personal and household transformations built around car ownership, new household appliances, stylish clothes, and an appetite for enjoyment. As Donald Horton, Dallas's main collaborator on the New York content study, noted, viewer engagement was reinforced by a faux intimacy of address, carefully designed to "give the illusion of face-to-face relationship" (Horton and Wohl 1956, 215). In contrast to the paternalist address, resonant of encounters with teachers, bureaucrats, and churchmen, that had characterized much public service programming, the personalities of the commercialized screen appeared as imaginary friends. This vision was domesticated within the early ITV system in the advertising magazines that littered the schedules. Fronted by well-known broadcast personalities these long form productions, often lasting for thirty minutes, promoted a range of brands in a variety of settings, from department stores to fictional pubs; but because they were classified as information or entertainment programming they evaded the limits placed on the amount of spot advertising allowed (Murdock 1992b). They were eventually discontinued, to be replaced by a concerted drive to incorporate promotional messages into programming through sponsorship and product placement.

This increasing installation of commodities at the heart of popular visual culture opened up two lines for critical inquiry, both of which led back to Marx's writings and to later theorists working in the Marxist tradition.

Marx's decision to begin the first volume of *Capital* with a discussion of commodities is animated by two considerations. By presenting the enticing array of finished goods on sale in shops as the outcome of exploitative labor relations he moves analysis from the dynamics of markets and exchange to the organization of production, from the operation of prices to the unequal possession of property. This decisive shift of focus is the fundamental basis of the critique of conventional political economy announced in the book's subtitle. But he also does something else. By comparing commodities to religious fetish objects, invested with magical properties, he underlines their power to mystify and obscure the conditions that produced them. By focusing attention relentlessly on the consumer's personal connection to the object, and the promised transformations in well-being it will deliver, the carefully orchestrated appearance of commodities conceals their true human and environmental costs. Pursuing these arguments leads in one direction to an extended analysis of labor and exploitation and in the other to an analysis of ideology. Both are essential for a comprehensive critique.

Dallas's model presents commercial television as an "invisible triangle" of relations between advertisers, broadcasters, and audiences (Smythe 1989, 134). It is "invisible" because the exploitative relations between audiences and advertisers are concealed by the apparently free gift of programming. In a crucial move he translates Marx's analysis of labor into the sphere of consumption, exposing the exploitation concealed by the appearance of free exchange. As he notes, the industry's economic viability depends on how much advertisers are prepared to pay to commandeer audience attention and receptiveness measured both "quantitatively and qualitatively, demographically and psychographically" (134). This "audience-power" constitutes the primary commodity the industry manufactures and is captured by programming that aims to be sufficiently engaging to stop people switching channels during the ad breaks. There is nothing in this account so far that could not be found in any standard industry manual.

What distinguished Dallas's analysis was his emphasis on the exploitative nature of this system. Audiences appear to get something for nothing since they don't pay for the programs they watch directly, but, he argues, borrowing a metaphor from A.J. Liebing's analysis of the press, programming is the equivalent of the "free lunches" that saloon proprietors offer to attract potential customers and entice them to spend money on drinks once inside. Like the bar owner, advertisers recoup the cost of promotion in the prices charged for their goods and services. As a result, Dallas argues, audiences are exploited twice over. Not only do they "work to market goods to themselves without being paid" through their attention to the ads and the time and effort expended on shopping, they "repay the advertiser for guiding them in that marketing work" (Smythe 1989, 135).

When Dallas sent me the draft of his original "Blindspot" article to comment, it was clear that he had provided an essential critical codification of the basic political economy of commercial broadcasting. But I felt strongly that in his drive to counter what he saw as the over attention paid to questions of ideology by European Marxists, he had avoided key issues raised in that debate. By then I had increasingly come to see advertising as the pivot of a promotional culture that, by relentlessly celebrating consumerism as a world view, was undermining the ideal of citizenship and eroding the ethos of mutuality, reciprocity and social care essential for democratic life. Attempting to understand this process of displacement led me to re-read Marx from a different vantage point to Dallas.

Marxist accounts of the operation of ideology can be read as a series of meditations on successive setbacks for movements aimed at implementing radical change. In the wake of defeat, analysis has repeatedly revolved around attempts to understand the persistence of established structures of power and the dynamics of popular compliance.

The first such moment follows the collapse of the insurgencies that swept across Europe in 1848. Disruption had been particularly intense in France, but the revolution of that year had "led not to Socialism or even universal suffrage, but the reign of Napoleon III" (Nasar 2011, 28). Marx's faith in the imminent overthrow of the established order was severely shaken. Observing the growing commercial power of Britain, where he had been forced to decamp after being expelled from Paris, he gloomily concluded, in an article of 1850, that it "seems to be the rock that breaks the revolutionary waves" (Nasar 2011, 38). His pessimism was confirmed when the Great Exhibition opened in Hyde Park the following year. Organized by another German émigré, Queen Victoria's husband Prince Albert, it showcased the vigour of British industry and the inventiveness and abundance of the commodities it produced under a spectacular glass and steel structure, nicknamed the "Crystal Palace." Four million domestic visitors and large numbers of tourists from overseas flocked to see it. To add insult to injured revolutionary pride, it opened on May the first, the day that the movement that Marx had drafted the *Communist Manifesto* for, had declared a day of international workers' solidarity. He responded to this reversal by devoting himself even more concertedly to the study of political economy in a search for the underlying mechanisms that sustained capitalism and the dynamics that might dismantle them (Adamson 1985). One key strand in this project was the analysis of fetishism sketched initially in the series of notebooks compiled in 1857 and 1858, now known as the *Grundrisse*, and developed more fully in the first volume of *Capital* published a decade later, in 1867. This placed people's relations to the commodities they consumed and the ways their attachment was secured by the orchestration of appearances, at the heart of his analysis. It was not the only account of ideology he offered, however. It coexisted in his writings with a model that emphasised the way consent was secured by the intentional promotion of narratives and images designed to naturalize prevailing structures of power and present the class interests of dominant groups as general interests. It was this second variant that was pursued most vigorously in the second major meditation on radical defeat, in the years following World War I.

The immediate aftermath of conflict fermented unrest and insurrection across Europe. There was a revolutionary rising in Berlin and short-lived Socialist republics declared in Munich and Budapest. But after the dust had settled, the only enduring shift in power had occurred in Russia, which many commentators of the time saw as the least promising site for an overthrow of the old order. Not only were socialist and communist movements turned back everywhere else, but within a few years Fascist governments had come to power in Germany and Italy. Accounting for this spectacular reversal became a major preoccupation among Marxist theorists living to the west of the borders established by the new Soviet Union. Their search for explanations, led by Adorno and Horkheimer and other members of the "Frankfurt School" in Germany and by Antonio Gramsci in Italy, spread out in a number of directions, but the analysis of ideology and popular compliance increasingly assumed a key role. In exile in the United States, key members of the

Frankfurt group devoted themselves to analysing the rhetorical devices employed in reactionary appeals and uncovering the social and psychological roots of everyday authoritarianism. In prison in Italy, Gramsci reflected on the construction of the way diverse power centres coalesced to form "historic blocs" and the construction of populist appeals grounded in the mobilization of common sense understandings. These foci were dominant but not exclusive. Other Marxist-influenced writers were pursuing the seductions of consumption. Sigfried Kracauer explored how the sumptuous sets of movies enticed shop girls to imagine themselves sharing the lifestyles they sold to others, and in his compendious but unfinished archaeology of the Paris shopping arcades Walter Benjamin embarked on a sustained excavation of the origins and dynamics of consumer society. This move to locate media audiences within the wider complex of commodity culture played a central role in my own intellectual formation (Murdock 2000).

I came of age politically in the 1960s, firstly as an undergraduate at the London School of Economics, which saw the first sustained student occupation of a British university in my final year, and then as a graduate student at Sussex University in 1968, a year of turbulence and unrest across the advanced capitalist countries. In France, demonstrations and strikes shook the state to its foundations. At the end of May, the President, Charles De Gaulle, who had briefly left the country at the height of the protests, called an election. Well over half a million of his supporters marched through the centre of the city, and on polling day, the 23rd of June, he and his party won a crushing victory. Seven days earlier the police had retaken the Sorbonne, the university's central building, which had been occupied by students and acted as the iconic centre of protest and debate. This pattern of reversal was repeated elsewhere, and in a potent echo of 1848 and 1918, order was restored in Britain in a conservative and reactionary form, with the installation of neoliberalism as the dominant working ideology.

The year after the initial "Blindspot" debate appeared saw the election of a radical conservative government, led by Margaret Thatcher, dedicated to a root and branch restructuring of the economy based on a market fundamentalist vision. Attempts to unpack the appeal of her double mantra, that "There is no alternative" (to rampant market forces) and "No society. Only Individuals and their families," led again in two main directions. The first, developed by Stuart Hall and his colleagues, drew on Gramsci and explored "Thatcherism" as a particularly potent new variant of populism. Its power lay in its ability to build a new structure of popular feeling that welded together demonization (of unions, welfare claimants, and cultural liberals), resentment (against migrants), disillusion (with public services), and aspirations (for greater personal choice). It was a brilliant and essential analysis. My own growing concern with the eclipse of citizenship led me to focus on what I saw as its principle antithesis, the celebration of consumer culture as the primary site of self-expression and realisation (Murdock 1992a). This consumerist worldview found concrete expression in the presentation of high taxes as an unwarranted imposition, largely wasted on the underserving, and the elevation of increased personal spending as a new civic duty. Policies that dismantled the welfare system and eroded the rights and opportunities of the poor were presented as putting more money "in your pocket" and increasing opportunities for consumption. This exhortation was underwritten by a massive extension of consumer credit designed to address the crisis of accumulation that, by then, was all too evident. Capital's response to stagnation in the markets for the standardised consumer goods generated by mass production was to introduce finer distinctions of style and presentation, generating "hitherto unknown opportunities for the individualized expression of social identities" (Streck 2012, 33).

This dynamic was particularly visible in the proliferating range of youth subcultures. Radicals, scanning the social landscape for signs of fracture, were tempted to interpret these truculent, in-your-face gestures as promising signals of non-compliance. This essentially romantic perspective was written into the title of one of the most influential books to come out of the Birmingham Centre of Contemporary Cultural Studies, *Resistance Through Rituals* (Hall and Jefferson 1976). I contributed a chapter based on a wider exploration of the way rhetorics around "the generation gap" were obscuring the persistence of class divisions in access to opportunities. At the same time, in collaboration with Peter Golding, I was also working on developing a critical political economy of corporate structures and strategies in the media industries. This work suggested that innovations in style could all too easily be incorporated into business plans searching for saleable product differentiations. As George Melly, essayist, jazz singer and acute observer of the emerging "pop culture" noted in his 1972 book, *Revolt into Style,* aesthetic gestures of anger and refusal were being translated into new promotional campaigns and product designs. Subcultural activists' customising and cut-and-paste alterations to standard consumer objects offered a cost source of corporate research and development. The spaces of self and collective expression on the streets were also sites of corporate exploitation.

The unbroken ascendency of market ideology initiated by the Thatcher governments has persisted to the present and has marked every area of institutional life, including broadcasting. Over this period British television has moved ever closer to the U.S. model of the "invisible triangle" anatomized by Dallas. Rupert Murdoch's Sky operation has emerged as the country's sole national satellite operator. The formerly separate, regionally based operators with the ITV system have merged to form a single national company and successfully lobbied to be released from a number of their former public service obligations and to be allowed to carry product placements. In 1998 Channel 4, which was established in 1982 expressly to address social constituencies that had been underrepresented on the mainstream channels, assumed responsibility for selling its own advertising time and moved towards a more commercially populist strategy, launching the English version of the iconic reality television show *Big Brother*. In this increasingly inhospitable environment the BBC came under mounting pressure. There were more frequent calls for the Corporation to be wholly or partly privatized or for its ring-fenced funding to be opened to contest. In response, I joined commentators rallying to its defense. My support rested on three main grounds. Firstly, the continuing prohibition on advertising exempted it from the exploitative relations at the heart of Dallas's "invisible triangle." Secondly, despite its organisational and operational shortcomings, its commitment to public service remained the best guarantor of cultural resources for citizenship that the commercial sector would not provide. Thirdly, it had responded to competition from ITV and the radical commitments of some of its producers by becoming more plural and open in its approach to questions of representation and participation. It was at this point that the Internet began to develop as an increasingly mass utility.

In his early work, Dallas outlined a "transactionist view of the relation between mass media content and audience members," arguing that viewers actively construct interpretations, utilizing "not only the explicit layer of meaning in the content but also the innumerable latent or contextual dimensions" (Smythe 1954, 144). By way of illustration he offers a range of possible interpretations of a televised professional wrestling match. Interestingly Roland Barthes's essay on wrestling is one of the most quoted pieces in his hugely influential book *Mythologies*, originally published in 1957, organized around magazine essays on French popular culture written in the early 1950s. He sees the associations that images arouse, rather than what they ostensibly show, as the key mechanism anchoring ideology in everyday experience and understanding. Advertisers had already grasped the marketing potential of latent layers of content. In an article published in 1944, Ernest Dichter, a Viennese émigré with a background on psychoanalysis, argued that the journey towards making a purchase mimicked the stages of getting to know someone." Every product we buy is at first a stranger to us, we have to warm up to it, get to know it better until after a while it becomes part of our personality" and an extension of ourselves. He describes this process as a "psychodramatic performance where products serve as co-actors" (Dichter 1944, 432). A critical reading of this formulation points to a way of combining a conception of ideology oriented around commodity fetishism with an analysis of audience agency, since it suggests that it is the invitation to interact with brands as familiar companions who can still surprise us that beckons and then binds us most securely into the consumer system. Dallas didn't follow this "transactionist" logic through into his later work since it would have cut across his contention that the concerted focus given to content in media studies was an "idealist" error that directed attention away from the analysis of audience exploitation (Smythe 1989, 136). This is an unproductive dichotomy. Developing a structural account does not preclude an analysis of "transactional" agency. On the contrary, contemporary conditions require both.

The agency exercised by television viewers centred around the construction of personal interpretations which could be shared and negotiated in immediate social settings. Viewers might go on to create artefacts (compiling scrapbooks, for example) or display their preferences in more general social arenas (e.g., by attending fan conventions), but these were exceptions. The Internet has transformed the active audience into the interactive audience. Personal responses can be widely circulated in blogs and postings on social networking sites. Narratives and program elements can be re-edited, modified, parodied, and released for general consumption on file sharing sites. But as a number of commentators have pointed out, this extended field of user agency is inserted into a remodeled version of Dallas's "invisible triangle." In return for access to Internet platforms that are free at the point of use, "netizens" sign over all the information their online interactions provide about their likes, tastes, and social connection, and in many cases, the intellectual property rights to everything they post or contribute as well. Not only is every trace of user interaction deposited in the pool of "Big Data" and mined to devise promotional appeals tailored more precisely to individual consumer profiles, website owners go beyond the exploitation of users' time and attention to appropriate their creativity, knowledge, expertise and friendships, encouraging them to talk up products in their conversations, both online and offline, and contribute ideas and energy to developing new products and modifying existing ones.

Unpaid activity is simultaneously the primary commodity produced by the web and a potential source of innovations to the commodities that users later purchase. Dallas's work continues to offer an essential route into analyzing these new extensions of exploitation. But for a full account we also need to retrieve his earlier sketch of transactional agency and reengage with analyses of fetishism to explore the continuing seductions of commodity culture.

If we follow the development of the consumer system, from Marx down to the present, and explore the interlocking and successive innovations in media and retailing systems, we can trace a clear trajectory in which promotional rhetorics and spaces become progressively both more pervasive and more immersive (Murdock 2013b). By promising people that it is possible to buy their way out of trouble with purchasable solutions to personal and social problems, the animating ideology of consumerism cancels the collective responsibilities traditionally embedded in the identities of the worker and the citizen. Understanding how this commercial enclosure of everyday activity and imagination operates through the promotion of commodities is an essential starting point for any effort to retrieve a conception of citizenship that speaks to the internal complexities of contemporary societies and the shared challenges that transcend national borders.

But there is a second task.

Alongside the explosive growth and consolidation of the commercialized Internet, we see public cultural institutions—museums, libraries, universities, galleries, archives—digitalizing their holdings and developing new forms of public involvement and collaborative activity. Providing an accessible and engaging point of entry into this emerging public cultural domain offers public service broadcasting a powerful new potential rationale (Murdock 2005). The BBC, for example, is currently taking the leading role in creating an online public space that networks together the digital holdings of a range of Britain's leading public institutions. It is also developing new forms of popular collaboration. The invitation to audience members to record everyday conversations for potential deposit in the national oral archive and broadcast on the radio, is an instructive case in point. Responding certainly involves labor but it is outside the circuit of commodification. It contributes to an openly accessible public resource rather than a commercial asset protected by intellectual property regimes.

As citizens, we are asked to contribute to the common good in two ways. We are required to pay taxes and urged to participate in discussions on how best to spend them on extending shared resources and equalizing access, and we are encouraged to volunteer our time and expertise to specific projects designed to enhance the quality of collective life. Historically, the moral economies of public goods and gifting that underpin these two interventions have been embedded in separate institutional domains. Digitalization offers an opportunity to combine them in a new kind of cultural commons that can act as a counterweight to the pressures of commercial enclosure (Murdock 2013a).

The convergence that matters most to the future of democratic life is not, as Henry Jenkins and others have argued, between commercial corporations and creative netizens but between established public cultural institutions and bottom-up, grassroots, cultural origination. It has become fashionable to herald the transformation of consumers into "prosumers" and celebrate their increasing integration into the circuits of production. In offering a critical alternative it is not enough to reveal the exploitation at the heart of this process; we need to argue for an alternative grounded in a commitment to build a new cultural commons based on relations of reciprocity.

In making the case for non-commercial broadcasting at the beginning of the 1950s, Dallas appealed to an attitude which he saw installed "deep in our mores" that: "the responsibility of the agencies which serve our minds and our souls is on a much higher level than the responsibilities of the agencies which supply us soap and razor blades" (Smythe 1950, 463).

Defining what that responsibility is under present conditions and how it might be translated into practical institutional forms that foster a culture of participatory citizenship, in a communications landscape transformed by digitalisation and dominated by market fundamentalist assumptions of corporate privilege, is one of the central challenges now facing critical inquiry. Rereading the history of popular resistance to commercial enclosure reveals both a refusal of commodification and a defense of commoning based on shared responsibility for sustaining essential communal resources. A liveable future requires both.

CHAPTER SEVENTEEN

Value, the Audience Commodity, and Digital Prosumption: A Plea for Precision[1]

Edward Comor
University of Western Ontario

Not long after Dallas Smythe published his audience commodity thesis, Alvin Toffler predicted that people soon would be customizing the things they consume. Networked computers, he said, will enable production and consumption to be in the hands of everyday people, thus eclipsing the era of mass industrialization and mass consumption. He called the agent of this transformation in production and consumption the *prosumer* (Toffler 1980).

In recent years, both Smythe's thesis and Toffler's prosumer have re-emerged. Interest in both stems, most likely, from the rapid development of digital technologies, the commodification of online activities, and efforts to encourage various forms of direct participation in the development and marketing of commodities. More generally, post-Fordist capitalism and the neoliberal policies shaping it reflect and have generated a more competitive world economy in which various fractions of organized labor have been gutted. These developments, and probably the ascendancy of what is called autonomist Marxism, have contributed to a renewed interest in the prosumer as a new kind of exploited value-producing worker.

"With the diffusion of networking technologies," write Kozinets et al. (2008), "collective consumer innovation is taking on new forms that are transforming the nature of consumption and work and, with it, society..." (339). Like Toffler, Zwick et al. (2008) believe these developments to be transformational because "the productive value of social cooperation, communication, and affect ... represents a closing of the economic and ontological gap between consumption and production" (182). People, they believe, are being empowered to realize their potentials through prosumption's mobilization of "the collective know-how" (184). According to Pridmore and Zwick (2012), user-generated activities online "have moved practices of prosumption, the constant blurring of the production–consumption line, *to the center of economic value creation*" (109; emphasis added).

As intriguing as these claims are, they entail, I think, simplified, generalized, or incorrect readings of Marx's foundational conceptualization of value. Consequently, much has been written about supposedly new political economic relations emerging through capital's use of digital technologies, as if the way in which value is being created today is fundamentally different than it was in previous decades and centuries. In this chapter I respond to these claims by launching a friendly critique of the audience commodity thesis and a not-so-friendly challenge to its application by several analysts of digital prosumption. This will be done in four stages. The first involves an overview and assessment of how the audience commodity thesis is being applied to prosumption. The second relates these applications to several premises used by autonomist Marxists and others. The third entails a precise explication of Marx's value theory—an explication that is necessary if we are to move forward with the kind of analytical clarity that much of the current literature has been lacking. In the fourth stage, I briefly reassess the audience commodity thesis and digital prosumption in light of value theory and the methodological precision it requires.

Herein I argue that while Smythe's audience commodity thesis (and its revision by Jhally and Livant) is overly economistic, its resurrection and modification as a means of assessing digital prosumption involves a profound misunderstanding of Marx's economics. Indeed, the thesis, when applied too mechanically or used in ways that overstep its analytical capabilities, has facilitated some rather "vulgar" or totalizing conceptualizations while, today, in its application to supposedly new political economic processes, an idealist version of Smythe's work has emerged. My conclusion is that the audience commodity thesis, while still useful (at least as an analytical metaphor), has been limited due to an absence of rigor in its contemporary applications concerning the question of value.

The Audience Commodity and the Exploitation of Labor

A core postulate of Smythe's audience commodity thesis is that commercial broadcasters sell audiences to advertisers. What, asks Smythe, "is the principal product of the commercial mass media?" The answer, he says, is analogous to Marx's category labor power—"audience power" (Smythe 1981, 255). Smythe thus frames the dynamics of the media-audience-advertiser relationship in a way that centrally involves the development of something quite remarkable: the economic exploitation of audiences during their time away from paid work. To quote him directly,

> What mode of work is it which has the following characteristics: One is born into it and stays in it from infancy to the old folks home; one is not consulted as to the precise work to be done tomorrow; work tasks are presented and done; and lastly, one is unpaid? Answer? Slavery? Yes, and the audience too? (Smythe 1978, 125)

Jhally and Livant (1986) developed Smythe by arguing that it is the audience's "watching time" that is sold by media capitalists to advertisers. Unlike Marx, who insisted that the sale of labor power requires a formal transaction between capitalist and worker (a transaction involving a wage), for Jhally and Livant the remuneration for the audience's watching time is not directly monetized. Instead, broadcasters, "pay" audiences by providing them with what is shown in between the advertisements: programming. Furthermore, because it is the audience's watching time that is being sold, surplus watching time—manipulated quantitatively (i.e. absolute surplus value) and qualitatively (relative surplus value)—is the basis of the value produced through this relationship (Jhally 1987, 71–78). While Smythe argued that the audience works for advertisers to satisfy the latter's quest for surplus value, for Jhally surplus value instead is generated for the media capitalists.

As with Marx, who argued that the free and equal exchange of labor power for a wage is simultaneously *un*free and *un*equal, Jhally explains that although the participation of the audience appears to be voluntary, broadcasters, in fact, compel audiences to watch. The reason, he says, is that more than just the program as a kind of wage is involved in the transaction as the messages received are culturally and psychologically meaningful. Indeed, these messages constitute, in cultural practice, a necessary kind of labor (Jhally 1987, 69). Ultimately it is the need to comprehend and live within capitalism as a social formation and, more specifically, the audience's need to fit into its (commodified) social norms that makes the seemingly autonomous choice to take part (beyond one's choice to select this or that program) to be, in fact, no real choice at all. Indeed, commercial mass media, say Jhally and Livant, "are not characterized primarily by what they put into audiences (messages) but by what they take out (value)" (1986, 143).

An obvious question that stems from this work, and a question that contemporary audience commodity theorists rarely ask in relation to prosumption, is how can media capitalists economically exploit something they do not formally own (or even rent[2]): the time and labor of the audience or prosumer? And, for that matter, how can television viewers or online prosumers themselves sell what they do not formally own; their watching time, watching power or, more generally, their knowledge?[3]

For Marx, a fundamental condition for capital's ability to generate surplus value is the worker's legally institutionalized freedom (i.e. her autonomy) to sell her labor power as a commodity in return for a wage. This formal price-mediated exchange (through the wage labor contract) is essential if we are to understand the complex nature of value and why it takes the form it does in capitalist society. Why, precisely, this point is important will be addressed below. For now, we should note that most audience commodity theorists are not overly concerned about such specificities.

Arguably, today, the most prolific of these is Christian Fuchs. In assessing prosumptive relations, Fuchs understands digital media capitalists, such as the executives at Facebook, to be valorizing their online participants, not because Facebook users are paid a formal wage (as they are not) but, instead, because their attention, personal information, and online relationships are being commodified. According to Fuchs, here we have "an extreme form of exploitation, in which the produsers [i.e. prosumers] work completely for free and are therefore infinitely exploited" (Fuchs 2010a, 191).

Audiences, says Smythe, are exploited in their working lives as wage laborers and in their "free" time as the advertising industry's audience commodity (Smythe 1977, 3). For Jhally, in addition to their formal wage labor relationships, the audience labors for media capitalists. Thus, for both Smythe and Jhally, the audience/worker is the subject of a *double* exploitation (Lebowitz 1986, 167). For Fuchs our present situation is even worse. Now, with digital prosumption, exploitation has reached its zenith as the exploitation of practically everyone is becoming, to repeat the word he references, *infinite*. According to Fuchs, "knowledge workers" are paid or not paid to directly or indirectly "produce or reproduce the social conditions of the existence of capital and wage labor...necessary for the existence of society..." and, as such, we have all become the world's "exploited class" (Fuchs 2010b, 141). But here, as in other contemporary applications of the audience commodity thesis, questionable generalizations have been turned into premises—premises that then are used to construct some rather dubious arguments.

For example, in implicitly and correctly recognizing, as does Marx, that capitalist production is a *process* involving the inter-related moments of production, distribution, exchange, and consumption (Marx 1984), Fuchs incorrectly views exploitation to be an almost universal condition as "all knowledge workers, unpaid and paid, are part of an exploited class" (Fuchs 2010b, 149). With this, several assertions follow, including the view that through prosumption "capital exploits *all* members of society *except itself*" (ibid, 144; emphases added).

Once we accept this claim, there are analytical implications. If all the time that we spend engaged in digital prosumption or knowledge production more generally is exploited labor time, key categories in Marxist economics such as socially necessary labor time (the time needed to reproduce labor power) are rendered moot. Moreover, given the fact that a portion of the monies paid to wage earning (knowledge) workers must cover the costs of their reproduction, Fuchs' assertion that prosumers and other unwaged workers (including homeworkers, students, parents, and many others) produce "completely for free" (as "neither variable nor constant investment costs" are associated with their labor) raises additional questions. For one thing, how can this "extreme form of exploitation" (ibid, 148) take place in the absence of the remuneration needed to reproduce this form of labor? By conflating our "labor" as "knowledge workers" (itself involving the apparent expulsion of socially necessary labor time) with the formal economic process of exploiting waged workers, economic precision is lost and generalizations abound. Indeed, if everyone who directly or indirectly produces or reproduces the social conditions of capitalist production and its reproduction are exploited (at least in any sense of Marx's understanding of the economics of exploitation), who do we include or exclude as the exploited, and why? Surely, given their contributions to the production of the knowledge used to produce and reproduce capitalist relations, why should we exclude Rupert Murdoch, Larry Page, and Mark Zuckerberg from membership in this exploited class? Saying that everyone *except capitalists* are exploited simply will not do.[4]

According to Fuchs,

> The users who google data, upload or watch videos on YouTube, upload or browse personal images on Flickr, or accumulate friends with whom they exchange content or communicate online via social networking platforms ... constitute an audience commodity that is sold to advertisers. *The difference between the audience commodity on traditional mass media and on the Internet is that in the latter case the users are also content producers* ... [I]n the case of the Internet the audience commodity is a produsage/prosumer commodity. The category of the produsage/prosumer commodity ... [constitutes] *the total commodification of human creativity*. (2010a, 191–92; emphasis added)

What Fuchs is saying, if we are to take him literally, is that prosumers are themselves commodities because their time, attention, and information about their lives are utilized by advertisers and marketers while their knowledge is monetized, sold, and used to further capitalist interests. In this totalization of exploitative and value-creating relationships, however, the historical position of the prosumer as a participant in or contributor to our contemporary political economy and culture is occluded. In the absence of a quantitatively or qualitatively precise use of Marx's concepts and categories, it is unclear, for instance, why digital prosumption is substantively new and different. Critical academics (not to mention marketers and advertisers) have known for decades that audiences of all kinds are active in the production process (van Dijck 2009), while Smythe and Jhally's (pre-digital) versions of the audience commodity thesis also recognize this fact as marketers and advertisers have, for at least a century, consciously collected information and tapped people for their creative ideas (Beniger 1986, esp. ch. 8; Huws 2003, 140–42).

Another proponent of the seemingly revolutionary role played by digital prosumption in terms of the transformation of how capital extracts value is Eran Fisher (2012). "The value of Facebook," he argues, "is derived from Facebook's unprecedented ability to have access to information, store, own, process, and analyze it, and deliver it to its customers. … What is new and unique about Facebook," he continues, is that much of this information "emanates from the very practice of using it, from being a media of communication and sociability." As such, with Facebook at least, the "means of communication (media) and a means of production (technology) converge." Through social media sites, Fisher adds, information is not merely harvested, it "gets articulated within the specific context of social networks, i.e., that of communication and sociality" (176). Facebook, he concludes, "epitomizes a new form of production relations, where value is created not primarily by workers of the company, but by the audience" (177).

Historically, of course, digital prosumption itself is new but does it really constitute a new way of creating value? Unless we believe the mainstream illusion that individuals are inherently autonomous, all 'personal' information and knowledge surely is the outcome of commercial and non-commercial social relations and mediations. A human being, as Marx put it, is "an animal which can individualize himself only within society. Production by an isolated individual outside society … is just as preposterous as is the development of speech without individuals who live *together* and talk to one another" (Marx 1987, 125; emphasis in original). More to the point, accumulating and modifying ideas and inclinations that are socially and individually produced—the process of taking what people produce in the absence of a wage to develop commodities and re-shape ideas in ways that evoke purchases—has long been part of business.

In the 1920s, for instance, Edward Bernays initiated, for his client Procter & Gamble, a series of contests to promote one of the company's brands, Ivory Soap. This involved millions of school children being solicited to express their creativity by carving sculptures out of the soap—creations, once submitted, that were judged by P&G executives. The best sculptures subsequently were exhibited in museums across the United States (Tye 1998, 56). Parental (and corporate-promulgated) concerns about cleanliness were combined with the artistic expressions of children to produce a marketing campaign mediated through inter-personal and mass communications. As a contemporary prosumption theorist might put it, the knowledge of unpaid prosumers—American children and their parents—was exploited in the task of producing consumers for a brand of soap.

In this example and many others,[5] it is hard to see why Google or Facebook or YouTube executives are doing much that is fundamentally different. To repeat, there is no such thing as individual knowledge or creativity that does not involve various dimensions of social intercourse. The implication of this for Fuchs, Fisher, and others (e.g. Rey 2012) is that it is not enough to say that Marx's theory of value has been proven wrong by "new" prosumer-facilitated processes as similar forms of prosumption have been utilized by commercial interests for a very long time. An alternative argument is that Marx's value theory has been extended to virtually all areas of human interaction but, as discussed above, this yields some extraordinary analytical generalizations and, as developed below, it undermines value theory itself. A final alternative is that Marx's theory of value was *always* wrong. If, however, Marx was correct and the dynamics underlying how value is produced in capitalist political economies has not fundamentally changed, it is more accurate to say that whether intellectual "labor" is used to develop and sell Ivory Soap or contemporary commodities using social media, all such activities facilitate the creation of value in the *production process*. This, however, is not the same thing as producing value through *production itself.*

The Influence of Autonomist Marxism

The turn away from or confusion concerning Marx's value theory appears to stem (directly or indirectly) from the ascent of autonomist Marxism.[6] According to Maurizio Lazzarato, a "great transformation" in capitalism took place in the 1970s. Through post-Fordism, "immaterial" forms of labor emerged that entailed "a social labor power that is independent and able to organize both its own work and its relations with business entities" (Lazzarato 2006, 137). Hardt and Negri later defined immaterial labor as labor that "creates immaterial products, such as knowledge, information, communication, a relationship, or an emotional response" (Hardt and Negri

2004, 108). In every era (such as Fordism or post-Fordism) they say that one form of labor is, in effect, hegemonic. Immaterial labor has become dominant in this way not only because, quantitatively, such jobs seem to be proliferating but, also, because all kinds of contemporary work activities require the use of "communication mechanisms, information, knowledges, and affect" (115).

Hardt and Negri argue that during the industrial era there was a clear boundary between work time and other time. But now, "an idea or an image comes to you not only in the office but also in the shower or in your dreams" (111–12). For our purposes, the key shift that Hardt and Negri identify is that, in conjunction with "social production" taking place both "inside and outside the factory walls, so too it takes place equally inside and outside the wage relationship" (135). Marx's theory of value, which they assume is the exploitation of the proletariat's labor power as quantified in terms of labor time, thus is being eradicated given that the division between working time and non-working time is dissipating. Moreover, Marx's understanding that surplus value is derived from private exploitation has been replaced by "the private appropriation of part or all of the value that has been produced *as common*" (Hardt and Negri 2004, 150; emphasis added). It is from this point that Fuchs, Fisher, and others (Rey 2012; Cova et al. 2011) develop their shared conceptualization that digital prosumption reflects and institutionalizes the exploitation of virtually everyone in the task of creating value.[7]

To repeat, a core building block of this departure from past Marxist thought is the notion that immaterial labor not only is distinct in relation to other forms of labor (a questionable assertion, I think, in light of the mental and physical properties of all kinds of labor) but also that one type of labor can be qualitatively hegemonic. To make this assertion Marx's focus on *the specificities of the labor process* is replaced by *what workers are producing*. While, according to Camfield, this constitutes a fetishistic approach (33), in my mind it also confuses the neoliberal zeitgeist of the post-Fordist "information age" with the dynamics underlying it. More to the point, the very idea that any type of labor can be hegemonic (apart from Gramsci's understanding that the hegemony of a given class, including the proletariat, constitutes an extensive political, economic and cultural project) reflects a blatant misreading of Marx. Hardt and Negri, in fact, cite *The Grundrisse* where Marx comments that "In all forms of society there is one specific kind of production which predominates over the rest, whose relations thus assign rank and influence to the others. It is a general illumination which bathes all the other colours and modifies their particularity" (quoted in Hardt and Negri 2000, 40). But here Marx is referring to a society's ruling class (and thus its institutionalization of production processes that reproduce *its historical dominance*) and not the fastest-growing or "cutting-edge" component of the labor force (Camfield 2007, 36).

Another misinterpretation of Marx's work is apparent in their assertion that the eroding division between formal working time and "time of life" (Hardt and Negri 2004, 145) reflects the diminishing importance of formal wage labor relations. To quote Camfield (2007) on this,

> The fact that some workers may come up with ideas related to their paid employment while they are in the shower or dreaming rather than the office does not signal the erasure of a temporal division which remains extremely important in the everyday lives of hundreds of millions of people. Rather, it demonstrates how today 'work time' casts its shadow over the rest of life for researchers, designers and other such workers, whose employers, like other employers, are demanding higher levels of productivity… (45)

Capital almost always tries to get more for less out of paid labor, even to the (contradictory) point of eliminating paid labor altogether. This is not because waged labor is becoming less important but, instead, waged labor remains central in the process of creating surplus value. The fact that digital prosumption facilitates capital's use of unpaid labor, far from being something new, is an elaboration of the ongoing drive to develop productive forces (such as information and communications technologies) which, in turn, evoke modifications in relations of production, including the relationship between labor and capital. Although productive relations also may induce modifications in the production process, these relations are not themselves productive forces. Marx, in fact, was quite definitive that laboring activities (or as I will call these later, "concrete" activities) are not productive forces. Instead, it is labor power, as a commodity owned by the worker and sold to the capitalist, that constitutes such a force. As G.A. Cohen (2000) explains,

> Marx attaches high importance to the distinction between labouring and labour power, regarding it, indeed, as his crucial conceptual innovation. … For as long as the distinction remains unmade it is, in Marx's view, impossible to explain … how it is that the worker receives less than the value of what he produces… What has value is not labouring but labour power… This, and not labour, is what the proletarian sells to the capitalist, who pays less for it than the value of what he is able to make it produce. … But if labour power is what the proletarian sells, it follows that labour power is what he owns. Neither he nor anyone else owns labouring… It is his ownership of his labour power which fits the proletarian into the economic structure of society. (43)

Autonomist Marxists, Fuchs, Rey, and others believe that the ascent of digital prosumption somehow refutes or extends Marx on the different yet inter-related roles played by waged and unwaged workers and the importance of maintaining an analytical distinction between the two. Even in the absence of a careful examination of value theory, a cursory review of recent capitalist history demonstrates that something isn't quite right with this rejection or radical modification of Marx. Take, for example, the annual U.S. Pillsbury Bake-Off competition, that began in 1949, in which recipes using the company's products are submitted for judgment and their potential inclusion in an annually published collection (Young and Young 2010, 269). Here the unpaid prosumer consensually provides a corporation with her creative labor in the form of a recipe. Prosumption and the knowledge it yields then are used to market Pillsbury commodities. American homemakers thus labor to facilitate the production process, reducing the costs of finding new ways to use Pillsbury products and to provide the company with an effective marketing vehicle. Importantly, the surplus value (and not, of course, the profit) of what Pillsbury makes remains a matter of getting more for less out of the waged workers it employs to produce it.

This drive to get more for less out of waged workers and the role of prosumption in obtaining new ideas and stimulating sales seems straightforward enough yet, for some, this is not a sufficient explanation for how value is being generated in a supposedly new kind of capitalism. Beyond conflating formally waged work time with other time, Hardt and Negri also meld labor in general (or non-commodified "concrete" labor) with labor power (the abstract commodity that workers sell through the wage labor contract). I now examine Marx's value theory directly in order to clarify how significant this is for those who want to accurately assess the audience commodity thesis in light of contemporary developments involving digital prosumption.

Marx on Value

The value of a commodity, says Marx, is the "objectified or materialized" expression of the amount of labor that has gone into its creation (Marx 1976, 129). How, specifically, is this labor objectified or materialized? If one factory owner employs workers to make chairs and these workers are slow, the time it takes to manufacture them will be longer than the time it takes for relatively fast workers in a competing factory to do the same job. Thus, if the value of a commodity is the sum of the time it takes to make it, the identical chair made in the more time-consuming manner would be more valuable. Clearly, as Marx recognized, this is incorrect. But if this time-based quantitative measurement is not the basis of value, what is? Perhaps the answer instead lies in the fact that commodities are exchanged in accordance with relative demand—a demand that reflects a particular commodity's use value to consumers? This, however, is precisely the mainstream political economy approach that Marx sought to correct.

If we say, instead, that the value of one commodity (such as a chair) is the outcome of its use value in relation to another (and thus another commodity's use value) we find ourselves embedded in a circular explanation. In our minds, as a consequence of everyday experiences, this understanding of value makes sense as we routinely weigh the relative use values of virtually everything we purchase (should I spend my money on the chair or a jacket, or something else?). But if our chair is indeed valued in relation to another commodity, the value of either one, and all commodities for that matter, logically should add up to the sum total of such use value comparisons. Marx knew that this could not be true as, empirically, capitalism constitutes a system in which value *grows*.

Marx moves forward by recognizing that the value of a particular commodity purchased by the capitalist—*labor power*—is unique. The creation of value in capitalism, he argues, directly involves the process of modifying everyday or "concrete" labor into the commodity called labor power. Concrete labor, which is qualitatively heterogeneous and embedded in the production of all sorts of knowledge, is not easily quantifiable (we can measure a ton of steel, but how do we measure a ton of knowledge?). Comparing what I do and think about each day with what someone else does and thinks about is, at best, an amorphous pursuit. This is why Marx recognizes the analytical importance of *labor power as an abstraction*—a legal and economic category that is expressed and materialized through the wage labor contract. Unlike concrete labor, abstract labor power is quantifiable, primarily in terms of its exchange value as represented through prices. It is this kind of quantification that enables capitalists to measure and formally exploit something that is, essentially, *un*measurable and, formally speaking in terms of creating surplus value, *un*exploitable: concrete labor. By abstracting labor into the commodity called labor power, Marx was able to explain value in an altogether novel way.

To take this point to a more transhistorical level, in contemporary writings on digital prosumption, a relation or activity, including the concrete production or use of knowledge, too often is confused with what Marx classified generally as productive forces. For Marx, productive forces directly enable the production of valuable commodities and, as such, knowledge itself may or may not constitute such a force. When assessing capitalism specifically, Marx was careful to avoid conflating productive relations and knowledge activities. In relation to this he was trying to clarify Adam Smith's distinction between productive and unproductive labor. For example, a worker's religious beliefs may well facilitate the production process at a particular time and in a specific place, and a religious teacher may, in extraordinary circumstances, be a productive worker, but this should not be confused with religious knowledge generally or the labor of a religious instructor specifically as necessary for the production of the value in a commodity. In his chapter "Theories of Productive and Unproductive Labour" in *Theories of Surplus Value* (1863), Marx criticizes Henri Storch for treating "spiritual labour" (generally defined) in ways that many contemporary analysts of prosumption might well recognize:

> According to Storch, the physician produces health (but also illness), professors and writers produce enlightenment (but also obscurantism), poets, painters, etc., produce good taste (but also bad taste), moralists, etc., produce morals, preachers religion ... and so on... It can just as well be said that illness produces physicians, stupidity produces professors and writers, lack of taste poets and painters, immorality moralists, superstition preachers... This way of saying in fact that all these activities, these services, produce a real or imaginary use-value is repeated by later writers in order to [incorrectly] prove that they are productive workers ... , that is to say, that they directly produce ... products of material labour and consequently immediate wealth. (ibid)

In his response to another political economist, Nassau Senior, who argued that because the military protects farmers, soldiers should be viewed as productive workers in the agricultural sector, Marx asks whether or not, in their absence, the material conditions for the production of corn (to name just one relevant commodity) would remain intact. Marx concluded that while in some circumstances *social* conditions may necessitate this kind of explicit protection, the labor provided by the soldier is not *formally* necessary. Furthermore, in the Addenda to Part I of *Theories of Surplus-Value*, he makes it clear that activities that merely *stimulate* production should not be seen to be materially productive. He cites criminal activities, as an example, and extrapolates on it:

> The effects of the criminal on the development of productive power can be shown in detail. Would locks ever have reached their present degree of excellence had there been no thieves? Would the making of bank-notes have reached its present perfection had there been no forgers? ... Crime, through its constantly new methods of attack on property, constantly calls into being new methods of defence, and so is as productive as strikes for the invention of machines. And if one leaves the sphere of private crime: would the world-market ever have come into being but for national crime? Indeed, would even the nations have arisen? And hasn't the Tree of Sin been at the same time the Tree of Knowledge ever since the time of Adam? (ibid)

A theory of value thus requires us to make a distinction between productive and unproductive labor.[8] It also compels us to quantify value in abstract monetary terms, just as the capitalist does in relation to price measurements and the worker when assessing the sale of her labor power. Marx, however, recognized that such quantitative measurements are by themselves inadequate given the largely hidden dynamics shaping value. One way to think about the complexity of this approach is to recall that abstract (waged) labor is crucial for the formal creation of value at the point of production, while the more general concept of concrete labor is fundamental to the reproduction of the broader production process. This dual understanding involving labor's abstract application as a commodity through contracts and wages recognizes that value itself involves both measurable and unmeasurable activities and relations.[9]

The primary question that needs to be answered is not *what is value?* but, instead, *what determines its forms?* "The essential difference between the various economic forms of society," writes Marx, "between for instance, a society based on slave labour and one based on waged labour, lies only in the form in which … surplus labour is in each case extracted from the actual producer, the labourer" (Marx 1976, 217). Unlike capitalist political economies, where the exploitation of labor power is the basis of producing surplus value, in slave or feudal society there is no such sale of labor power. The extraction of a surplus instead is based on explicit forms of domination, often involving the use or threat of violence. The proletarian, unlike the slave or serf, is free to withhold his labor but, of course, he is also compelled to enter into some form of wage labor relation given his need for money to survive. Money, it should be underlined, plays a crucial role in obscuring the human (labor) relations that are "objectified or materialized" in the commodity form itself. Money, after all, also is a unique commodity. While it is produced like any other commodity, it also serves the logistical function of being the economy's "universal equivalent" (Marx 1976, esp. ch. 2) and, in this capacity, it constitutes "the socially recognized incarnation of human labour in the abstract" (Harvey 2006, 241). In effect, the money commodity is relatively autonomous in that its supply is alterable (through state policies, individual hoarding, etc.) while its exchange value is not necessarily locked into any direct relationship with demand.

For Marx, *only money concretely demonstrates and mediates the value process*. To quote Michael Williams on this point,

> specific labours cannot count as abstract labour prior to exchange, so that their contribution to production cannot count as value: value cannot pre-exist exchange, but is, in part, constituted by it. … [V]alue under developed capitalism is the form which social evaluation of the worth of an activity or object tendentially takes. In the abstract, value *is* pure, quantitative form, which finds its autonomous physical expression only in money. (Williams 1998, 191; emphasis in original)

Fuchs, it should be pointed out, has rightly criticized others, such as Adam Arvidsson, for rejecting the audience commodity thesis in a way that may be mistaken for the position I am articulating in this chapter. According to Arvidsson,

> the labor theory of value only holds if labor has a price, if it has been transformed into a commodity that can in some way be bought and sold on a market. It is clear already at this point that it is difficult to apply the labor theory of value to productive practices that do not have a given price, that unfold outside of the wage relation. (2011, 265)

Fuchs responds by rejecting Arvidsson's assertion that exploitation requires the labor in question to be sold for a price. Fuchs is both correct and incorrect on this point but in ways that he does not appear to understand. Rather than grasping the dual nature of value—value creation as a process involving both the mobilization of concrete and abstract labor (as distinct yet inter-related)—he, instead, chides Arvidsson on grounds that are politically laudable but analytically weak. "Arvidsson's false assumption that exploitation is only present if a wage is paid," says Fuchs, "downplays the horrors of exploitation and implies also that classical slaves and houseworkers are not exploited. His assumption has therefore problematic implications in the context of racist modes of production and patriarchy" (2012, 732).

Perhaps in the guise of a polemic this position is useful but I do not think it is helpful in a debate concerning the economics of value and exploitation.[10] In sum, *value cannot be understood in the absence of its association with money, yet money is not value*. Neither the labor process itself nor money-mediated relations (as expressions of commensurable use values) explain value. Instead, *each defines the other* and, as such, value in capitalism stems from the labor process and monetary relations being both independent and interdependent. One implication is that labor power itself cannot produce value. Labor power is essential, of course, in the production of commodities that have value—a value that is objectified through price—but it is not itself definable merely through references to price. Similarly, concrete labor cannot (in and of itself) yield value. It is, in fact, the form in which labor power is used to produce other commodities that yields value and this form is comprehensible only through the use of money. Unwaged concrete prosumer labor therefore cannot *itself* produce surplus value and, as such, the digital prosumer's assumed economic exploitation also is false.

Given Marx's complex, dialectical and, indeed, materialist understanding of value, Fuchs and others might consider a different interpretation. While surplus value remains the outcome of capital's exploitation of waged labor, technological developments, capitalism's enormous dependency on credit to sustain investment and consumption, neoliberal trade agreements enabling capital to access labor and produce goods all over the world, not to mention an almost universal curbing of organized labor through state activities, all have facilitated the suppression of wages and the extension of corporate capabilities in their extraction of surplus value. It is this extraordinary ability to produce value on the backs of waged workers, coupled with the inarguable ascent of power in the hands of capitalists engaged in distribution and sales (including financiers, shippers, marketers, retailers, telecommunications companies, as well as internet services and social media corporations) that, I think, best explains a world characterized by wage suppression and the ongoing transference of wealth to the already wealthy. Facebook, of course, gathers, organizes, and sells user generated content *but* the work it formally exploits—and thus the value it directly produces—is the result of its processing and packaging of this content for its primary clients: advertisers and marketers.

As this chapter illustrates, an assessment of value that paradoxically reflects the empiricist illusions Marx sought to redress persists. Smythe and Jhally embraced this empiricism, while Fuchs' modification of value theory has generated yet another version of idealism. To quote Marx directly, "If we say that, as values, commodities are simply congealed quantities of human labour [i.e. the empiricist understanding of value] our analysis reduces them … to the level of *abstract value…*" If, however, value is understood in the absence of its objectified (monetary) form (as in Fuchs and others), the value produced in capitalist political economies is devoid of "a form of value [that is] distinct from their natural forms" (Marx 1976, 141; emphases in original).

In capitalism, for a commodity to have an objectively discernable value, the use of both labor power and money are required. It is in this specific process that the relationship between use and exchange value ceases to be dichotomous as the value of, for example, the chair addressed earlier can only transcend its use value (and thus its tautological 'value' in relation to other use values) when it is objectively related to other commodities (such as the jacket) through the use of prices (Arthur 1979, 75). In the absence of this conceptualization, audience commodity theorists tend to marry seemingly analogous but different concepts (audience power with labor power, for example) or, in rejecting or radically revising Marx's value theory, they are free to view digital prosumption as the basis of some kind of universal exploitation. Not only is the absence of an explicit money or wage-mediated transaction problematic (as money is the means of objectifying value), equating audience or prosumer participation with exploited labor is logically and empirically dubious, at least in the context of Marx's theory.

Toward a Precise Analysis of Digital Prosumption

Implicit in my argument is the view that the audience commodity thesis should be applied more metaphorically than literally. Furthermore, applications of the thesis to digital prosumption, particularly in the context of autonomist Marxist premises concerning a new form of capitalism, are questionable.

For Marx, generally speaking, time spent during each moment of the production process needs to be minimized. Temporal inefficiencies and gaps between the moments of production and consumption are costly to capital, particularly in relation to the efforts of competitors to accelerate this same process (which is why monopoly conditions often limit the drive towards efficiency). Advertisements of various kinds are almost as old as capitalism itself and resources allocated to advertising have been expenditures justified in light of the need to increase and accelerate sales while also minimizing the costs associated with circulation time. Commercial media thus developed (and continue to develop) largely in response to (or in conjunction with) improved production efficiencies. Today, advertising and marketing, as specialized service industries, are evolving in relation to the needs of all kinds of commodity producers to sell what the workers they hire produce, thus enabling them to realize the surplus values generated in production. Media capitalists, as in the past, are paid a share of the surplus produced and materialized in the form of money.

For Smythe, advertisers somehow commodify and buy audiences. For Jhally, media capitalists somehow commodify and buy audience watching time. But, to step back from these debatable perspectives, at least one economic fact is clear: *audiences are important because they are potential consumers*. If, after all, audiences buy one brand of beer instead of another, the advertising company's client—a brewery—willingly pays for this service (paying both the ad firm and the media capitalist for facilitating the realization of its product's value).

Methodologically, what is important is our comprehension of how the concrete and the abstract are inter-related and how relations and mediations shape what is perceived. After all, one of Marx's most important epistemological insights is that through the political economic relations that structure and mediate capitalism, dominant ideas and conceptions often are quite upside-down. And as Lebowitz writes, the pre-digital audience commodity literature reflects such illusions as its "starting-point…is the inverted concept of the sale of audiences and audience-time" to capital (1986, 170).

A more careful use of the audience commodity thesis is found in Vincent Manzerolle's work. According to him,

> The audience commodity and audience work do not *actually* produce surplus value directly. Rather, the conservation and realization of surplus value in the sphere of circulation occurs through the intervention of capital in the materialization of social communicative, co-operative and cognitive capacities of audiences; it is in this sense that the audience can be said to actually 'work.' (Manzerolle 2010, 460)

For Manzerolle, the audience or, for our purposes, the prosumer is active in contributing to the production process and the facilitation of the circulation of capital but, in the economic creation of value itself, her "labor" is unproductive. Here the reader might think I am "splitting hairs" but, as I have argued from the outset, for the concepts and categories we use to have analytical meaning it is important that we resist the tendency to over-extend or totalize them, for when we do it becomes impossible to specify "what, in particular circumstances, would count as the other of it" (Eagleton 2007, 7). For example, if "free" time or "living" time (in the absence of a contract-mediated waged relationship) has, through prosumption, become some kind of commodified labor time, what then is the analytical meaning of "free" time or "living" time or, for that matter, "labor" time?

Reality is not reducible to empirical observation or abstract analysis. What Marx developed, instead, is what he referred to as "science"—a methodology that impels us to go beyond the surface-level of things through abstract thought *and* the process of testing this reasoning using the past and present as empirical resources. Most past and present audience commodity theorists instead have constructed conceptual universes that entail, to use E.P. Thompson's phrase, their "own ideality upon the phenomena of material and social existence, rather than engaging in continual dialogue with these" (Thompson 1978, 205). While verifiable facts regarding digital prosumption are attainable—such as the demonstrable utilization of creativity and knowledge by capital—these do not and cannot independently disclose the dynamics and potentials of these activities. To use an analogy, we cannot characterize an army by referring to the soldiers who belong to it because, to do so, we would need to identify them *as soldiers* and, to do this, we would have to refer back to the original concept, the army (ibid, 222). Similarly, claims concerning the audience commodity and digital prosumption cannot be made without interrogating the parameters of what constitutes the audience commodity and digital prosumption. If this is not done correctly, as with the questionable assumptions informing claims concerning the dawning of a new kind of capitalism and new means of creating value, what results is a *reductio ad absurdum*.

Conclusions

Rather than extending the audience commodity thesis into an all-encompassing formula, instead I suggest we assess it in terms of Marx's conceptualization of production as a holistic process. To his credit, Fuchs does precisely this by recognizing that almost all forms of knowledge play a role in this process. Cova et al. (2011) also understand that production and consumption are "two sides of the same coin" but then over generalize and overstate the point. Through the Internet, production and consumption are being melded together and, as such, prosumption (or, as they prefer, "co-creation") is becoming central to "value creation" (232). But, to repeat, such pronouncements, which extrapolate from a general comprehension of the production process, in effect, dissolve value theory itself.

There is no doubt that we are living through a period in which the drive to harness and monetize the creativity and knowledge of prosumers is extraordinary. Moreover, as Napoli understands, what is qualitatively different today is the "enhanced ability to access an audience" and online media organizations' engagement with this audience's "creative expression"—a creativity that "they in turn monetize" (Napoli 2010, 513). But, then, to leap to conclusions such as surplus value now is being created directly through the exploitation of unpaid prosumer labor and, as such, value theory itself needs to be radically re-thought or rejected is, I think, more than a leap of logic; it is more akin to leaping off an analytical cliff.

For those of us trying to understand digital prosumption, the audience commodity thesis *is* relevant. In its application to commercial broadcasting, the thesis, when not used literally, enables us to better comprehend the dynamics shaping the form and content of these activities. Moreover, Smythe's "free lunch" analogy was and remains insightful as it situates commercial media in terms of their economic, historical and cultural roles vis-à-vis more general production process developments. As Smythe puts it,

> The information, entertainment and 'educational' material transmitted to the audience is an inducement (gift, bribe or 'free lunch') … [It] consists of materials which wet the prospective audience members' appetites and thus (1) attract and keep them attending to the programme, newspaper or magazine, and (2) cultivate a mood conducive to favourable reaction to the explicit and implicit advertisers' messages. (1977, 5)

If the aim of a television program is to deliver a particular 'demographic' to paying advertisers in the context of what else is on (i.e. what competition is present) and the technologies being used (a remote control, a DVR, etc.), the audience commodity thesis illuminates the dynamics shaping the pacing, the effects, and the information that is (or is not) provided on commercial as well as non-commercial services. With digital prosumption, the "free lunch" is the informative, fun, creative, social, and often ego-enhancing activities that impel online participation, and these efforts to engage us as audiences/consumers/citizens directly inform our analyses of the environments and activities at hand. When, however, this approach is extended to postulate that audiences or prosumers themselves constitute a specific commodity—one, in the absence of a formal wage labor contract, that is exploited—we enter into the simplistic realm of economism (distilling the complexities of production and reproduction into some kind of empirical calculus). As Nicholas Garnham (1979) put it in his response to Smythe, such applications misunderstand "the function of the commodity form as an abstraction within Marxist economic theory" (132).

Ursula Huws (2003), in her critique of academic and popular conceptualizations of our now "weightless" economy, argues that online mediations have facilitated the "notion" that "anything can now be done by anyone, anywhere: that the entire population of the globe has become a potential virtual workforce" (146). However, in recognizing this condition, Huws also understands that value in our capitalist political economy still must be objectified through abstract relationships and monetary measurements. This complex duality, linking capitalist dynamics with the upside-down illusions they generate, needs to be kept in mind as we move forward in the task of assessing, with precision, the development of relations and realities involving online communications. Contemporary analyses of value, the audience commodity, and digital prosumption should do the same.

Notes

1 Des Freedman deserves many thanks for his insightful critique of an earlier draft of this chapter. I am also thankful to Hamilton Bean for pointing me towards some of the empirical examples used herein.

2 In his critique of an alternative to the audience commodity thesis—"broadcasting rent"—Jhally writes that "human activity (watching) as a power simply disappears. Instead we have the audience as a raw material, being worked upon and processed by the agents of capital" (120). There is an either/or quality to this response that suggests, I think, the limitations of the audience commodity thesis and how value theory itself is represented through it.

3 Meehan's (1984) alternative thesis should be noted here as she recognized that ratings firms accumulate and organize data about audiences—data they process into what she calls the "ratings commodity" (223). Of course information about or knowledge derived from audience activities is not the same thing as the audience itself entering into a

contract involving their direct remuneration in return for their labor power. Instead, this kind of audience commodity is more analogous to the services provided by a firm that monitors workers and sells information about their behavior (whether to assist a corporation in its efforts to improve workplace efficiencies or to help it screen prospective employees by selling dossiers concerning past job performances).

4 It can be argued that Murdoch, Page, Zuckerberg and others are not exploited because they steer the workers who are exploited but, to repeat, at what point is the knowledge used to steer also or instead the knowledge implemented to produce or reproduce the exploitative production process? For Fuchs, there is no analytically or logically coherent answer to this question.

5 Applied research on consumer behavior, involving techniques for gathering and processing preferences and lifestyles, has been part of marketing and public relations since the 1950s (Peterson and Merino 2003, 101). In the 1970s, the first electronic consumer databases were deployed using consumer credit information and geographic location to produce a "powerful predictor of all manner of consumption practices" (Burrows and Gane 2006, 795). In the 1980s, one of America's biggest marketing firms, Western Wats, integrated telephones with computers to significantly expand the scale and precision of applying individual knowledge to the capitalist production process (McFarland 2008, 139).

6 Arguments made in this section are directly influenced by David Camfield (2007). His work is recommended for those interested in a more extensive critique of autonomist Marxism.

7 Having identified this commonality, it is important to recognize that Fuchs critiques Negri for his rejection of Marx's value theory (Fuchs 2012f, 645) but, as discussed below, his alternative generates another form of idealism. Social media, argues Fuchs, enables capital to exploit both paid and unpaid forms of labor. Facebook users, he says, "create use values" that "take on the form of commodities that are sold to advertising clients so that the concrete [unwaged] Facebook labour that satisfies human and social needs is at the same time an abstract, value-generating activity" (Fuchs 2012f, 637).

8 In Meehan's alternative to Smythe and Jhally—referencing the "ratings commodity"—rather than the audience, productive labor is performed by the contracted employees of Nielsens and other such firms who accumulate, interpret, and produce a specific kind of product for broadcasters, advertisers, and marketers. Thus the workers employed by Nielsen's are "productive" while the television audience is not.

9 I am not saying that Marx's theory is without problems. One concept that remains controversial is socially necessary labor time. For Marx, it constitutes a measurement, expressed as a monetary cost, of how much of the average worker's wage necessarily goes into reproducing the labor power needed in relation to a commodity's production. It is this average that (loosely) determines, among other things, the minimal wage required to produce a particular product in a specific place and time. Both Marxists (Sweezy 1942) and non-Marxists (Sraffa 1960) initiated critiques of this concept as, among other things, a complex of factors that are not readily apparent, nor measurable, fall outside the parameters of such monetary calculations. But here, again, Marx is not endorsing an economistic approach to measuring the costs of labor and, thus, the essence of value.

Rather than value being reducible to labor's exchange value (its price) or concluding that such quantitative measurements are dismissible illusions, Marx is trying to comprehend why labor and value take the forms they take (Elson 1979, 123). Also see Steedman et al. (1981).

10 Fuchs stresses the temporal aspects of labor time in his treatment of value as he recognizes value to be the objectification of this time. In doing this, however, he divorces this objectification from its inter-relationship with money (i.e. price). He concludes that exploitation takes place before the commodity produced is sold and that "[e]ven if a commodity is not sold…labour has been exploited" (Fuchs 2012f, 634). This, in effect, frames value in terms of a labor time or price either/orism. It also enables Fuchs to concur with many autonomist Marxists that, today, both paid and "unpaid workers are productive" as "they create the commons that are exploited by capital, and are therefore part of the proletariat" (ibid).

CHAPTER EIGHTEEN

Dallas Smythe Reloaded: Critical Media and Communication Studies Today

Christian Fuchs
University of Westminster

The new capitalist crisis has resulted in a new interest in the works of Karl Marx. We take this as opportunity for discussing foundations of Marxist Media and Communication Studies and which role Dallas Smythe's works can play in this context. First, I discuss the relevance of Marx and Marxism today. Second, I give a short overview of the relevance of some elements of Dallas Smythe's work for Marxist Media and Communication Studies. Dallas Smythe reminds us of the importance of engagement with Marx's works for studying the media in capitalism critically. Third, I engage with the relationship of Critical Political Economy and Critical Theory in Media and Communication Studies. Both Critical Theory and Critical Political Economy of the Media and Communication have been criticized for being one-sided. Such interpretations are mainly based on selective readings. They ignore that in both approaches there has been with different weightings a focus on aspects of media commodification, audiences, ideology, and alternatives. Critical Theory and Critical Political Economy are complementary and should be combined in Critical Media and Communication studies today. Finally, I draw some conclusions.

Introduction

- "Marx makes a comeback" (*Svenska Dagbladet*. Oct 17, 2008)
- "Crunch resurrects Marx" (*The Independent*. Oct 17, 2008)
- "Crisis allows us to reconsider left-wing ideas" (*The Irish Times*. Oct 18, 2008)

- "Marx exhumed, capitalism buried" (*Sydney Morning Herald*. Oct 23, 2008)
- "Marx Renaissance" (*Korea Times*. Jan 1, 2009)
- "Was Marx Right All Along?" (*The Evening Standard*. March 30, 2009)

These news clippings indicate a renewed interest in Karl Marx's works concomitant with the new global crisis of capitalism. This chapter poses the following questions: how have the crisis and the Marxist resurgence impacted Media and Communication Studies and what can we learn from Dallas Smythe in the contemporary situation?

The next section deals with the disappearance and return of Marx; the third section focuses on Dallas Smythe's importance for Marxist Media and Communication Studies today; the fourth section discusses the relationship between Critical Political Economy and Critical Theory; and I draw some conclusions in the last section.

The Disappearance and Return of Marx

In 1977, Dallas Smythe published his seminal article "Communications: Blindspot of Western Marxism," in which he argued that Western Marxism had not given enough attention to the complex role of communications in capitalism. Over the past 35 years, the rise of neoliberalism resulted in a turn away from an interest in social class and capitalism. Instead, it became fashionable to speak of globalization (e.g., Beck 1999), postmodernism (Lyotard 1984), and, with the fall of Communism, even the end of history (Fukuyama 1992). In essence, Marxism became the blindspot of all social sciences. The combination of neoliberalism with postmodernism, late-modernism, culturalism and new conservatism was anything but anti-capitalist, and all the more anti-Marxist. As a consequence, Marxist academics were marginalized, structurally disadvantaged, institutionally discriminated against and it was increasingly career threatening to take an explicitly Marxist approach to social analysis.

The declining interest in Marx and Marxism is visualized in Figure 1 showing the number of articles in the Social Sciences Citation Index that contain one of the keywords "Marx," "Marxist," or "Marxism" in the article topic description and were published in the five time periods 1968–1977, 1978–1987, 1988–1997, 1998–2007, 2008–2012.[1] These periods are chosen for two reasons: (1) to determine if there has been a change since the start of the new capitalist crisis in 2008; and (2) because social upheavals in 1968 marked a break that also transformed academia.

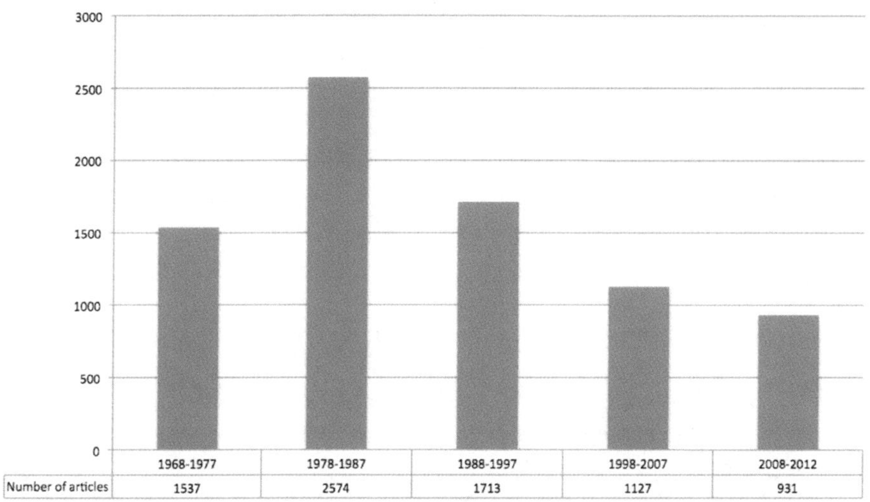

Figure 1: Articles published about Marx and Marxism that are listed in the Social Sciences Citation Index, January 22, 2013.

Figure 1 shows that there was a relatively large output of academic articles about Marx in the period 1978–1987 (2574). Given that the number of articles published increases historically, interest in the period 1968–1977 also seems to have been high. One can observe a clear contraction in the output of articles focusing on Marx, in the periods 1988–1997 (1713) and 1998–2007 (1127).

There are multiple reasons for the disappearance of Marx:

- The rise of neoliberal and neoliberal class struggle from above.
- The commodification of everything, including the commons and public universities.
- The rise of postmodernism in social science and culture.
- The lack of trust in alternatives to capitalism.
- The relatively low presence and intensity of economic struggles.
- Conducting Marxist studies was not conducive for an academic career or academic reputation in a climate of conservative backlash and commodification of academia.

"Monetary crises, independent of real crises or as an intensification of them, are unavoidable" in capitalism (Marx 1894, 649). For Marx, financial crises are not avoidable by regulating financial markets or moral rules that limit greed. Greed is, for him, a structural feature of capitalism that derives from the necessity of capitalists to accumulate ever more capital and to increase profit rates or to perish. Competition between capitals and the need to expand accumulation result in attempts to create "financial innovations" that have a high risk, but can yield very high short-run revenues. The fictitious value signified by commercial papers stands in no direct relation to the actual value created by companies. Financial bubbles are the effect, i.e. share prices that do not reflect actual profitability and which fall rapidly once a burst of the financial bubble is triggered by events that destroy the investors' expectations for high future returns. The new global economic crisis that started in 2008 is the most obvious reason for a resurgence of interest in Marx.

This shift is, however, multidimensional and has multiple causes:

- The new global economic crisis has resulted in an increasing interest in the dynamics and contradictions of capitalism and the notion of crisis itself.
- Neoliberalism and the precarization of work and life are consequences of class society, exploitation and commodification.
- New social movements (the anti-corporate movement, global justice movement, Occupy movement) have an interest in questions of class.
- The financialization of the economy can be analysed with categories such as the new imperialism or fictitious capital.
- New global wars bring about an interest in the category of imperialism.
- Contemporary revolutions and rebellions (such as the Arab spring) give attention to the relevance of revolution, emancipation, and liberation.
- The globalization discourse has been accompanied by discussions about global capitalism.
- The role of mediatization, ICTs, and knowledge work in contemporary capitalism was anticipated by Marx' focus on the General Intellect.
- Many precariously working university scholars and students have a logical interest in Marxian theory.

Indicative of an increased interest in capitalism as an object of study in Media and Communication Studies, several special issues have focused on the role of communication, media, and culture in the recent capitalist crisis.[2]

Some Cultural Studies scholars have, in light of the crisis, admitted indirectly a lack of focus on capitalism, class, and the economy in their field. For example, in the special issue on "The Economic Crisis and After" published by the journal *Cultural Studies*, Lawrence Grossberg wrote that "it is true that the challenge of finding better ways of incorporating economic analysis into the conjuncturalist project of cultural studies has become more urgent and more visible in recent years" (Grossberg 2010a, 295). Mark Hayward wrote in the introduction that the special issue should "remind scholars working in cultural studies that the economy is, and must remain, a site of constant engagement and experimentation" (Hayward 2010, 289). Grossberg (2012, 320) says that it has become "necessary [...] that people who write about culture are taking questions of economics seriously. [...] We don't do the work of taking what's been written about economics seriously within the discipline [of Cultural Studies]—and that includes neoclassical, but it also includes a wide range of heterodox forms of economics. It means that it also entails doing empirical work." An easier and more appropriate way to formulate these "reminders" and "challenges" is that Cultural Studies has to turn into or at least take up aspects of Critical Political Economy in order to adequately understand contemporary capitalism and the role of media and communication. The tradition of Critical Political Economy of Media, Communication, Information and Culture has given attention to the commodification of content and audiences, labor spatialization, class, gender, race, social movements, hegemony and ideology (Mosco 2009). Although a contemporary challenge is that "labor remains a blind spot of western communication studies, including the political economy tradition" (Mosco 2011, 358) some recent work has helped to overcome this blindspot (see: McKercher and Mosco 2007; Mosco and McKercher 2008; Burston, Dyer-Witheford, and Hearn 2010; Mosco, McKercher and Huws 2010). The latter tradition has arguably made critical-theoretical and empirical efforts to come to grips with the relationship between communication and a capitalist political economy, whereas Cultural Studies has to, as Toby Miller (2012, 322) argues, "rethink the anti-Marxism" because it is the "wrong target."

When Grossberg (2010b, 318) says that "cultural studies does need to take questions of economics more seriously, especially because of the specific realities, relationships, and forces of the contemporary conjuncture," then one wonders if economics was really *ever* unimportant for the study of culture, the media, and communication? Has it been unimportant for studying these phenomena during the time of neoliberal capitalism since the 1970s? Has it been unimportant during the time of Fordist mass production and the rise of consumer culture? Has it been unimportant during the time of the war economy in the late 1930s and 1940s? Has it been unimportant during the stock market crashes dotting the twentieth century? Has it been unimportant during the time of the rise of imperialism? Has it been unimportant during the time of the rise of industrial capitalism? The answer to all of these questions is that the economy has always been relevant for the study of culture, communication, the media, and information, but that Cultural Studies has increasingly ignored the economy, not out of societal necessity, but because of its turn away from Marxism and towards postmodernism. The problem of Cultural Studies is, as Robert Babe says, that its "poststructuralist turn [...] instigated the separation" (Babe 2009, 9) from economics. A reintegration requires first and foremost "setting aside poststructuralist cultural studies" (Babe 2009, 196) and seriously engaging with political economy. Specifically, I recommend Marxist studies, as exemplified in the work of Dallas Smythe.

Dallas Smythe and Marxist Media and Communication Studies Today

In the article "On the Political Economy of Communications," Smythe (1960) defined the "central purpose of the study of the political economy of communications" as the evaluation of "the effects of communication agencies in terms of the policies by which they are organized and operated" and the analysis of "the structure and policies of these communication agencies in their social settings" (Smythe 1960, 564). Whereas there are foundations of a general political economy in this paper, there are no traces of Marx in it. Janet Wasko (2004, 311) argues that although "Smythe's discussion at this point did not employ radical or Marxist terminology, it was a major departure from the kind of research that dominated the study of mass communications at that time." Wasko (2004, 312) points out that it was in the "1970s that the political economy of media and communications (PE/C) was explicitly defined again but this time within a more explicitly Marxist framework." She mentions in this context the works of Nicholas Garnham, Peter Golding, Armand Mattelart, Graham Murdock, Dallas Smythe, as well as the Blindspot Debate (Wasko 2004, 312–313).

Smythe (1977, 1) later argued that "western Marxist analyses have neglected the economic and political significance of mass communications systems." Robin Mansell (1995, 51) argues that Smythe was engaged in establishing a Critical Media and Communication Studies that "had at its core the need to interrogate the systemic character of capitalism as it was expressed through the means of structures of communication." Smythe's focus, according to Mansell, was on exposing "through critical research the articulation of political and economic power relations as they were expressed in the institutional relations embedded in technology and the content of communication in all its forms" (Mansell 1995, 47). Smythe was not interested in developing a general political economy of communication, but a

"Marxist theory of communication" (Smythe 1994c, 258) and argued that critical theory means "Marxist or quasi-Marxist" theory (Smythe 1994c, 256). I therefore think that it is consequent and important to characterize Smythe's approach not just as Critical Communication Research, but as Marxist Communication Studies, which means a unity of theoretical/ philosophical, empirical, and ethical studies of media and communication. Such an approach focuses on the analysis of contradictions, structures and practices of domination, exploitation, struggles, ideologies, and alternatives to capitalism in relation to media and communication. One should not split off the importance of Marx and Marxism from Smythe's approach and reduce him to having established a critical empirical research methodology. Janet Wasko stresses that Marx's 11th Feuerbach thesis ("The philosophers have only interpreted the world, in various ways; the point is to change it") applied to the work and life of Dallas Smythe: "Analyzing and understanding the role of communications in the modern world might be enough for most communication scholars. But Dallas Smythe also sought to change the world, not only by his extensive research and teaching in academia, but in his work in the public sector, and through his life as a social activist" (Wasko 1993, 1).

Smythe (1981, xvi–xviii) identified eight core aspects of a Marxist political economy of communications: materiality, monopoly capitalism, audience commodification and advertising, media communication as part of the base of capitalism, labor power, critique of technological determinism, consciousness, arts, and learning. Smythe's works can today remind us of the importance of the engagement with Marx's works for studying the media in capitalism critically. Marx developed a Critique of the Political Economy of Capitalism, which means that it is: a) an analysis and critique of capitalism, b) a critique of liberal ideology, thought and academia, and c) transformative practice.

Karl Marx (1976) titled his magnum opus not *Capital: A Political Economy*, but rather *Capital: A Critique of Political Economy*. Political economy is a broad field, incorporating traditions of thinking grounded in classical liberal economic thought. Marx studied and was highly critical of such writers as Malthus, Mill, Petty, Ricardo, Say, Smith, Ure, and others. His main criticism of bourgeois political economy is that it fetishizes capitalism. Its thinkers "confine themselves to systematizing in a pedantic way, and proclaiming for everlasting truths, the banal and complacent notions held by the bourgeois agents of production about their own world, which is to them the best possible one" (Marx 1976, 175). They postulate that categories like commodities, money, exchange value, capital, markets, or competition are anthropological features of all societies, thereby ignoring the categories' historical character and enmeshment within class struggles. Marx showed the contradictions of political economic thought and took classical political economy as a starting point for a critique of capitalism that considers "every historically developed form as being in a fluid state, in motion" and analyzes how "the movement of capitalist society is full of contradictions" (Marx 1976, 103). Such analysis calls for the "development of the contradictions of a given historical form" by political practice (619) and means that Marx's approach is "in its very essence critical and revolutionary" (Marx 1976, 103).

There are different forms of the political economy of media and communication. Vincent Mosco (2009) distinguishes between neoconservative, institutional, Marxian, feminist, and environmental approaches in political economy. Dwayne Winseck (2011) speaks of political economies of media and identifies a conservative/neoclassical approach, a radical approach, a Schumpeterian institutional approach, and the cultural industries approach. Applying Marx's distinction between political economy and a critique of political economy, one can say that there are certainly Political Economies of Media and Communication, but only one Critique of the Political Economy of Media and Communication: the one grounded in Marx's works and Marxian analysis.

Social scientists in the first half of the twentieth century mapped research paradigms for communications and media. Paul Lazarsfeld (1941/2004) differentiated between traditional and critical research; Max Horkheimer (1937/2002) similarly distinguished between traditional and critical theory. Smythe took up the task of further elaborating this research agenda along relatively critical or administrative ambitions (Smythe 1981, chapter 11).[3]

According to Smythe, "The basis for distinguishing critical and administrative theory and research is in: (1) the kinds of problems chosen for study; (2) the kinds of research methods used in the study; and (3) the ideological predisposition of the researcher either to criticize and try to change the existing politico-economic order or to defend and strengthen it" (Smythe 1984, 205). The second type of theory would be "dialectical, historical, and materialistic" (Smythe 1984, 206).

Smythe (1994c, 256ff) distinguished between administrative and critical theory and between administrative researchable problems and critical problems:

> By administrative theory, I refer to the applications of neopositivistic, behavioral theory. By critical theory, I refer to applications of critical (Marxist or quasi-Marxist) theory. By administrative researchable problems, I mean how to market goods, how to improve the efficiency of media operations, etc. By critical problems, I mean research addressed to macro institutional structure and policies. (Smythe 1994c, 256)

The distinction between administrative/traditional and critical research certainly is still of crucial relevance today. Several decades of neoliberal transformations have weakened the conditions for conducting critical research in the social sciences and humanities as well as in Media and Communication Studies. If one takes a comparative look at the special issues of the *Journal of Communication* that have reflected on the status of the field in 1983 (Volume 33, Issue 3) and 2008 (Volume 58, Issue 4), then these developments become strikingly clear. Whereas in 1983 there were, besides overall positivistic outlooks on the field, also papers with titles like "Emancipation or Domestication: Toward a Utopian Science of Communication", "The Debate over Critical vs. Administrative Research: Circularity or Challenge", "On Critical and Administrative Research: A New Critical Analysis", "The Political and Epistemological Constituents of Critical Communication Research", "The Critical Researcher's Dilemma", "Critical Research and the Role of Labor", "Critical Research in the Information Age", "Power and Knowledge: Toward a New Critical Synthesis", "The Importance of Being Critical—In One's Own Fashion." The 2008 issue in contrast shows cleansing of engagement with critical approaches and instead features articles with titles like "Empirical Intersections in Communication Research: Replication, Multiple Quantitative Methods, and Bridging the Quantitative-Qualitative Divide," "The Evolution of Organizational Communication," "Transdisciplinary Science: The Nexus Between Communication and Public Health," or "The Intersection of Communication and Social Psychology: Points of Contact and Points of Difference." My hope is that in 2020 a special journal issue that reflects on the status of the field will hold the title "The New Rise of Marxism." Smythe's stress on the distinction between conventional and critical-Marxist research reminds us of the fronts of contemporary intellectual struggles.

Besides the reminder of the importance of a Marxist approach for studying communication and the stress on the distinction between administrative and critical research, a third aspect of the relevance of Dallas Smythe's works today is the renewal of the audience commodity concept in debates about digital labor.

Janet Wasko (2005b, 29) argues that "with the increasing spread of privatized, advertiser-supported media, the audience commodity concept has been accepted by many political economists, as well as other communication theorists." In recent years, there has been a revival of interest in Dallas Smythe's works, especially in relation to the question: are users of commercial social media *workers* and are they exploited? Critical conferences have helped to advance the discourse on digital labor[4]. The audience commodity concept has in this context played a crucial role (see e.g., Fuchs 2011a;

2011b; 2011d; 2010b; 2009; Cohen 2008, Fisher 2012; Kang and McAllister 2011; Lee 2011; Manzerolle 2010; McStay 2011; Napoli 2010; Prodnik 2012; Sandoval 2012).

The digital labor debate is not the subject of this article, but I want to briefly point out one of its dimensions that relates to Smythe. Smythe (1981, 47) argued: "For the great majority of the population [...] 24 hours a day is work time" (Smythe 1981, 47). Sut Jhally (1987) has argued that due to the rise of the audience commodity, the living room has become a factory. Mario Tronti (cited in: Cleaver 1992, 137) has taken this idea one step further by arguing that society has become a factory, and that the boundaries of the factory extend spaces for the exploitation of wage labor. Nick Dyer-Witheford (2010, 485) speaks in this context of the emergence of the "factory planet." The exploitation of user labor on commercial Internet platforms like Facebook and Google is indicative of a phase of capitalism, in which we find an all-ubiquitous factory that is a space of the exploitation of labor. Social media and the mobile Internet make the audience commodity ubiquitous and the factory not limited to your living room and your typical space wage labor—the factory and work place surveillance are omnipresent. *The entire planet is today a capitalist factory*. The digital labor debate not only opens up a connection to contemporary debates about Dallas Smythe's concept of the audience commodity, but also a theoretical connection between Smythe and Autonomist Marxism.

Critical Political Economy and Critical Theory of Media and Communication

Smythe argued that Gramsci and the Frankfurt School advanced the concepts of ideology, consciousness, and hegemony as areas "saturated with subjectivism and positivism" (Smythe 1981, xvii). They would have advanced an "idealist theory of the communications commodity" (Smythe 1977, 2) that situates the media only on the superstructure of capitalism and forgets to ask what economic functions they serve. For Smythe (1977), the material aspect of communications is that audiences are exploited and sold as commodities to advertisers. He was more interested in aspects of surplus value generation within the media than the ideological effects of message content. Smythe called for analyzing the media more in terms of surplus value and exploitation and less in terms of manipulation. Nicholas Garnham (1990, 30) shares with Smythe the insight that the Political Economy of Communication should "shift attention away from the conception of the mass media as ideological apparatuses" and focus on the analysis of their "economic role" in surplus value generation and advertising. The analysis of media as "vehicles for ideological domination" is for Garnham (2004, 94) "a busted flush" that is not needed for explaining "the relatively smooth reproduction of capitalism."

Given the analyses of Smythe and Garnham, the impression can be created that Frankfurt School Critical Theory focuses on ideology critique, whereas the Political Economy of Media/Communication focuses on the analysis of capital accumulation by and with the help of the media. Both Smythe and Garnham criticise ideology critique and point out the need for a commodity- and surplus value-centred analysis of the capitalist media. But political economy and ideology critique are not mutually exclusive, but rather require each other. Although widely read works of the Frankfurt School focused on ideology (Adorno, Frenkel-Brunswik; Levinson and Sanford 1950; Horkheimer and Adorno 2002; Marcuse 1964), other books in its book series *Frankfurter Beiträge zur Soziologie* dealt with the changes of accumulation in what was termed late capitalism or monopoly capitalism (Pollock 1956; Friedmann 1959). The Marxist political economist Henryk Grossmann was one of the most important members of the *Institut für Sozialforschung* in the 1920s and wrote his main work at the Institute (Grossmann 1929). Although only a few will today agree with Grossmann's theory of capitalist breakdown, it remains a fact that Marxist political economy was an element of the *Institut für Sozialforschung* right from its beginning and had, in Pollock and Grossmann, two important representatives. After Horkheimer had become director of the Institute in 1930, he formulated an interdisciplinary research program that aimed at bringing together philosophers and scholars from a broad range of disciplines, including economics (Horkheimer 1931). When formulating their general concepts of critical theory, both Horkheimer (Horkheimer and Adorno 2002, 244) and Marcuse (1941) had a combination of philosophy and Marx's Critique of the Political Economy in mind.

Just as Critical Political Economy was not alien to the Frankfurt School, ideology critique has also not been alien to the approach of the Critical Political Economy of the Media and Communication. For Murdock and Golding (1974, 4), the media are organizations that "produce and distribute commodities," means for distributing advertisements, and they also have an "ideological dimension" by disseminating "ideas about economic and political structures." Murdock (1978, 469) stressed in the Blindspot Debate that there are non-advertising based culture industries (like popular culture) that sell "explanations of social order and structured inequality" and "work with and through ideology—selling the system" (see also: Artz 2008, 64). Murdock also argued in the Debate that Smythe did not sufficiently acknowledge Western Marxism in Europe and that one needs a balance between ideology critique and political economy for analyzing the media in capitalism. Smythe also acknowledged the importance of ideology when talking about the "Consciousness Industry" (Smythe 1981, 4–9, 270–299). In contrast to the Frankfurt School, he does not understand ideology as false consciousness, but as a "system of beliefs, attitudes, and ideas" (Smythe 1981, 171). The task of the Consciousness Industry is, for Smythe, to make people buy commodities and pay taxes (Smythe 1994c, 250). Its further task is to promote values that favour capitalism and the private property system (Smythe 1994c, 251–253). One role of the capitalist media would be the "pervasive reinforcement of the ideological basis of the capitalist system." For example, assumptions like "human nature is necessarily selfish and possessive. It has always been this way: You can't change human nature" (Smythe 1994c, 251). So while Smythe criticized the Frankfurt School, he did advance and confirm the importance of ideology critique. Robert Babe argues in this context that although Smythe stressed the need for a materialist theory of culture that sees audience power "as the media's main output" (Babe 2000, 133f), his concept of the Consciousness Industry "is 'idealist' in Smythe's sense of the term" (Babe 2000, 134).

A difference between Critical Political Economy of the Media and Critical Theory is that the first is strongly rooted in economic theory and the second in philosophy and social theory. Dallas Smythe acknowledged this difference: "While the cutting edge of critical theory lies in political economy, critical theory in communications has the transdisciplinary scope of the social sciences, humanities, and arts" (Smythe 1984, 211). Smythe defined Critical Theory broadly as "criticism of the contradictory aspects of the phenomena in their systemic context" (Smythe and Van Dihn 1983, 123) and therefore concluded that Critical Theory is not necessarily Marxist. The historical Critical Theory of the Frankfurt School has its roots in Marxist political philosophy, so the question is if one should really have a broad definition of the term "critical," as Smythe suggests, that does not focus on a systemic critique of capitalism.

The approaches of the Frankfurt School and of the Critique of the Political Economy of Media and Communication should be understood as being complementary. There has been a stronger focus on ideology critique in the Frankfurt School approach for historical reasons. For Horkheimer and Adorno, the rise of German fascism, the Stalinist praxis and American consumer capitalism showed the defeat of the revolutionary potentials of the working class (Habermas 1984, 366f). They wanted to explain why the revolutionary German working class followed Hitler, which engendered interest in the analysis of the authoritarian personality and media propaganda.

The Anglo-American approach of the Political Economy of Media and Communication was developed by people in countries that did not experience fascism, which might be one of the factors that explain the differences in emphasis on ideology and capital accumulation. Whereas North American capitalism was, after 1945, based on pure liberal ideology, anti-communism, and a strong consumer culture, German post-war capitalism was built on the legacy of National Socialism and a strong persistence of fascist thinking.

Horkheimer's (1947) notion of instrumental reason and Marcuse's (1964) notion of technological rationality open up connections between the two approaches. Horkheimer and Marcuse stressed that in capitalism there is a tendency that freedom of action is replaced by instrumental decision making on the part of capital and the state so that the individual is expected to only react and not to act. The two concepts are grounded in Georg Lukács' (1923/1972) notion of reification, which is a reformulation of Marx's (1976) concept of fetishism. Reification means "that a relation between people takes on the character of a thing and thus acquires 'phantom objectivity', an autonomy that seems so strictly rational and all-embracing as to conceal every trace of its fundamental nature: the relation between people" (Lukács 1923/1972, 83). The media in capitalism are modes of reification in multiple senses. First, they reduce humans to the status of consumers of advertisements. Second, culture is in capitalism to a large degree connected to the commodity form, in the form of cultural commodities that are bought by consumers and in the form of audience and user commodities that media consumers/Internet prosumers become themselves. So citizens, on the one hand, buy newspapers, magazines, DVDs, music, computers, mobile phones, laptops, tablets etc., and, on the other hand, do not have to pay for the access to Google, Facebook, YouTube and Twitter because their personal usage and social relations data are sold as commodities to advertisers. Third, in order to reproduce its existence, capitalism has to present itself as the best possible (or only possible) system, and it makes use of the media in order to try to keep this message (in all its differentiated forms) hegemonic. The first and the second dimension constitute the economic dimension of instrumental reason, the third dimension is the ideological form of instrumental reason. Capitalist media are necessarily means of advertising and commodification and spaces of ideology. Advertising and cultural commodification make humans instruments for economic accumulation. Ideology aims at instilling the belief in the system of capital and commodities into human's subjectivity. The goal is that human thoughts and actions do not go beyond capitalism, do not question and revolt against this system and thereby play the role of instruments for the perpetuation of capitalism. It is of course an important question to which extent ideology is always successful and to which degree it is questioned and resisted; but the crucial aspect about ideology is that it encompasses strategies and attempts to make human subjects instrumental in the reproduction of domination and exploitation.

"The wealth of societies in which the capitalist mode of production prevails appears as an 'immense collection of commodities'; the individual commodity appears as its elementary form" (Marx 1976, 125). Marx begins the analysis of capitalism with the analysis of the commodity: its use value, exchange value, value, the labor embodied in it, and the value forms of the commodity, including the money form (x commodity A = y amount of money). Next, Marx turns to the analysis of ideology as an immanent feature of the commodity. The "mysterious character of the commodity-form" is that human social relations that create commodities are not visible in the commodity, but appear as "the socio-natural properties of these things." "The definite social relation between men themselves [take in ideologies] [...] the fantastic form of a relation between things" (Marx 1976, 165). Ideologies legitimatize various phenomena, such as wage labor, by creating the impression that the latter exist always and naturally and by ignoring the historical and social character of things.

Smythe said that the "starting point for a general Marxist theory of communications is [...] the theory of commodity exchange" (Smythe 1994c, 259). Adorno acknowledged that "the concept of exchange is [...] the hinge connecting the conception of a critical theory of society to the construction of the concept of society as a totality" (Adorno 2000, 32). Commodity and commodity exchange are crucial concepts for Critical Political Economy and Critical Theory. As the commodity concept is connected to both capital accumulation and ideology, both approaches should start simultaneously with the value aspects and the ideology aspects of media commodities.

Accumulation and ideology go hand in hand. An example: "social media." After the dot.com crisis in 2000, there was a need for establishing new capital accumulation strategies for the capitalist Internet economy. The discourse on "social media" assumed this task. At the same time, investors were reluctant to invest finance capital after the crisis. Nobody knew if the users were interested in microblogs, social networking sites, etc. The rise of social media as a new capital accumulation model was accompanied by a social media ideology: that social media are new ("Web 2.0"), pose new opportunities for participation, will bring about an "economic democracy," enable new forms of political struggle ("Twitter revolution"), more democracy ("participatory culture"), etc. The rise of new media was accompanied by a techno-deterministic, techno-optimistic ideology. This ideology was necessary for convincing investors and users to support the social media capital accumulation model. The political economy of surplus value generation on social media heavily interacted with ideology in order to enable the economic and discursive rise of "social media."

Cultural Studies scholars who tend to say that the Frankfurt School and the Critical Political Economy of Media and Communication are pessimistic, elitist, and neglect audiences, have a simplified understanding of these two approaches (see for example: Hall 1986, 1988; Grossberg 1995/1998). They say that the concept of ideology as false consciousness makes "both the masses and the capitalists look like judgemental dopes" (Hall 1986, 33). Hall (1988, 44) criticizes Lukács, whose works have been one of the main influences on the Frankfurt School. He says that the false consciousness theorem is simplistic because it assumes that "vast numbers of ordinary people, mentally equipped in much the same way as you or I, can simply be thoroughly and systematically duped into misrecognizing entirely where their real interests lie." Cultural Studies scholar Lawrence Grossberg (1995/1998) argued that both the Frankfurt School and Political Economy have a simple "model of domination in which people are seen as passively manipulated 'cultural dupes'" (616) and that for them "culture matters only as a commodity and an ideological tool of manipulation" (618).

In contrast to such claims, Dallas Smythe actually had a very balanced view of the audience: capital would attempt to control audiences, but they would have potentials to resist.

> People are subject to relentless pressures from Consciousness Industry; they are besieged with an avalanche of consumer goods and services; they are themselves produced as (audience) commodities; they reproduce their own lives and energies as damaged and in commodity form. But people are by no means passive or powerless. People do resist the powerful and manifold pressures of capital as best they can. (Smythe 1981, 270)

Likewise, Adorno, who is vilified by many Cultural Studies scholars as the prototypical cultural pessimist and elitist, had a positive vision for a medium like TV. The German word for television, *Fernsehen,* literally means to watch into the distance. For television "to keep the promise still resonating within the word, it must emancipate itself from everything within which it [...] refutes its own principle and betrays the idea of Good Fortune for the smaller fortunes of the department store" (Adorno 2005, 57). This is indirectly a call for the creation of alternative media that question the status quo. Adorno did not despise popular culture. He was, for example, a fan of Charlie Chaplin and he pointed out the critical role of the clown in popular culture (Adorno 1996). Even in the "Culture Industry" chapter of the *Dialectic of the Enlightenment*, the positive elements of popular culture are visible. For example, Adorno writes that "traces of something better persist in those features of the culture industry by which it resembles the circus" (Horkheimer and Adorno 2002, 114). In his essay *Erziehung nach Auschwitz* (*Education after Auschwitz*), Adorno (1977, 680) wrote about the positive role that TV could play in anti-fascist education in Germany after Auschwitz. If one goes beyond a superficial and selective reading of Adorno, then one will find his deep belief in the possibility of emancipation and in the role that culture can play in it. English translations of Horkheimer's and Adorno's works are imprecise because the language of the two philosophers is complex and not easily translatable. But besides the problem non-German speakers face when reading Horkheimer and Adorno, there seems to be a certain reluctance in Cultural Studies to engage thoroughly with the Frankfurt School's and Critical Political Economy's origins in order to set up a straw man.

David Hesmondhalgh (2010, 280) claims that "Smythe's account is crude, reductionist and functionalist, totally underestimating contradiction and struggle in capitalism" and that it "has totally lost its connection to pragmatic political struggle." Similarly, in a contemporary critique of Smythe's audience commodity theory and its application to digital media, Brett Caraway (2011) argues that "Smythe's theory represents a one-sided class analysis which devalues working-class subjectivity" (696), gives "no discussion of wage struggles, product boycotts, or consumer safety" (700), and thereby conducts "audience commodity fetishism" in which "we are all now merely cogs in the capitalist machine" (700). Caraway's criticism of political economy coincides with his celebration of the "creative energy residing in the new media environment" (706), which sets his analysis on par with social media determinists like Henry Jenkins, who argue that "the Web has become a site of consumer participation" (Jenkins 2006, 133) and that media are today a locus of "participatory culture" (Jenkins 2006). These criticisms are based on uninformed or deliberately selective readings of Smythe that ignore his focus on alternative media as counterpart to audience commodification. Unlike certain cultural studies scholars, Smythe does not celebrate audiences as always rebelling and does not argue for social-democratic reformism that tolerates exploitation and misery; his analysis rather implies the need for the overthrow of capitalism in order to humanize society and the overthrow of the capitalist media system in order to humanize the media.

Dallas Smythe did not ignore the ability of humans to create alternative futures, which is shown by the fact that he engaged with the idea of an alternative communication system. For Smythe, subjectivity is revolutionary subjectivity that aims at fundamentally transforming society and establishing an alternative media system. Critics like Hesmondhalgh and Caraway overlook this aspect of Smythe's approach. Mao wrote in 1957 about big-character posters (Dazibao, Tatsepao): "We should put up big-character posters and hold forums."[5]

When Dallas Smythe wrote in the early 1970s about communication in China in his article, "After Bicycles, What?" (Smythe 1973/1994, 230–244), he took up Mao's idea of the big-character posters for thinking about how to democratically organize the broadcasting system. He spoke of a "two-way system in which each receiver would have the capability to provide either a voice or voice-and-picture response. [...] a two-way TV system would be like an electronic tatzupao system" (Smythe 1973/1994, 231f). These thoughts paralleled the ideas of Hans Magnus Enzensberger's (1970) concept of emancipatory media use, Walter Benjamin's (1934, 1936/2006) idea of the reader/writer, and Bertolt Brecht's (1932/2000) notion of an alternative radio in his radio theory.

Mao had the idea of a media system that is controlled by the people in grassroots processes, and Smythe applied this idea in formulating a concept of alternative electronic media. Yuezhi Zhao (2011) points out the relevance of Smythe's article and his ideas of a non-capitalist communication system for China. Given a world dominated by the logic of neoliberal capitalism (both in the West and China), she stresses, inspired by Smythe, the importance of establishing communications and societies that are based on non-capitalist logic.

Dallas Smythe was fundamentally concerned with processes of commodification, which is reflected in his creation of the audience commodity category. Although he was critical of some other Marxist theories of culture, important elements of ideology critique and alternative media accompany his focus on the audience commodity. He was furthermore deeply concerned about social struggles for a better world and democratic communications. Smythe's work was connected to politics, e.g., he worked with unions for improving the working conditions of communications workers, gave testimonies and conducted studies in favour of public ownership of satellites, public service broadcasting and affordable universal access to telecommunications, and against corporate media control and monopolization (Yao 2010). He also was involved in debates about the establishment of a New World Information and Communication Order and acted as public intellectual. The claim that Smythe had no connection to political struggles is false and ideological.

Conclusion

I have stressed in this chapter that Dallas Smythe's works are helpful in at least three ways today:

- Smythe reminds us of the importance of Marxism and Marx for critically studying media and communication in capitalism.
- Smythe stressed the distinction between administrative and critical research that is a crucial line of struggle in the time of neoliberalism and the new capitalist crisis.
- Smythe's audience commodity concept informs what is called the digital labor debate.

I also discussed the relationship of Critical Theory and Critical Political Economy of Media and Communication Studies. A combination of these two approaches is fruitful and important for Marxist and Critical Media and Communication Studies today. I have shown that commonly held objections against both approaches are wrong and that the two approaches can be complementary, although there are also historical and theoretical differences.

The task for a Critical Theory and the Critique of the Political Economy of Communication, Culture, Information and the Media is to focus on:

a) processes of capital accumulation (including the analysis of capital, markets, commodity logic, competition, exchange value, the antagonisms of the mode of production, productive forces, crises, advertising, etc.),
b) class relations (with a focus on work, labor and the mode of the exploitation of surplus value),
c) domination in general,
d) ideology (both in academia and everyday life).

This task of critical research further includes the analysis of and engagement in struggles against the dominant order, which are enacted in part through social movement struggles aided by social media that aim at the establishment of a democratic socialist society—one based on communication commons as part of structures of commonly-owned means of production (Fuchs 2011c). The approach thereby realizes that in capitalism all forms of domination are connected to forms of exploitation (Fuchs 2011c).

Marxist scholarship has historically had to deal with attempts of repression. Marxist scholars are often facing surveillance, overt or hidden repression, and discrimination (in terms of promotion, applications, acceptance of publications, appointments and firing, resources, research funding, and so on). This has also affected and continues to affect the conditions of conducting critical communication scholarship. Dallas Smythe (1907–1992) reported in his autobiographic memories about the surveillance he was undergoing: "It is possible to distinguish five periods when I was under FBI investigation and/or surveillance. [...] an unidentified female FCC employee visited J. Edgar Hoover's office and told his top deputy that I, along with [others] [...] were pinkos" (Smythe 1994a, 41). From Smythe's FBI file: "It is further recommended that the Springfield Office [of the FBI] be permitted to make limited discreet inquiries [...] to keep abreast of Smythe's Communist activities" (42). Smythe argued that he found out from his FBI file that while he was at the University of Illinois, Willbur Schramm, who is considered by some as a crucial founding figure of Media and Communication Studies, was an anti-communist FBI-informant who made claims about the political attitudes of Smythe to the Bureau (Lent 1995; Smythe and Guback 1994).

In the late 1960s, right-wing groups demanded that Herbert Marcuse (1898–1979) should be fired from his professorship at UC San Diego. In August 1968, the American Legion wanted to buy his university contract for $20,000 USD. A Marcuse puppet carrying a sign that said "Marxist Marcuse" was hoisted with a rope around its neck on a pole in front of San Diego City Hall on January 15, 1969, by anonymous Marcuse haters. This was not only a murder threat, but also a genuinely fascist action with symbolic value, given that Marcuse had to emigrate from anti-Semitic National Socialist Germany in the 1930s because, as a Marxist coming from a Jewish family, he would have probably been killed by the Hitler regime. Then-U.S. Vice President Spiro Agnew said that Marcuse is "literally poisoning a lot of young minds."[6]

Horst Holzer (1935–2000), a German Critical Political Economist of Media and Communication, was one of the most prominent victims of the German *Berufsverbote* (occupational bans) for members of the DKP (German Communist Party). He was appointed to the Chair in Communication and Aesthetics at the University of Bremen by the appointment committee in 1971, but the SPD (Social Democratic Party of Germany)-dominated Senate of Bremen denied him the position because of his DKP membership. He was denied appointment at the University of Oldenbourg in 1972, at the Pädagogische Hochschule Berlin in 1973, the University of Marburg in 1973, and faced a denial of tenure and suspension at the University of Munich in 1974 (see Bönkost 2011).

The logic of repression against Marxists can work by the denial of resources, exclusion from decision-making, ideology, or overt violence. The latter culminates in the appeal to kill Marxists. The contemporary logic of repression against Marxists has most directly been expressed in Anders Breivik's "A European Declaration of Independence," in which one can find 1,112 occurrences of the terms Marx/Marxist/Marxism. He describes "Cultural Marxist profiles" and says that the following intellectuals are the main representatives of the cultural Marxist-worldview that he considers as his main enemy: Georg Lukács, Antonio Gramsci, Wilhelm Reich, Erich Fromm, Herbert Marcuse, and Theodor Adorno. Breivik writes in his declaration that "armed resistance against the cultural Marxist/multiculturalist regimes of Western Europe is the only rational approach." He calls for killing Marxists. The logic of repressing Marxism and Marxists leads in the last instance to this very thought, namely that culture and society are threatened by Marxism and that its representatives should therefore be annihilated. Breivik's arguments show at the same time that the logic of anti-Marxism thought to the end and in the very last instance leaves democratic grounds and enters the realm of fascism. If fascism considers Marxism as its principle antagonist, then it might precisely be the case that in order to resist fascism one has to turn towards Marxism, which illustrates the importance of being a Marxist today.

Although Marxist scholarship is facing repression, it is not impossible to conduct research based on this approach and it is in fact theoretically, academically, and politically necessary. Further support is given to conducting Marxist research of media and the Internet by the circumstance that there is a critical mass of scholars, who know each other and can support each other. Giving the existing anti-Marxist biases, Marxists have to work harder and more in order to be able to build institutional grounds for their works. But this work is worth pursuing—for political and ethical reasons.

Michael Burawoy reflects on the question of when is the right time for conducting critical/Marxist social science: "How often have I heard faculty advise their students to leave public sociology until after tenure—not realizing (or realizing all too well?) that public sociology is what keeps sociological passion alive. [...] Once they have tenure, they [...] may have lost all interest in public sociology, preferring the more lucrative policy world of consultants or a niche in professional sociology. Better to indulge the commitment to public sociology from the beginning, and that way ignite the torch of professional sociology" (Burawoy 2007, 40).

When is the right time for Marxist social science, Marxist Media and Communication Studies, and Marxist Internet Studies? Do we have to wait? We cannot wait. All times are the right times as long as injustice exists in the world. Critical social science requires networks, passion, courage, commitment, and solidarity. What Marxist communication scholars need to do is to keep up the struggle, to build research networks, operate journals, organize conferences, practice mutual aid, engage in visible debates as public intellectuals, and connect academic struggles to other struggles.

Notes

1 İrfan Erdogan (2012) has analyzed 210 articles that mentioned Marx and that were published in 77 selected media and communication journals between January 2007 and June 2011. He found that "Mainstream studies ignore and liberal-democrats generally appreciate Marx," whereas the main criticisms of Marx come from "so-called 'critical' or 'alternative' approaches," whose "'alternatives' are 'alternatives to Marx'" and critical in the sense of a "criticism directed against Marx" (Erdogan 2012, 382). At the same time as there are sustained attempts to downplay the importance of Marx for the study of society, media, and communication, there are indicators of a certain degree of new engagement with Marx. One of them is the special issue of tripleC (http://www.triple-c.at) "Marx is Back—The Importance of Marxist Theory and Research for Critical Communication Studies Today" (Fuchs and Mosco 2012) that features 29 original articles engaging with Marxist thought. Another one was the conference "Critique, Democracy and Philosophy in 21st Century Information Society. Towards Critical Theories of Social Media," at which a sustained engagement with Marx and communication today took place, especially by and among PhD students (see Fuchs 2012d).

2 *tripleC—Journal for a Global Sustainable Information Society: Capitalist Crisis, Communication & Culture* (2009, Vol. 8, No. 2, pp. 193–309, edited by Christian Fuchs, Matthias Schafranek, David Hakken, Marcus Breen). * *International Journal of Communication: Global Financial Crisis* (2010, Vol., edited by Paula Chakravartty and John D.H. Downing) *Cultural Studies: The Economic Crisis and After (2010, Vol. 24, No. 3, pp. 283–444.

3 "By 'critical' researchable problems we mean how to reshape or invent institutions to meet the collective needs of the relevant social community [...] By 'critical' tools, we refer to historical, materialist analysis of the contradictory process in the real world. By 'administrative' ideology, we mean the linking of administrative-type problems and tools, with interpretation of results that supports, or does not seriously disturb, the status quo. By 'critical' ideology, we refer to the linking of 'critical' researchable problems and critical tools with interpretations that involve radical changes in the established order" (Smythe and Van Dihn 1983, 118).

4 Examples of conferences include, "Digital Labour: Workers, Authors, Citizens" (University of Western Ontario 2009), "The Internet as Playground and Factory" (New School 2009).

5 In 1958, Mao wrote: "The Tatsepao, or big-character poster, is [a] powerful new weapon, a means of criticism and self-criticism which was created by the masses during the rectification movement; at the same time it is used to expose and attack the enemy. It is also a powerful weapon for conducting debate and education in accordance with the broadest mass democracy. People write down their views, suggestions or exposures and criticisms of others in big characters on large sheets of paper and put them up in

conspicuous places for people to read." http://www.marxists.org/reference/archive/mao/selectedworks/volume5/mswv5_65.htm

6 The Ku Klux Klan wrote a letter to Marcuse, saying: "Marcuse, you are a very dirty communist dog. […] 72 hours more Marcuse, and we will kill you. Ku Klux Klan" Source: *Herbert's Hippopotamus. Marcuse in Paradise*. A Film by Paul Alexander Juutilainen. http://video.google.com/videoplay?docid=-5311625903124176509.

Bibliography

Abzug, R. 2011. "The Future of Television Audience Research: Changes and Challenges." http://www.learcenter.org/pdf/futureoftvaudience.pdf.

Adamson, W.L. 1985. *Marx and the Disillusionment of Marxism*. Berkley, CA: University of California Press.

Adorno, T. W. 1977. *Kulturkritik und Gesellschaft II*. Frankfurt am Main: Suhrkamp.

———. 2005. "Prologue to Television." In *Critical Models*, 49–57. New York: Columbia University Press.

———. 1996. "Chaplin Times Two." *Yale Journal of Criticism* 9 (1): 57–61.

———. 2000. *Introduction to Sociology*. Cambridge, UK: Polity.

Adorno, T. W., E. Frenkel-Brunswik, D. Levinson and N. Sanford. 1950. *The Authoritarian Personality*. New York: Harper & Row.

Albrechtslund, A. 2008. "Online Social Networking as Participatory Surveillance." *First Monday* 13 (3). http://firstmonday.org/htbin/cgiwrap/bin/ojs/index.php/fm/article/view/2142 /1949/.

Allor, M. 1988. "Relocating the Site of the Audience." *Critical Studies in Mass Communication* 5 (3): 217–33.

Amin, S. 1974a. "In Praise of Socialism." *Monthly Review* 29(4): 1–16 September.

———. 1974b. *Accumulation of a World Scale*. New York: Monthly Review Press.

———. 1975. "Toward a Structural Crisis of World Capitalism." *Socialist Revolution* April: 9–44.

Ampofo, L. 2011. "The Social Life of Real-Time Social Media Monitoring." *Participations* 8 (1): 21–47.

Anderson, C. 2006. *The Long Tail: How the Future of Business Is Selling Less of More*. New York: Hyperion.

Anderson, P. 1976. *Considerations on Western Marxism*. London: New Left Books.

Andrejevic, M. 2002. "The Work of Being Watched: Interactive Media and the Exploitation of Self-Disclosure." *Critical Studies in Media Communication* 19 (2): 230–48.

———. 2004. *Reality TV: The Work of Being Watched*. Lanham, MD: Rowman and Littlefield.

———. 2007. *iSpy: Surveillance and Power in the Interactive Era*. Lawrence, KS: University Press of Kansas.

———. 2008. "Watching Television Without Pity: The Productivity of Online Fans." *Television and New Media* 9 (1): 24–46.

———. 2009a. "Exploiting YouTube: Contradictions of User-generated Labor." In *The YouTube Reader*, edited by P. Snickars and P. Vonderan, 406–23. Stockholm: National Library of Sweden.

———. 2009b. "The Twenty-First Century Telescreen." In *Television Studies After TV: Understanding Television in the Post-Broadcast Era*, ed. G. Turner, and J. Tay, 31–40. New York: Routledge.

———. 2011. "Surveillance and Alienation in the Online Economy." *Surveillance & Society* 8 (3): 278–87.

———. 2013. *Infoglut: How Too Much Information is Changing the Way We Think and Know*. New York: Routledge.

Andrews, K., and P. M. Napoli. 2006. "Changing Market Information Regimes: A Case Study of the Transition to the BookScan Audience Measurement System in the U.S. Book Publishing Industry." *Journal of Media Economics* 19 (1): 33–54.

Ang, I. 1991. *Desperately Seeking the Audience*. London: Routledge.

"API versus Protocol." http://c2.com/cgi/wiki?ApiVsProtocol. Accessed February 19, 2013.

Araujo, L., J. Finch, and H. Kjellberg. 2010. *Reconnecting Marketing to Markets*. Oxford: Oxford University Press.

Arminen, I. 2009. "Intensification of Time-Space Geography in the Mobile Era." In *The Reconstruction of Space and Time: Mobile Communication Practices*, edited by R. Seyler Ling and S. W. Campbell, 89–108. New Brunswick, N.J.: Transaction Publishers.

Arthur, C. 1979. "Dialectic of the Value-Form." In *Value: The Representation of Labour in Capitalism*, edited by D. Elson, 67–81. London: CSE Books.

Artz, L. 2008. Media Relations and Media Product: Audience as Commodity. *Democratic Communiqué* 22 (1): 60–74.

Arvidsson, A. 2005. "Brands: A Critical Perspective." *Journal of Consumer Culture* 5 (2): 235—58.

———. 2006. *Brands: Meaning and Value in Media Culture*. New York: Routledge.

———. 2009. "The Ethical Economy: Towards a Post-Capitalist Theory of Value." *Capital & Class* 33 (1): 13–29.

———. 2011. "Ethics and Value in Customer Co-Production." *Marketing Theory* 11 (3): 261–78.

———. 2013. "The Potential of Consumer Publics." *Ephemera* 13 (2): 367–91.

Arvidsson, A., and E. Colleoni. 2012. "Value in Informational Capitalism and on the Internet." *The Information Society* 28 (3): 135–150.

Aspinall, A. 1949. *Politics and the Press, C. 1780–1985*. London: Home & Van Thal.

Asquith, K. 2009. "A Critical Analysis of the Children's Food and Beverage Advertising Self-Regulatory Initiatives." *Democratic Communiqué* 23 (2): 41–60.

Babe, R.E. 1975. *Cable Television and Telecommunications in Canada: An Economic Analysis*. East Lansing, MI: Michigan State University Press.

———. 1990. *Telecommunications in Canada: Technology, Industry, and Government*. Toronto: University of Toronto Press.

———. 1996. "Paeans to Dallas Smythe." *Journal of Communication* 46 (1): 179–82.

———. 2000. *Canadian Communication Thought: Ten Foundational Writers*. Toronto: University of Toronto Press.

———. 2006a. *Culture of Ecology: Reconciling Economics and Environment*. Toronto: University of Toronto Press.

———. 2006b. "The Political Economy of Knowledge: Neglecting Political Economy in the Age of Fast Capitalism (as Before)." *Fast Capitalism* 2(1). http://www.uta.edu/huma/agger/fastcapitalism/2_1/babe.html

———. 2009. *Cultural Studies and Political Economy: Toward a New Integration*. Lanham, MD: Lexington Books.

———. 2011. "Convergence and Divergence: Telecommunications, Old and New." In *Media, Structures, and Power: The Robert E. Babe Collection*, edited by E. A. Comor, 134–51. Toronto: University of Toronto Press

Baker, C. E. 2002. *Media, Markets, and Democracy*. Cambridge: University of Cambridge Press.

Balas, G. R. 2011. "Eavesdropping at Allerton: The Recovery of Paul Lazarsfeld's Progressive Critique of Educational Broadcasting." *Democratic Communiqué* 24: 1–16.

Balnaves, M., and T. O'Regan. 2010. "The Politics and Practice of Television Ratings Conventions: Australian and American Approaches to Broadcast Ratings." *Continuum: Journal of Media & Cultural Studies* 24 (3): 461–74.

Banet-Weiser, S. 2007. *Kids Rule!: Nickelodeon and Consumer Citizenship.* Duke University Press.

Bar, F., and C. Sandvig. 2008. "US Communication Policy After Convergence." *Media, Culture & Society* 30 (4): 531–50.

Baran, P., and P. Sweezy. 1966. *Monopoly Capitalism.* London: Pelican.

Barney, D. 2000. *Prometheus Wired: The Hope for Democracy in the Age of Network Technology.* Vancouver: UBC Press.

Barnouw, E. 1978. *The Sponsor.* New York: Oxford University Press.

Bauman, Z. 2007. *Consuming Life.* Cambridge, UK: Polity Press.

Beck, U. 1999. *What Is Globalization?* Cambridge: Polity Press.

Becker, K., and F. Stalder, eds. 2009. *Deep Search: The Politics of Search Beyond Google.* Innsbruck, Austria: StudienVerlag.

Beckett, A. 2012. "Governing the Consumer: Technologies of Consumption." *Consumption, Markets & Culture* 15 (1): 1–18.

Benda, C.G. 1979. "State Organization and Policy Information: The 1970 Reorganization of the Post Office Department." *Politics and Society* 9 (2): 123–151.

Beniger, J.R. 1986. *The Control Revolution: Technological and Economic Origins of the Information Society.* Cambridge, MA: Harvard University Press.

Benjamin, W. 1934. Der Autor als Produzent. In *Medienästhetische Schriften,* 231–47. Frankfurt am Main: Suhrkamp.

———. 1936/2006. "The Work of Art in the Age of Mechanical Reproduction." In *Media and Cultural Studies: KeyWorks*, edited by M.G. Durham and D.M. Kellner, 18–40. Malden, MA: Blackwell.

Bennett, T. 1982a. "Media, 'Reality,' Signification." In *Culture, Society, and the Media*, edited by M. Gurevitch, T. Bennett, J. Curran, and J. Woollacott, 287–308. London: Metheun.

———. 1982b. "Theories of media, theories of society." In *Culture, Society, and the Media,* edited by M. Gurevitch, T. Bennett, J. Curran, and J. Woollacott, 30–55. London: Metheun.

Berger, P.L., and T. Luckmann. 1966. *The Social Construction of Reality.* New York: Anchor Books.

Bergreen, L. 1980. *Look Now, Pay Later: The Rise of Network Broadcasting.* Garden City, NY: Doubleday.

Bermejo, F. 2007. *The Internet Audience: Constitution and Measurement.* New York: Peter Lang.

———. 2009. "Audience Manufacture in Historical Perspective: From Broadcasting to Google." *New Media & Society* 11 (1/2): 133–54.

Boddy, W. 2004. "Interactive Television and Advertising Form in Contemporary U.S. Television." In *Television After TV: Essays on a Medium in Transition*, edited by L. Spigel, and J. Olsson, 113–32. Durham, NC: Duke University Press.

Bolin, G. 2009. "Symbolic Production and Value in Media Industries." *Journal of Cultural Economy* 2 (3): 345–61.

Bönkost, J. 2011. "Im Schatten des Aufbruchs. Das erste Berufsverbot für Horst Holzer und die Uni Bremen." *Grundrisse* 39: 29–37.

Bonnington, C. 2012. "Congress Queries App Developers on Their Data Privacy Practices." *Wired,* March 23. http://www.wired.com/gadgetlab/2012/03/congress-app-data-storage/

Bonsu, S.K., & Darmody, A. 2008. "Co-creating Second Life: Consumer Cooperation in Contemporary Economy." *Journal of Macromarketing* 28 (4): 355–68.

Bottomore, T. 1975. *Marxist Sociology*. London: Macmillan.

Bourdieu, P. 1973. "Cultural Reproduction and Social Reproduction." In *Knowledge, Education and Cultural Change*, edited by R. Brown. London: Tavistock Publications.

Bourdon, J., and C. Meadel. 2011. "Inside Television Audience Measurement: Deconstructing the Ratings Machine." *Media, Culture & Society* 33 (5): 791–800.

boyd, d., and E. Hargittai. 2010. "Facebook privacy settings: Who cares?" *First Monday* 15 (8). http://firstmonday.org/htbin/cgiwrap/bin/ojs/index.php/fm/article/view/3086/2589.

boyd, d., and K. Crawford. 2012. "Critical Questions for Big Data." *Information, Communication and Society* 15 (5): 662–79.

Brady, R.A. 1933. *The Rationalization Movement in German Industry*. Berkeley, CA: University of California Press.

———. 1943. *Business as a System of Power*. New York: Columbia University Press.

Brand, S. 1986. *The Essential Whole Earth Catalog: Access to Tools and Ideas*. Garden City, NY: Doubleday.

Braverman, H. 1974. *Labour and Monopoly Capitalism*. New York: Monthly Review Press.

Brecht, B. 1932/2000. "The Radio as a Communications Apparatus." In *Bertolt Brecht on Film & Radio*, edited by M. Silberman, 41–46. London: Methuen.

Brown, B., and A. Quan-Haase. 2012. "'A Workers' Inquiry 2.0': An Ethnographic Method for the Study of Produsage in Social Media Contexts." *tripleC* 10 (2): 488–508.

Brown, L. 1971. *Television: The Business Behind the Box*. New York: Harcourt Brace Jovanovich.

Brooks, T., S. Gray, and J. Dennison. 2010. *The State of Set-top Box Viewing Data as of December, 2009*. New York: Council on Research Excellence.

Bunz, M. 2013. "As You Like It: Critique in the Era of Affirmative Discourse." In *Unlike Us Reader: Social Media Monopolies and Their Alternatives*, edited by G. Lovink and M. Rasch, 137–45. Amsterdam: Institute of Network Cultures.

Burawoy, M. 2007. "For Public Sociology." In *Public Sociology*, edited by D. Clawson, R. Zussman, J. Misra, N. Gerstel, R. Stokes, D.L. Anderton, and M. Burawoy, 23–64. Berkeley, CA: University of California Press.

Burrows, R., and N. Gane. 2006. "Geo-demographics, Software and Class." *Sociology* 40 (5): 793–812.

Burston, J., N. Dyer-Witheford, and A. Hearn, eds. 2010. "Digital labour. Special issue." *Ephemera* 10 (3/4): 214–539.

Buxton, W.J. 2012. "The Rise of McLuhanism, The Loss of Innis-sense: Rethinking the Origins of the Toronto School of Communication." *Candian Journal of Communication* 37 (4): 577–93.

Buzzard, K. S. 2002. "The Peoplemeter Wars: A Case Study of Technological Innovation and Diffusion in the Ratings Industry." *Journal of Media Economics* 15 (4): 273–91.

Callon, M., C. Meadel, and V. Rabeharosoa. 2002. "The Economy of Qualities." *Economy and Society* 31(2): 194–217.

Camfield, D. 2007. "The Multitude and the Kangaroo: A Critique of Hardt and Negri's Theory of Immaterial Labour." *Historical Materialism* 15 (2): 21–52.

Caraway, B. 2011. "Audience Labour in the New Media Environment: A Marxian Revisiting of the Audience Commodity." *Media, Culture, and Society* 33 (5): 693–708.

Carey, J.W. 1969. "The Communications Revolution and the Professional Communicator." *Sociological Review Monograph* 13: 23–38.

———. 2009. *Communication as Culture: Essays on Media and Society*, Revised edition. New York: Routledge.

Carey, J., and L. Grossberg. 2006. “From New England to Illinois: The Invention of (American) Cultrual Studies.” In *Thinking with James Carey: Essays on Communications, Transportation, History*, edited by J. Packer and C. Roberston, 11–28. New York: Peter Lang.

Carlson, M. 2006. “Tapping into TiVo: Digital Video Recorders and the Transition from Schedules to Surveillance in Television.” *New Media & Society* 8 (1): 97–115.

Carr, N. 2008. “Is Google Making Us Stupid?” *The Atlantic*, July/August, 56–58, 60, 62–63.

———. 2010a. *The Shallows: How the Internet Is Changing the Way We Think, Read and Remember*. London: Atlantic.

———. 2010b. “Author Nicholas Carr: The Web Shatters Focus, Rewires Brains.” *Wired*, May 24. http://www.wired.com/magazine/2010/05/ff_nicholas_carr/.

Castañeda, M. 2007. “The Complicated Transition to Broadcast Digital Television in the United States.” *Television & New Media* 8 (2): 91–106.

Cavoukian, A. 2012. “Privacy by Design and the Emerging Personal Data Ecosystem.” In *Privacy By Design*. Toronto, ON: Information and Privacy Commissioner.

Chakravartty, P., and Y. Zhao, eds. 2008. *Global Communications: Toward a Transcultural Political Economy*. Lanham, MD: Rowman and Littlefield.

Chamberlin, E. H. 1931. *The Theory of Monopolistic Competition*. Cambridge, MA: Harvard University Press.

Chao, L. 2009. “Something Borrowed… Chinese Companies Succeed by Taking an Existing Technology and Then Tweaking it for a Local Audience.” *Wall Street Journal*, November 16. http://online.wsj.com/article/SB10001424052748704222704574501434029141444.html.

———. 2011a. “Baidu Net Profit Nearly Doubles on Ad Growth.” *Wall Street Journal*, October 28. http://online.wsj.com/article/SB1000142405297020368750457700267284501862.html.

———. 2011b. “China Summons Internet Executives.” *Wall Street Journal*, November 4.

———. 2013. “Chinese Tech Titans Eye Brazil.” *Wall Street Journal*, January 8. http://online.wsj.com/ article/SB10001424127887323401904578159370572501456.html.

Chao, L., and A. Back. 2010. “China’s Baidu Weighs Life After Google.” *Wall Street Journal*, January 22. http://online.wsj.com/article/SB1000142405274870442320457501679148528 2632.html.

Chen, B. X. 2013. “Smartphones Becomes Life’s Remote Control.” *New York Times*, January 11. http://www.nytimes.com/2013/01/12/technology/smartphones-can-now-run-consume rs-lives.html?_r=0.

Cheng, J. 2011. "Senators Press Apple, Google for Answers About Location Tracking." *Ars Technica*, May 10. http://arstechnica.com/apple/2011/05/senators-press-apple-google-for-answers-about-location-tracking/.

Chung, G., and S. Grimes. 2005. “Data Mining the Kids: Surveillance and Market Research Strategies in Children’s Online Games.” *Canadian Journal of Communication* 30 (4): 527–48.

Cisco. 2013. “Cisco Visual Networking Index: Global Mobile Data Traffic Forecast Update, 2012–2017.” http://www.cisco.com/en/US/solutions/collateral/ns341/ns525/ns537/ns705/ns827/white_paper_c11-520862.html.

Citi Research. 2013. *2013 Retail Technology Deep Dive*. Citigroup Global Markets.

Cleaver, H. 1992. "The Inversion of Class Perspective in Marxian Theory: From Valorisation to Self-Valorisation." In *Open Marxism, Vol. 2*, edited by W. Bonefeld, R. Gunn and K. Psychopedis, 106–44. London: Pluto.

Clement, A. 1992. "Electronic Workplace Surveillance: Sweatshops and Fishbowls." *Canadian Journal of Information Science* 17 (4): 18-45.

Clement, A., and L.R. Shade. 2000. "The Access Rainbow: Conceptualizing Universal Access to the Information/Communications Infrastructure." In *Community Informatics: Enabling Communities with Information and Communications Technologies*, edited by M. Gurstein, 1–20. Hershey, PA: Idea Group.

Coase, R.H. 1950. *British Broadcasting: A Study in Monopoly*. London: Longmans, Green and Co.

———. 2004. "My Evolution as an Economist." In *Lives of the Laureates: Eighteen Nobel Economists*, 4th ed., edited by W. Breit and B.T. Hirsch, 189–207. Cambridge, MA: MIT Press.

Cochoy, F. 2007. "A Brief Theory of the 'Captation' of Publics: Understanding the Market with Little Red Riding Hood." *Theory, Culture & Society* 24 (7/8): 203–23.

Cohen, G.A. 2000. *Karl Marx's Theory of History: A Defence*. Princeton, NJ: Princeton University Press.

Cohen, N. 2008. "The Valorization of Surveillance: Towards a Political Economy of Facebook." *Democratic Communiqué* 22 (1): 5–22.

Collier, R.B., and D. Collier. 2002. *Shaping the Political Arena: Critical Junctures, the Labor Movement, and Regime Dynamics in Latin America*. South Bend, IN: Notre Dame University Press.

Comor, E. 1994. "Harold Innis's Dialectical Triad." *Journal of Canadian Studies* 29 (2): 111–27.

———, ed. 1996. *The Global Political Economy of Communication: Hegemony, Telecommunication, and the Information Economy*. New York: St. Martin's Press.

———. 1998. *Communication, Commerce, and Power: The Political Economy of America and the Direct Broadcast Satellite, 1960–2000*. New York: MacMillan.

———. 2008. *Consumption and the Globalization Project: International Hegemony and the Annihilation of Time*. New York: Palgrave MacMillan.

———. 2011. "Contextualizing and Critiquing the Fantastic Prosumer: Power, Alienation and Hegemony." *Critical Sociology* 37 (3): 309–27.

Coté, M., and J. Pybus. 2007. "Learning to Immaterial Labour 2.0: MySpace and Social Networks." *Ephemera* 7 (1): 88–106.

Cova, B., and D. Dalli. 2009. "Working Consumers: The Next Step in Marketing Theory?" *Marketing Theory* 9 (3): 315–39.

Cova, B., D. Dalli, and D. Zwick. 2011. "Critical Perspectives on Consumers' Role as 'Producers': Broadening the Debate on Value Co-Creation in Marketing Processes." *Marketing Theory* 11 (3): 231–41.

Crawford, S. 2013. *Captive Audience: The Telecom Industry and Monopoly Power in the New Gilded Age*. New Haven, CT: Yale University Press.

Crupi, A. 2012. "CBS' Poltrack Unveils New Media Planning Tool." *Adweek*, October, 3. http://www.adweek.com/news/advertising-branding/cbs-poltrack-unveils-new-media-planning-tool-144160.

Dahlgren, P. 1998. "Critique: Elusive audiences." In *Approaches to Audiences: A Reader*, edited by R. Dickinson, R. Harindranath, O. Linne, 298–310. London: Arnold.

Davenport, T.H., and J.C. Beck. 2001. *The Attention Economy: Understanding the New Currency of Business*. Boston, MA: Harvard Business School Press.

Davidow, B. 2012. "Exploiting the Neuroscience of Internet Addiction." *The Atlantic*, July 18. http://www.theatlantic.com/health/archive/2012/07/exploiting-the-neuroscience-of-internet-addiction/259820/.

DeNardis, L. 2012. "Hidden Levers of Internet Control: An Infrastructure-based Theory of Internet Governance." *Information, Communication & Society* 15 (5): 720–38.

Draper, N. 2012. "Group Power: Discourses of Consumer Power and Surveillance in Group Buying Websites." *Surveillance & Society* 9 (4): 394–407.

Dean, J. 2010. *Blog Theory: Feedback and Capture in the Circuits of Drive*. Cambridge, MA: Polity.

Danielian, N.R. 1939. *AT&T*. New York: Vanguard Press.

Delaney, K.J. 2006. "Google Looks to Boost Ads with YouTube." *Wall Street Journal*, October 10 October.

Dichter, E. 1944. "Psychodramatic Research Project on Commodities as Intersocial Media." *Socioemtry* 7 (4): 432

Donaton, S. 2006. "How to Thrive in the New World of User-Created Content: Let Go." *Advertising Age*, May 1. http://adage.com/columns/article?article_id=108884.

Duffy, B.E. 2010. "Empowerment Through Endorsement? Polysemic Meaning in Dove's User-Generated Advertising." *Communication, Culture & Critique* 3 (1): 26–43

Dyer-Witheford, N. 1999. *Cyber-Marx: Cycles and Circuits of Struggle in High Technology Capitalism*. Urbana, IL: University of Illinois Press.

———. 2009. "The Circulation of the Common." http://www.globalproject.info/it/in_movimento/Nick-Dyer-Witheford-the-Circulation-of-the-Common/4797.

———. 2010. "Digital labour, Species-Becoming and the Global Worker." *Ephemera* 10 (3/4): 484–503.

Eagleton, T. 2007. *Ideology: An Introduction.* London: Verso.

Eckersley, P. 2011. "Some Facts About Carrier IQ." https://www.eff.org/deeplinks/2011/12/carrier-iq-architecture

Edgecliffe-Johnson, A. 2010. "Online Sites Alter View on TV Ratings. *Financial Times*, May 21. http://www.ft.com/cms/s/0/2778ce34-6508-11df-b648-00144feab49a.html.

Efrati, A. 2010. "Online-ad Rebound Drives Net at Google." *Wall Street Journal*, October 15. http://online.wsj.com/article/SB20001424052748704361504575552461758109040.html.

———. 2012a. "Google's Revenue Begins to Slow." *Wall Street Journal*, January 20.

———. 2012b. "New Display Ad Push Adds to Bag of Tricks." *Wall Street Journal*, January 20.

Egan, J. 2008. *Relationship Marketing: Exploring Relational Strategies in Marketing*. Harlow, UK: Pearson Education Limited.

Elmer, G. 2004. *Profiling Machines: Mapping the Personal Information Economy.* Cambridge, MA: MIT Press.

Elson, D. 1979. "Marx's Value Theory of Labour." In *Value: The Representation of Labour in Capitalism*, edited by D. Elson, 115–80. London: CSE Books.

Ely, R. 1886. *Labor Movement in America.* New York: Crowell & Company.

Engels, F. 1891. "Preface." In K. Marx, *Wage Labor and Capital: Value, Price and Profit*. New York: International Publishers.

Enzensberger, H.M. 1970. "Baukasten zu einer Theorie der Medien." In *Kursbuch Medienkultur*, edited by L. Engell, O. Fahle, B. Neitzel, J. Vogel and C. Pias, 264–78. Stuttgart, Germany: DVA.

Erdogan, İ. 2012. "Missing Marx: The Place of Marx in Current Communication Research and the Place of Communication in Marx's Work." *tripleC* 10 (2): 349–91.

Ettema, J.S., and D.C. Whitney. 1994. "The Money Arrow: An Introduction to Audiencemaking." In *Audiencemaking: How the Media Create the Audience*, edited by J.S. Ettema and D.C. Whitney, 1–18. Thousand Oaks, CA: SAGE.

Ewen, S. 1976. *Captains of Consciousness*. New York: McGraw Hill.

Falkinger, J. 2005. "Limited Attention as the Scarce Resource in an Information-Rich Economy." *IZA Discussion Papers* 1538. Bonn, DE: Institute for the Study of Labor.

Federal Trade Commission. 2012. *A Review of Food Marketing to Children and Adolescents: Follow-Up Report*. Washington, DC.

Fejes, F., and J. Schwoch. 1987. "A Competing Ideology for the Information Age: A Two Sector Model for the New Information Society." In *The Ideology of the Information Age*, edited by J.D. Slack and F. Fejes. Norwood, NJ: Ablex.

Fiber, B. 1984. "Tuning Out Ads a Growing Trend." *The Globe and Mail*, October 31.

Fine, B. 2010. *Theories of Social Capital: Researchers Behaving Badly*. London: Pluto.

Fisher, E. 2012. "How Less Alienation Creates More Exploitation? Audience Labour on Social Networking Sites." *tripleC* 10 (2): 171–83.

Fitch, D. 2011. *Who's still afraid of the DVR?* Millward Brown: Point of View. http://www.millwardbrown.com/Libraries/MB_POV_Downloads/MillwardBrown_POV_Still_Afraid_of_DVRs.sflb.ashx.

Fletcher, O. 2010. "Boss talk: Baidu's CEO Pursues Long-term Growth." *Wall Street Journal*, August 4.

Fortunati, L. 2002. "The Mobile Phone: Towards New Categories and Social Relations." *Information, Communication, and Society* 5 (4): 513–28.

Foster, R.J. 2007. "The Work of the ''New Economy'': Consumers, Brands, and Value Creation." *Cultural Anthropology* 22 (4): 707–31.

———. 2011. "The Uses of Use Value: Marketing, Value Creation, and the Exigencies of Consumption Work." In *Inside Marketing: Practices, Ideologies, Devices*, edited by D. Zwick and J. Cayla, 42–57. Oxford and New York: Oxford University Press.

Foucault, M. 1991. "Governmentality." In *The Foucualt Effect*, edited by G. Burchell, C. Gordon and P. Miller, 87–104. Chicago, IL: The University of Chicago Press.

Franklin, S. 2012. "Cloud Control, or the Network as Medium." *Cultural Politics* 8 (3): 442–64.

Friedman, T.. 2007. *The World is Flat*, 3rd ed. New York: Picador.

Friedman, W. 2012a. "Nielsen Buys SocialGuide, Extends Social TV Data Reach." *MediaDailyNews*, November 13. http://www.mediapost.com/publications/article/187068/nielsen-buys-socialguide-extends-social-tv-data-r.html?edition=53451#axzz2CASfFwaH.

———. 2012b. "Social-Media TV Biz Forecast to Hit 256 Bil by 2017." *MediaDailyNews*, October 12. http://www.mediapost.com/publications/article/185060/social-media-tv-biz-forecast-to-hit-256-bil-by-20.html.

———. 2012c. "Social Media Buzz Ups Traditional TV Ratings." *MediaDailyNews*, March 14. http://www.mediapost.com/publications/article/170201/social-media-buzz-ups-trad-tv-ratings.html.

———. 2012d. "Stronger Measures: Viewers' Attachment to Shows, Categories." *MediaDailyNews*, October 1. http://www.mediapost.com/publications/article/184232/stronger-measurements-viewers-attachment-to-show.html.

Friedmann, G. 1959. *Grenzen der Arbeitsteilung*. Frankfurter Beiträge zur Soziologie, Volume 7. Frankfurt am Main: Europäische Verlagsanstalt.

Fuchs, C. 2008. *Internet and Society: Social Theory in the Information Age*. New York: Routledge.

———. 2009. "Information and Communication Technologies and Society A Contribution to the Critique of the Political Economy of the Internet." *European Journal of Communication* 24 (1): 69–87.

———. 2010a. "Class, Knowledge and New Media." *Media, Culture & Society* 32 (1): 141–50.

———. 2010b. "Labor in Informational Capitalism and on the Internet." *The Information Society* 26: 179–96.

———. 2011a. "A Contribution to the Critique of the Political Economy of Google." *Fast Capitalism* 8 (1). http://www.fastcapitalism.com/.

———. 2011b. "Critique of the Political Economy of Web 2.0 Surveillance." In *Internet and Surveillance. The Challenges of Web 2.0 and Social Media*, edited by Christian Fuchs, Kees Boersma, Anders Albrechtslund and Marisol Sandoval, 31–70. New York: Routledge.

———. 2011c. *Foundations of Critical Media and Information Studies*. London: Routledge.

———. 2011d. "The Contemporary World Wide Web: Social Medium or New Space of Accumulation?" In *The Political Economies of Media: The Transformation of the Global Media Industries*, edited by D. Winseck and D. Y. Jin, 201–20. London: Bloomsbury.

———. 2011e. "Web 2.0, Prosumption, and Surveillance." *Surveillance & Society* 8 (3): 288–309.

———. 2012a. "Dallas Smythe Today—The Audience Commodity, the Digital Labour Debate, Marxist Political Economy and Critical Theory. Prolegomena to a Digital Labour Theory of Value." *triple-C* 10 (2): 692–740.

———. 2012b. "Google Capitalism." *tripleC* 10 (1): 42–48.

———. 2012c. "Implications of Deep Packet Inspection (DPI) Internet Surveillance for Society." *The Privacy & Security Research Paper Series*. http://www.projectpact.eu/documents-1/%231_Privacy_and_Security_Research_Paper_ Series.pdf.

———. 2012d. "New Marxian times! Reflections on the 4th ICTs and Society Conference 'Critique, Democracy and Philosophy in 21st Century Information Society. Towards Critical Theories of Social Media.'" *tripleC* 10 (1): 114–21.

———. 2012e. "The Political Economy of Privacy on Facebook." *Television & New Media* 13 (2): 139–59.

———. 2012f. "With or Without Marx? With or Without Capitalism? A Rejoinder to Adam Arvidsson and Eleanor Colleoni." *tripleC* 10 (2): 633–45.

Fuchs, C., and D. Winseck. 2011. "Critical Media and Communication Studies Today: A Conversation." *tripleC* 9 (2): 247–71.

Fuchs, C., and V. Mosco, eds. 2012. "Marx is back—The importance of Marxist theory and research for Critical Communication Studies today." *tripleC* 10 (2): 127–632.

Fukuyama, F. 1992. *The End of History and the Last Man.* New York: Free Press.

Futurescape. 2011a. "The Social TV Factor: How Social TV Impacts the TV Business." White Paper Report. http://www.futurescape.tv/social-tv-white-paper.html.

———. 2011b. *Social TV*, 2nd ed. www.futurescape.tv

Gabriel, Y., and T. Lang. 1995. *The Unmanageable Consumer: Contemporary Consumption and its Fragmentations*. London and Thousand Oaks, CA: Sage.

Galbraith, J.K. 1967. *The New Industrial State*. Boston, MA: Houghton Mifflin.

———. 1969. *The Affluent Society*, 2nd edition. Boston, MA: Houghton Mifflin.

Gandy, O.H., Jr. 1983a. *Audience Segmentation—Targeting Information Subsidies*. Paper presented at the national conference of the Speech Communication Association, Washington, DC, November.

———. 1983b. *Beyond Agenda Setting: Information Subsidies and Public Policy*. Norwood, NJ: Ablex.

———. 1993. *The Panoptic Sort: A Political Economy of Personal Information*. Boulder, CO: Westview.

———. 1995. "Tracking the Audience." In *Questioning the Media: A Critical Introduction*, edited by John Downing and Ali Mohammadi, 166–79. Thousand Oaks, CA: SAGE.

———. 2009. *Coming to Terms wtth Chance: Engaging Rational Discrimination and Cumulative Disadvantage*. Farnham, UK: Ashgate.

Garnham, N. 1979. "Contribution to a Political Economy of Communication." *Media, Culture & Society* 1 (2): 123–146.

———. 1990. *Capitalism and Communication*. London: SAGE.

———. 2004. "Class Analysis and the Information Society as Mode of Production."*Javnost* 11 (3): 93–104.

Gartner. 2012. "Gartner Says Worldwide It Spending on Pace to Surpass $3.6 Trillion in 2012." Stamford, CT: Gartner Research.

Gaudin, S. 2012. "Smartphones and Tablets May be Making You Sleepless, Fat and Sick." *Computerworld*, September 17.

Gergen, Kenneth. 2008. "Mobile Communication and the Transformation of Democratic Process." In *Handbook of Mobile Communications Studies*, edited by James Katz, 297–309. Cambridge, MA: MIT Press.

Gibson, J. 1979. *The Ecological Approach to Visual Perception*. Boston, MA: Houghton Mifflin.

Gibson, E. 2011. "Smartphone Dependency: A Growing Obsession with Gadgets." *USA Today*, July 27.

Gill, R., and A. Pratt. 2008. "In the Social Factory? Immaterial Labour, Precariousness and Cultural Work." *Theory, Culture & Society* 25 (7/8): 1–30.

Gitlin, T. 1980. *The whole World Is Watching*. Berkeley: University of California Press.

———. 1983. *Inside Prime Time*. New York: Pantheon Books.

Glantz, M. 2012. "The Shifting Role of GRPs in Media Buying." Cambridge, MA: Forrester Research.

Goel, V. 2013. "Twitter Buy a Referee in the Fight Over Online TV Chatter." *New York Times*, August 28, 2013. http://bits.blogs.nytimes.com/2013/08/28/twitter-buys-a-referee-in-the-fight-over-online-tv-chatter/?_r=0

Gold, D. A., C. Y.H. Lo, and E. O. Wright. 1975. "Recent Developments in Marxist Theories of the Capitalist State." *Monthly Review* 27 (2), October: 29–43.

Goldhaber, M. 2006. "The Value of Openness in an Attention Economy." *First Monday* 11 (6). http://firstmonday.org/ojs/index.php/fm/article/view/1334/1254

Goodman, A. 2011. "Eli Pariser on The Filter Bubble: What the Internet is Hiding from You." *Democracy Now!* (radio interview), 27 May. Transcript available online at: http://www.democracynow.org/2011/5/27/eli_pariser_on_the_filter_bubble.

Google. 2012. *Annual Report*. Mountain View, CA: Google.

"Google to Help Broker Video Ads." 2011. *Wall Street Journal*, January 11.

Graham, P. 2000. "Hypercapitalism: A Political Economy of Informational Idealism. *New Media & Society* 2 (2): 131–56.

Grimes, S. 2006. "Online Multiplayer Gaming: A Virtual Space for Intellectual Property Debates?" *New Media & Society* 8 (6): 969–90.

Gross, R., and A. Acquisiti. 2005. "Information Revelation and Privacy in Online Social." *Proceedings of the 2005 ACM Workshop on Privacy in the Electronic Society*, November 07, Alexandria, VA, USA.

Grossberg, L. 1995/1998. "Political Economy and Cultural Studies: Reconciliation or Divorce?" In *Cultural Theory and Popular Culture*, edited by John Storey, 600–12. Harlow, UK: Pearson.

———. 2010a. "Modernity and Commensuration. A Reading of a Contemporary (Economic) Crisis." *Cultural Studies* 24 (3): 295–332.

———. 2010b. "Standing on a Bridge. Rescuing Economies from Economists." *Journal of Communication Inquiry* 34 (4): 316–36.

———. 2012. "Interview." *European Journal of Cultural Studies* 15 (3): 302–26.

Grossmann, H. 1929. *Das Akkumulations- und Zusammenbruchsgesetz des kapitalistischen Systems*. Leipzig: C. L. Hirschfeld.

Guback, T. 1979. "Theatrical Film." In *Who Owns the Media?*, edited by B.M. Compaine, 199–298. New York: Harmony Books.

———. 1983. *Capital, labor power, and the identity of film*. Paper presented at the conference on Culture and Communication, Philadelphia, PA.

Guback, T., and S. Douglas. 1983. *Production and Technology in the Communications/ Information Revolution*. Paper presented at the national conference of the Speech Communication Association, Washington, DC, November.

Gurevitch, M., T. Bennett, J. Curran, and J. Woollacott, eds. 1982. *Culture, Society and the Media*. London: Methuen.

Habermas, J. 1984. *Theory of Communicative Action. Volume 1*. Boston, MA: Beacon Press.

Hagen, I., and J. Wasko, eds. 2000. *Consuming Audiences? Production and Reception in Media Research*. Cresskill, NJ: Hampton Press.

Halavais, Alexander. 2009. *Search Engine Society*. Cambridge: Polity.

Hall, S., and T. Jefferson. 1976. *Resistance Through Rituals: Youth Subcultures in Post-War Britain*. London: Hutchinson in association with the Centre for Contemporary Cultural Studies, University of Birmingham.

Hall, S. 1981/1988. "Notes on Deconstructing the Popular." In *Cultural Theory and Popular Culture: A Reader*, edited by J. Storey, 442–53. Hemel Hempstead: Prentice Hall.

———. 1982. "The Rediscovery of 'ideology': The Return of the Repressed in Media Studies." In *Culture, Society, and the Media*, edited by M. Gurevitch, T. Bennett, J. Curran, and J. Woollacott, 56–90. London: Metheun.

———. 1986. "The Problem of Ideology—Marxism Without Guarantees." *Journal of Communication Inquiry* 10 (2): 28–44.

———. 1988. "The Toad in the Garden. Thatcherism Among the Theorists." In *Marxism and the Interpretation of Culture*, edited by C. Nelson and L. Grossberg, 35–73. Urbana, IL: University of Illinois Press.

Hall, S., C. Critcher, T. Jefferson, J. Clarke, and B. Roberts. 1978. *Policing the Crisis: Mugging, the State, and Law and Order*. New York: Holmes & Meier.

Hardt, M. 1999. "Affective Labor." *Boundary 2* 26 (2): 89–100.

Hardt, M, and A. Negri. 2000. *Empire*. Cambridge, MA: Harvard University Press.

———. 2004. *Multitude: War and Democracy in the Age of Empire*. New York: Penguin.

———. 2009. *Commonwealth*. Cambridge, MA: Harvard University Press.

Harvey, D. 1982. *The Limits to Capital*. Chicago: University of Chicago Press.

———. 1990. *The Condition of Postmodernity*. Oxford: Blackwell.
———. 2010. *A Companion to Marx's Capital*. New York: Verso.
———. 2006. *Limits to Capital*. Oxford: Verso.
Havens, T., A.D. Lotz, and S. Tinic. 2009. "Critical Media Industry Studies: A Research Approach." *Communication, Culture, Critique* 2 (2): 234–53.
Hayward, M. 2010. "The Economic Crisis and After." *Cultural Studies* 24 (3): 283–94.
Hazlett, T. W, D. Porter, and V. Smith. 2011. "Radio Spectrum and the Disruptive Clarity of Ronald Coase." *Journal of Law and Economics* 54 (4): S125–S165.
Hearn, A. 2008. "Meat, Mask, Burden: Probing the Contours of the Branded Self." *Journal of Consumer Culture* 8 (2): 197–217.
Hedges, C. 2009. *Empire of Illusion: The End of Literacy and the Triumph of Spectacle*. Toronto: Vintage Canada.
Heidegger, M. 1977. *The Question Concerning Technology and Other Essays*. New York: Harper & Row.
Herzel, L. 1951. "'Public Interest' and the Market in Color Television Regulation." *University of Chicago Law Review* 18: 802–16.
Herzel, L. 1998. "My 1951 Colour Television Article." *Journal of Law and Economics* 41 (S2): 523–27.
Hesmondhalgh, D. 2010. "User-Generated Content, Free Labour and the Cultural Industries." *Ephemera* 10 (3/4): 267–84.
Heydebrand, W. 2003. "The Time Dimension in Marxian Social Theory." *Time & Society* 12 (2/30): 147–88.
———. 2002. *Critical Theory*. New York: Continuum.
Hill, J.A. 2011. "Endangered Childhoods: How Consumerism is Impacting Child and Youth Identity." *Media, Culture & Society* 33 (3): 347–62.
Horkheimer, M. 1931. "The State of Contemporary Social Philosophy and the Tasks of an Institute for Social Research." In *Critical Theory and Society: A reader*, edited by S. E. Bronner and D. Kellner, 25–36. New York: Routledge.
———. 1937/2002. "Traditional and Critical Theory." In *Critical Theory*, 188–252. New York: Continuum.
———. 1947. *Eclipse of Reason*. New York: Continuum.
Horkheimer, M., and T. W. Adorno. 2002. *Dialectic of Enlightenment*. Stanford, CA: Stanford University Press.
Horton, D., and R. Wohl. 1956. "Mass Communication and Parasocial Interaction: Observations on Intimacy at a Distance." *Psychiatry* 19: 215–29.
Humphrey, M. 2011. "Social TV: What Casey Anthony and Jersey Shore Teach About a New Metric." *Forbes*, August 15. http://www.forbes.com/sites/michaelhumphrey/2011/08/15/social-tv-what-casey-anthony-and-jersey-shore-teach-about-a-new-metric/3/.
Humphreys, L. 2011. "Who's Watching Whom? A Study of Interactive Technology and Surveillance." *Journal of Communication* 61 (4): 575–95.
Hurwitz, D. 1984. "Broadcast Ratings: The Missing Dimension." *Critical Studies in Mass Communication* 1 (2): 205–15.
Hussey, M. 2012. "Why Social Audience Measurement Hasn't Delivered Yet." *iMedia Connection*, March 1. http://www.imediaconnection.com/content/31100.asp.
Huws, U. 2003. *The Making of a Cybertariat: Virtual Work in a Real World.* London: Merlin.
Innis, H.A. 1950. *Empire and Communication.*Toronto, ON: University of Toronto Press.
———. 1964. *The Bias of Communication.* Toronto, ON: University of Toronto Press.

Jacobson, M. F., & Mazur, L. A. 1995. *Marketing Madness: A Survival Guide for a Consumer Society*. Boulder, CO: Westview Press.

Jackson, N. 2011. "Infographic: What Would You Give up to Keep Your Smartphone?" *The Atlantic*, August 9. http://www.techrepublic.com/blog/smartphones/infographic-americans-would-give-up-sex-before-smartphones/6714

Jaffe, J. 2005. *Life After the 30-Second Spot: Energize Your Brand with a Bold Mix of Alternatives to Traditional Advertising*. Hoboken, NJ: Wiley.

Jenkins, H. 2006. *Convergence Culture: Where Old and New Media Collide*. New York: New York University Press.

Jenkins, H., S. Ford and J. Green. 2013. *Spreadable Media: Creating Value and Meaning in a Networked Culture*. New York: New York University Press.

Jessop, B. 1977. "Remarks on Some Recent Theories of the Capitalist State." Unpublished paper, University of Cambridge.

Jhally, S. 1982. "Probing the Blindspot: The Audience Commodity." *Canadian Journal of Political and Social Theory* 6 (1–2): 204–10.

———. 1987. *The Codes of Advertising: Fetishism and the Political Economy of Meaning in the Consumer Society*. London: Pinter.

———. 1990. *The Codes of Advertising: Fetishism and the Political Economy of Meaning in the Consumer Society*. New York: Routledge.

———. 1993. "Communications and the Materialist Conception of History: Marx, Innis, and Technology." *Continuum* 7 (1): 161–82.

———. 2000. Advertising at the Edge of the Apocalypse. In *Critical Studies in Media Commercialism*, edited by R. Andersen and L. Strate, 27–39. New York: Oxford University Press.

———. 2007. *The Factory in the Living Room: How Television Exploits its Audience*. Distinguished Faculty Lecture, University of Massachusetts, March 8.

Jhally, S., and B. Livant. 1986. "Watching as Working: The Valorization of Audience Consciousness." *Journal of Communication* 36 (3): 124–43.

Johnson, T. 2011. "FCC Approves Comcast-NBC U Merger." *Variety*, January 18. http://www.variety.com/article/VR1118030437.

Johnson, W. 1971. *Super Spectator and the Electric Lilliputians*. Toronto: Little, Brown.

Kang, H., and M. P. McAllister. 2011. "Selling You and Your Clicks. Examining the Audience Commodification of Google." *tripleC* 9 (2): 141–53.

Katz, E. 2009. "The End of Television?" *The Annals of the American Academy of Political and Social Science* 625 (September): 6–18.

Kelly, K. 1998. New Rules for the New Economy: 10 Radical Strategies for a Connected World. New York: Viking.

———. 2009. "The New Socialism: Global Collectivist Society Is Coming Online." *Wired*, May 22. http://www.wired.com/culture/culturereviews/magazine/17-06/nep_newsocialism.

———. 2010. *What Technology Wants*. New York: Viking Adult.

Kelty, C. M. 2008. *Two Bits: The Cultural Significance of Free Software*. Durham, NC: Duke University Press.

Kim, P. 2001. "New Media, Old Ideas: The Organizing Ideology of Interactive TV." *Journal of Communication Inquiry* 25 (1): 72–88.

King, R. 2013. "Ibm Vp Describes the 'Empowered Consumer Era' at Sugarcon 2013." *ZDNet*. http://www.zdnet.com/ibm-vp-describes-the-empowered-consumer-era-at-sugarcon-2013-7000013739/.

Klein, N. 2000. *No Logo: Taking Aim at the Brand Bullies*. Toronto: Vintage Canada.

Knight, M.M., H.E. Barnes and F. Flugel. 1928. *Economic History of Europe*. New York: Houghton Mifflin.

Kotler, P. 1972. *Marketing Management,* 2nd edition. Englewood Cliffs, NJ: Prentice Hall.

Kozinets, R., A. Hemetsberger and H. Jensen Schau. 2008. "The Wisdom of Consumer Crowds." *Journal of Macromarketing* 28 (4): 339–54.

Kraidy, M. 2005. *Hybridity, Or the Cultural Logic of Globalization*. Philadelphia, PA: Temple University Press.

Kreiss, D., M. Finn and F. Turner. 2011. "The Limits of Peer Production: Some Reminders from Max Weber for the Network Society." *New Media & Society* 13 (2): 243–59.

Lace, S. ed. 2005. *The Glass Consumer*. Bristol: The Policy Press.

Lagace, M. 2004. *Your Customers: Use Them or Lose Them*. Harvard Business School Working Knowledge Series. Cambridge, MA. http://hbswk.hbs.edu/item/4267.html>.

Lanham, R. 2006. *The Economics of Attention*. Chicago: University of Chicago Press.

Lawler, R. 2011. "How Social Is Your Favorite TV Show? Bluefin Labs Knows." *Gigaom*, July 6. http://gigaom.com/video/bluefin-labs/.

Lazarsfeld, P. F. 1941/2004. "Administrative and Critical Communications Research." In *Mass communication and American social thought: Key texts, 1919–1968*, edited by J. D. Peters and P. Simonson, 166–73. Lanham, MD: Rowman & Littlefield.

———. 1947. "Audience Research in the Movie Field." *The Annals of the American Academy of Political and Social Science* 254: 160–68.

Lazzarato, M. 2006. "Immaterial Labor." In *Radical Thought in Italy: A Potential Politics*, edited by Paolo Virno and Michael Hardt, 142–57. Minneapolis, MN: University of Minnesota Press.

Leavitt, A. 2011. "Watching with the World: Television Audiences and Online Social Networks." *Convergence Culture Consortium*, March 9. http://www.convergenceculture.org/research/c3-watchingworld-full.pdf.

Leber, J. 2013. "How Wireless Carriers Are Monetizing Your Movements." *MIT Technology Review*, April 12. http://www.technologyreview.com/news/513016/how-wireless-carriers-are-monetizing-your-movements/

Lebowitz, M. 1986. "Too Many Blindspots on the Media." *Studies in Political Economy* 21 (Autumn): 165–73.

Lee, M. 2010a. *Free Information? The Case Against Google*. Champaign, IL: Common Ground.

———. 2010b. "A Political Economic Critique of Google Maps and Google Earth." *Information, Communication, and Society* 13 (6): 909–28.

———. 2010c. "Revisiting the 'Google in China' Question from a Political Economic Perspective." *China Media Research* 6 (2): 15–24.

———. 2011. "Google Ads and the Blindspot Debate." *Media, Culture, and Society* 33 (3): 433–47.

———. 2012. "Time and the Political Economy of Financial Television." *Journal of Communication Inquiry* 36 (4): 322–39.

———. 2013. "Information and Finance Capital." *Information, Communication & Society* 16 (7): 1139–1156.

Leiss, W., S. Kline, and S. Jhally. 1990. *Social Communication in Advertising*. Scarborough, ON: Nelson Canada.

Lent, J. A. 1995. "Interview with Dallas W. Smythe." In *A Different Road Taken*, edited by J. A. Lent, 21–42. Boulder, CO: Westview Press.

Lessing, L.P. 1956. *Man of High Fidelity: Edwin Howard Armstrong, a Biography*. Philadelphia, PA: Lippincott.
Letzing, J. 2012. "For Google, All Eyes on Costs Per Click." *Wall Street Journal*, April 9.
Levitt, T.N. 1976. "The Industrialization of Service." *Harvard Business Review* September: 63–75.
Lewerenz, S., and B. Nicolosi, eds. 2005. *Behind the Screen: Hollywood Insiders on Faith, Film, and Culture*. Grand Rapids, MI: Baker Books.
Lewis, E. H. 1968. *Marketing Channels*. New York: McGraw Hill.
Li, C., and J. Bernoff. 2008. *Groundswell: Winning in a World Transformed by Social Technologies*. Boston. MA: Harvard Business Press.
Liebling A.J. 1961. *The Press*. New York: Ballantine.
Linder, S. B. 1970. *The Harried Leisure Class*. New York: Columbia University Press.
Livant, B. 1979. "The Audience Commodity: On the 'Blindspot' Debate." *Canadian Journal of Political and Social Theory* 3 (1): 91–106.
———. 1982. "Working at Watching: A Reply to Sut Jhally." *Canadian Journal of Political and Social Theory* 6 (1-2): 211–15.
Livingstone, S. 2009. "Half a Century of Television in the Lives of Our Children." *The Annals of the American Academy of Political and Social Science* 625: 151–63.
Lo, B.W.N., and R.S. Sedhain. 2006. "How Reliable are Web Site Rankings? Implications for e-Business Advertising and Internet Research." *Issues in Information Systems* 7 (2): 233–38.
Lobet-Maris, C. 2009. "From Trust to Tracks: A Technology Assessment Perspective Revisited." In *Deep search: The Politics of Search Beyond Google*, edited by K. Becker and F. Stalder, 73–84. Innsbruck, Austria: StudienVerlag.
Lotz, A.D. 2007. *The Television Will Be Revolutionized.* New York: New York University Press.
Lukács, G. 1923/1972. *History and Class Consciousness*. Cambridge, MA: MIT Press.
Lury, C. 2004. *Brands: The Logos of the Global Economy*. New York: Routledge.
Lusch, R. F., S.L. Vargo. 2006. *The Service-Dominant Logic of Marketing: Dialog, Debate, and Directions*. Armonk, NY: M.E. Sharpe.
Lyon, D. 2001. *Surveillance Society: Monitoring Everyday Life*. Buckingham, UK: Open University Press.
Lyotard, J.F. 1984. *The Postmodern Condition: A Report on Knowledge*. Manchester, UK: Manchester University Press.
MacBean, J.R. 1975. *Film and Revolution.* Bloomington, IN: Indiana University Press.
MacKenzie, D. 2005. "Opening the Black Box of Global Finance." *Review of International Political Economy* 12 (4): 555–76.
Macpherson, C. B. 1964. *The Political Theory of Possessive Individualism: Hobbes to Locke*. London: Oxford University Press.
Magder, T. 1989. "Taking Culture Seriously: A Political Economy of Communications." In *The New Canadian Political Economy*, edited by W. Clement, and G. Williams, 278–96. Montreal, QC: McGill-Queens University Press.
———. 2009. "Television 2.0: The Business of American Television in Transition." In *Reality TV: Remaking Television Culture*, 2nd edition, edited by S. Murray, and L. Ouellette, 141–64. New York: New York University Press.
Mager, A. 2012. "Algorithmic Ideology: How Capitalist Society Shapes Search Engines." *Information, Communication and Society* 15 (5): 769–87.
Mander, J. 1978. Four Arguments for the Elimination of Television. New York: Morrow.

Mandese, J. 2012a. NBCU Taps Arbitron PPM for "Billion Dollar" Olympic lab, Does Not Disclose What it's Paying. *MediaDaily News*, March 12. http://www.mediapost.com/publications/article/169910/nbcu-taps-arbitron-ppm-for-billion-dollar-olympi.html.

———. 2012b. "comScore, Arbitron, Team for "5 Platform" Ratings, Covers Video, Audio, Display Across TV, Radio, PC, Phone, Tablet." *MediaPost*, September 13. http://www.mediapost.com/publications/article/182943/comscore-arbitron-team-for-5-platform-ratings.html.

Mangalindan M. 2003. "Seeking Growth, Search Engine Google Acts Like Ad Agency." *Wall Street Journal*, October 16.

Mansell, R. 1995. "Against the Flow. The Peculiar Opportunity of Social Scientists." In *A Different Road Taken. Profiles in Critical Communication*, edited by J. A. Lent, 43–66. Boulder, CO: Westview Press.

———. 2004. "Political Economy, Power and New Media." *New Media & Society* 6 (1): 74–83.

———. 2012. *Imagining the Internet: Communication, Innovation, and Governance*. Oxford: Oxford University Press.

Mansell, R., R. Samarajiva, and A. Mahan. 2002. *Networking Knowledge for Information Societies: Institutions and Interventions*. Delft, Netherlands: Delft University Press.

Manzerolle, V.R. 2010. "The Virtual Debt Factory: Towards an Analysis of Debt and Abstraction in the American Credit Crisis." *tripleC* 8 (2): 221–36.

———. 2011. "Mobilizing the Audience Commodity: Digital Labour in a Wireless World." *Ephemera* 10 (3/4): 455–69.

———. 2013. *Brave New Wireless World? Mapping the Rise of Ubiquitous Connectivity From Myth to Market*. PhD dissertation, University of Western Ontario.

Manzerolle, V., and A.M. Kjøsen. 2012. "The Communication of Capital: Digital Media and the Logic of Acceleration." *tripleC* 10 (2): 214–29.

———. 2014. "Dare et Capere: Virtuous Mesh / Targeting Diagram." In *The Imaginary App*, edited by S. Matviyenko and P. D. Miller. Cambridge, MA: MIT Press.

Manzerolle, V., and S. Smeltzer. 2011. "Consumer Databases and the Commercial Mediation of Identity: A Medium Theory Analysis." *Surveillance & Society* 8 (3): 323–37.

Marchand, R. 1985. *Advertising the American Dream: Making Way for Modernity, 1920–1940*. Berkley, CA: University of California Press.

Marcuse, H. 1941. *Reason and Revolution. Hegel and the Rise of Social Theory*, 2nd edition. London: Routledge.

———. 1964. *One-Dimensional Man*. Boston, MA: Beacon Press.

Marich, R. 2008. "Measuring Engagement: Audience Metric Exerts Increasing Influence on Ad Spending." Broadcasting & Cable, April 26. http://www.broadcastingcable.com/article/101689-COVER_STORY_Measuring_Engagement.php.

Marszalek, D. 2012. "Stations Up the Ante for Audience Research." *TVNewsCheck*, March 19. http://www.tvnewscheck.com/article/2012/03/19/58169/stations-up-the-ante-for-audience-research.

Martin, K., and I. Todorov. 2010. "How Will Digital Platforms Be Harnessed in 2010, and How Will They Change the Way People Interact With Brands?" *Journal of Interactive Advertising* 10 (2): 61–66.

Marx, K. 1844/1961. *Economic and Philosophic Manuscripts of 1844*. Translated by M. Milligan. Moscow: Foreign Languages Publishing House.

———. 1863. *Theories of Surplus-Value*. Available at http://www.marxists.org/archive/marx/works/1863/theories-surplus-value/

———. 1894. *Capital: Volume III*. Translated by D. Fernbach. London: Penguin.
———. 1973. *Grundrisse*. Translated by M. Nicolaus. London: Penguin.
———. 1976. *Capital: Volume I*. Translated by B. Fowkes. London: Penguin.
———. 1984. "Introduction to a Critique of Political Economy." In *The German Ideology*, edited by C.J. Arthur, 124–51. New York: International Publishers.
———. 1984. *The Poverty of Philosophy*. Translated by Harry Quelch. Moscow: Progress.
Marx, L. 1964. *The Machine in the Garden: Technology and the Pastoral Ideal in America*. New York: Oxford University Press.
Mattelart, A. 1991. *Advertising International: The Privatization of Public Space*. Trans. M. Chanan. New York: Routledge.
———. 2000. *Networking the World, 1794–2000*. Minneapolis, MN: University of Minnesota Press.
Maxwell, R. 1991. "The Image Is Gold: Value, The Audience Commodity, and Fetishism." *Journal of Film and Video* 43 (1/2): 29–45.
Maxwell, R., and T. Miller. 2012. *Greening the Media*. Oxford: Oxford University Press.
Mayer, V. 2011. *Below the Line: Producers and Production Studies in the New Television Economy*. Raleigh, NC: Duke University Press.
Mayr, O. 1971a. "Adam Smith and the Concept of the Feedback System: Economic Thought and Technology in 18th-Century Britain." *Technology and Culture* 12 (1): 1–22.
———. 1971b. "Maxwell and the Origins of Cybernetics." *Isis* 62 (4): 425–44.
McAllister, M. P. 2005. "Television Advertising as Textual and Economic System." In *A Companion to Television*, edited by J. Wasko, 217–237. Malden, MA: Blackwell.
McChesney, R. W. 1996. "The Internet and U.S. Communication Policy-Making in Historical and Critical Perspective." *Journal of Communication* 46 (1): 98–124.
———. 2004. *The Problem of the Media*. New York: Monthly Review Press.
———. 2007. *Communication Revolution: Critical Junctures and the Future of Media*. New York: The New Press.
———. 2013. *Digital Disconnect: How Capitalism is Turning the Internet Against Democracy*. New York: New Press.
McFarland, K. 2008. *The Breakthrough Company*. New York: Crown Business.
McGonigal, J. 2008. *Engagement Economy: The Future of Massively Scaled Collaboration and Participation*. Institute for the Future. http://www.iftf.org/node/2306.
McGuigan, J. 2000. "Sovereign Consumption." In *The Consumer Society Reader*, edited by M. J. Lee, 294–99. Malden, MA: Blackwell Publishing.
McGuigan, L. 2012a. "Consumers: The Commodity Product of Interactive Commercial Television, or, Is Dallas Smythe's Thesis More Germane than Ever?" *Journal of Communication Inquiry* 36 (4): 288–304.
———. 2012b. "Direct Marketing and the Productive Capacity of Commercial Television: T-commerce, Advanced Advertising, and the Audience Product." *Television & New Media* (December 11) DOI: 10.1177/1527476412467075.
McKercher, C., and V. Mosco, eds. 2007. *Knowledge Workers in the Information Society*. Lanham, MD: Lexington Books.
McLuhan, M. 1964. *Understanding Media: The Extensions of Man*. Cambridge, MA: MIT Press.
McMahon, D., and O. Fletcher. 2011. "Global Finance: China Studies Foreign IPOs—Tighter Regulation Could Reduce Options for Private Companies to Go Public." *Wall Street Journal*, September 21.

McStay, A. 2011. "Profiling Phorm: An Autopoietic Approach to the Audience-as-Commodity." *Surveillance and Society* 8 (3): 310–22.

McWilliam, G. 2000. "Building Stronger Brands through Online Communities." *Sloan Management Review* 41 (3): 43–55.

Meehan, E.R. 1983a. "Economics of Scale, Market Uncertainty, and Technical Innovation: Sources for Program Diversity?" Paper presented at the national conference of Speech Communication Association, Washington, DC, November.

———. 1983b. *Neither Heroes nor Villains: Towards a Political Economy of the Ratings Industry*. Ph.D. dissertation, University of Illinois-Urbana.

———. 1984. "Ratings and the Institutional Approach: A Third Answer to the Commodity Question." *Critical Studies in Mass Communication* 1 (2): 216–25.

———. 1986. "Conceptualizing Culture as Commodity." *Critical Studies in Mass Communication* 3 (4): 448–57.

———. 1993a. "Commodity Audience, Actual Audience: The Blindspot Debate." In *Illuminating the Blindspots: Essays Honouring Dallas W. Smythe*, edited by J. Wasko, V. Mosco, and M. Pendakur, 378–97. Norwood, NJ: Ablex.

———. 1993b. "Heads of Household and Ladies of the House: Gender, Genre, and Broadcast ratings, 1929–1990." In *Ruthless Criticism: New Perspectives in U.S. Communication History*, edited by W. S. Solomon and R. W. McChesney, 204–21. Minneapolis, MN: University of Minnesota Press.

———. 2000. Leisure or Labour? Fan Ethnography and Political Economy. In *Consuming Audiences? Production and Reception in Media Research*, edited by I. Hagen, and J. Wasko, 71–92. Cresskill, NJ: Hampton Press.

———. 2005. *Why TV Is Not Our Fault: Television Programming, Viewers, and Who's Really in Control*. Lanham, MD: Rowman and Littlefield.

Meehan, E.R., and J. Wasko. 2013. "In Defense of a Political Economy of the Media." *Javnost – The Public* 20 (1): 39–54

Melody, W. 1973. *Children's Television: The Economics of Exploitation*. New Haven, CN: Yale University Press.

———. 1993. "Dallas Smythe: A Lifetime at the Frontier of Communications." *Media, Culture & Society* 15 (2): 295–97.

———. 1994. "Dallas Smythe: Pioneer in the Political Economy of Communications." In *Counterclockwise: Perspectives on Communication*, edited by T. Guback, 1–6. Boulder, CO: Westview.

———. 1996. "Toward a Framework for Designing Information Society Policies." *Telecommunications Policy* 20 (4): 243–59.

Melody, W.H., L. Salter, and P. Heyer, eds. 1981. *Culture, Communication, and Dependency: The Tradition of H.A. Innis*. Norwood, NJ: Ablex.

Middleton, C. 2007. "Illusions of Balance and Control in an Always-on Environment: A Case Study of Blackberry Users." *Continuum: Journal of Media & Cultural Studies* 21 (2): 165–78.

Miller, P., and N. Rose. 1997. "Mobilizing the Consumer: Assembling the Subject of Consumption." *Theory, Culture & Society* 14: 1–36.

———. 2008. *Governing the Present: Administering Economic, Social and Personal Life*. Cambridge, UK: Polity.

Miller, T. 2007. *Cultural Citizenship: Cosmopolitanism, Consumers, and Television in a Neoliberal Age*. Philadelphia, PA: Temple University Press.

———. 2010. *Television Studies: The Basics*. New York: Routledge.

———. 2012. "Cultural Studies in an Indicative Mode." *Communication and Critical/ Cultural Studies* 8 (3): 319–22.

Moor, L. 2007. *The Rise of Brands*. Oxford: Berg.

Morley, D. 1993. Active Audience Theory: Pendulums and Pitfalls. *Journal of Communication* 43 (4): 13–19.

Mosco, V. 2004. *The Digital Sublime: Myth, Power, and Cyberspace*. Cambridge, MA: MIT Press.

———. 2009. *The Political Economy of Communication*, 2nd edition. Thousand Oaks, CA: SAGE.

———. 2011. "The Political Economy of Labor." In *The Handbook of Political Economy of Communications*, edited by J. Wasko, G. Murdock and H. Sousa, 358–80. Malden, MA: Wiley-Blackwell.

Mosco, V., and A. Herman. 1981. "Radical Social Theory and the Communications Revolution." In *Communication and Social Structure*, edited by E. McAnany, J. Schnitman, and N. Janus, 58–84. New York: Paeger.

Mosco, V., and C. McKercher. 2008. *The Laboring of Communication: Will Knowledge Workers of the World Unite?* Lanham, MD: Lexington Books.

Mosco, V., C. McKercher and U. Huws. 2010. "Getting the Message: Communications Workers and Global Value Chains." *Work Organisation, Labour & Globalisation* 4 (2): 1–9.

Mosco, V., and D. Schiller, eds. 2001. *Continental Order? Integrating North America for Cybercapitalism*. Lanham, MD: Rowman and Littlefield.

Mosco, V., and L. Kaye. 2000. "Questioning the Concept of the Audience." In *Consuming Audiences? Production and Reception in Media Research*, ed. I. Hagen, and J. Wasko, 31–46. Cresskill, NJ: Hampton Press.

Mosco, V., and J. Wasko, eds. 1983. *The Critical Communication Review, Volume I: Labor, the Working Class, and the Media*. Norwood, NJ: Ablex.

Mosco, V, J. Wasko, and M. Pendakur, eds. 1993. *Illuminating the Blindspots: Essays Honouring Dallas W. Smythe*. Norwood, NJ: Ablex.

Moss, David A., and Michael R. Fein. 2003. "Radio Regulation Revisited: Coase, the FCC, and the Public Interest." *Journal of Policy History* 15 (4): 389–416.

Murdock, G. 1978. "Blindspots About Western Marxism: A Reply to Dallas Smythe." *Canadian Journal of Political and Social Theory* 2: 109–19.

———. 1982. "Large Corporations and the Control of Communications Industries." In *Culture, Society, and the Media*, edited by M. Gurevitch, T. Bennett, J. Curran, and J. Woollacott, 118–150. London: Metheun.

———. 1992a. "Citizens, Consumers, and Public Culture." In *Media Cultures: Reappraising Transnational Media*, edited by M. Skovmand and K. C. Schroder, 17–41. London: Routledge.

———. 1992b. "Embedded Persuasions: The Fall and Rise of Integrated Advertising." In *Come on Down: Popular Media Culture in Post-War Britain*, edited by D. Strinati and Stephen Wagg, 202–31. London: Routledge.

———. 2000. "Peculiar Commodities: Audiences at Large in the World of Goods." In *Consuming Audiences? Production and Reception in Media Research*, edited by I. Hagen and J. Wasko, 47–70. Cresskill NJ: Hampton Press.

———. 2005. "Building the Digital Commons: Public Broadcasting in the Age of the Internet." In *Cultural Dilemmas of Public Service Broadcasting*, edited by P. Jauert and G. F Lowe, 213–30. Gotebourg: NORDICOM Goteborg University.

———. 2011. "Political Economies as Moral Economies: Commodities, Gifts and Public Goods." In *The Blackwell Handbook of the Political Economy of Communication*, edited by J. Wasko, G. Murdock and H. Sousa, 13–40. Oxford: Blackwell.
———. 2013a. "Communication in Common." *The International Journal of Communication* 7: 154–72.
———. 2013b. "Producing Consumerism: Commodities, Ideologies, Practices." In *Critique, Social Media and the Information Society*, edited by C. Fuchs and M. Sandoval, 125–43. London: Routledge.
Murdock, G., and P. Golding. 1974. "For a Political Economy of Mass Communications. In *The Political Economy of the Media I*, edited by P. Golding and G. Murdock, 3–32. Cheltenham: Edward Elgar.
Napoli, P.M. 1999. "Deconstructing the Diversity Principle." *Journal of Communication* 49 (4): 7–34.
———. 2003a. *Audience Economics: Media Institutions and the Audience Marketplace*. New York: Columbia University Press.
———. 2003b. "Environmental Cognitions in a Dual-Product Marketplace: A Participant-Observation Perspective on the U.S. Broadcast Television Industry." *International Journal on Media Management* 5 (2): 100–08.
———. 2005. "Audience Measurement and Media Policy: Audience Economics, the Diversity Principle, and the Local People Meter." *Communication Law & Policy* 10 (4): 349–82.
———. 2009. "Audience Measurement, the Diversity Principle, and the First Amendment Right to Construct the Audience." *St. John's Journal of Legal Commentary* 24 (2): 359–85.
———. 2010. "Revisiting 'Mass Communication' and the 'Work' of the Audience in the New Media Environment." *Media, Culture, and Society* 32 (3): 505–16.
———. 2011. *Audience Evolution: New Technologies and the Transformation of Media Audiences*. New York: Columbia University Press.
———. 2012a. "Program Value in the Evolving Television Audience Marketplace." Washington, DC: Time Warner Cable Research Program on Digital Communications.
———. 2012b. "Audience Evolution and the Future of Audience Research." *International Journal on Media Management* 14 (2): 79–97.
Nasar, S. 2011. *Grand Pursuit: The Story of Economic Genius*. New York: Simon & Schuster.
NBCUniversal. 2012. "NBCUniversal to Conduct Extensive Research Effort During 2012 London Olympics." February 13. http://www.nbcuni.com/corporate/newsroom/nbcuniversal-to-conduct-extensive-research-effort-during-2012-london-olympics/
Neff, J. 2011. "P&G Hikes Ad Spending to Record Levels." *Advertising Age*, August 5. http://adage.com/article/news/p-g-hikes-ad-spending-record-levels/229133/.
Neilson, B., and N. Rossiter. 2005. "From Precarity to Precariousness and Back Again: Labour, Life and Unstable Networks." *Fibreculture* 5.
Nichols, J., and R.W. McChesney. 2000. *It's the Media, Stupid.* New York: Seven Stories Press.
———, and ———. 2013. *Dollarocracy: How the Money and Media Election Complex is Destroying America.* New York: Nation Books.
Nielsen, A.C., Jr. 1966. "If Not the People…Who?" Presented to the Oklahoma City Advertising Club.
Nightingale, V., ed. 2011. *The Handbook of Media Audiences*. Malden, MA: Wiley-Blackwell.
Nixon, B. 2012. "Dialectical Method and the Critical Political Economy of Culture." *tripleC* 10 (2): 439—56.

Nordenstreng, K., and T. Varis. 1974. "Television Traffic—A One-Way Street?" *UNESCO Reports and Papers on Mass Communication*, 70. Paris: UNESCO.

O'Dwyer, R. 2013. "Spectre of the Commons: Spectrum Regulation in the Communism of Capital." *Ephemera* 13 (3): 497-526.

Østergaard, P., and J. Fitchett. 2012. "Relationship Marketing and the Order of Simulation." *Marketing Theory* 12 (3): 223–49.

Pariser, E. 2011. The Filter Bubble: What the Internet Is Hiding from You. New York: Penguin.

Parker, I. 1981. "Innis, Marx, and the Economics of Communication: A Theoretical Aspect of Canadian Political Economy." In *Culture, Communication, and Dependency: The Tradition of H.A. Innis*, edited by W. H. Melody, L. Salter, and P. Heyer, 127–43. Norwood, NJ: Ablex.

———. 1985. "Staples, Communications, and the Economics of Capacity, Overhead Costs, Rigidity, and Bias." In *Explorations in Canadian Economic History: Essays in Honour of Irene M. Spry*, edited by D. Cameron, 73–94. Ottawa: University of Ottawa Press.

Palamountain, J. C., Jr. 1969. "Vertical Conflict." In *Distribution Channels: Behavioral Dimensions*, edited by L. W. Stern. New York: Houghton Mifflin.

Palmer, B. 1974. "Class, Conceptions and Conflict: The Thrust for Efficiency, Managerial Views of Labour and the Working Class Rebellion, 1903–1922." *Review of Radical Political Economy*, 7 (2): 31–49.

Pasquinelli, M. 2009a. "Google Pagerank Algorithm: A Diagram of Cognitive Capitalism and the Rentier of the Common Intellect." In *Society of the Query Conference*. Amsterdam, The Netherlands.

———. 2009b. "Google's PageRank: Diagram of the Cognitive Capitalism and Rentier of the Common Intellect." In *Deep Search: The Politics of Search Beyond Google*, edited by K. Becker and F. Stalder, 152–62. Innsbruck, Austria: StudienVerlag.

Patel, K. 2011. "Will Social Media be the New Nielsen for TV Ad Buyers?" *Advertising Age*, October, 3. http://adage.com/article/digital/social-media-nielsen-tv-ad-buyers/230146/.

Pendakur, M. 2003. *Indian Cinema: Industry, Ideology, and Consciousness*. Cresskill, NJ: Hampton Press.

Pepitone, J. 2013. "Facebook Needs to Keep Mobile Momentum." *CNNMoney*, May 1.

Peters, J. D. 2009. "Calendar, Clock, Tower." Presentation to the *Media in Transition Conference* at MIT, April 24–26. http://web.mit.edu/comm-forum/mit6/papers/peters.pdf.

Peterson, R. A., and M. C. Merino. 2003. "Consumer Information Search Behavior and the Internet." *Psychology & Marketing* 20 (2): 99–121.

Petersen, S. M. 2008. "Loser Generation Content: From Participation to Exploitation." *First Monday* 13 (3). http://firstmonday.org/htbin/cgiwrap/bin/ojs/index.php/fm/article/view/2141/1948.

Pfanner, E. 2013. "French Tax Proposal Zeroes in on Web Giants' Data Harvest." *New York Times*, February 24.

Pickard, V. 2007. "Neoliberal Visions and Revisions in Global Communications Policy from NWICO to WSIS." *Journal of Communication Inquiry* 31 (2): 118–39.

———. 2010. "Reopening the Postwar Settlement for U.S. Media: The Origins and Implications of the Social Contract Between the Media, the State, and the Polity." *Communication, Culture & Critique* 3: 170–89.

———. 2011. "The Battle over the FCC Blue Book: Determining the Role of Broadcast Media in Democratic Society, 1945–8." *Media, Culture & Society* 33 (2): 171–91.

Polanyi, K. 2001. *The Great Transformation: The Political and Economic Origins of Our Time*, 2nd edition. Boston, MA: Beacon Press.

Pollock, F. 1956. *Automation. Materialien zur Beurteilung der ökonomischen und sozialen Folgen.* Frankfurter Beiträge zur Soziologie, Vol. 5. Frankfurt am Main: Europäische Verlagsanstalt.

Postman, N. 1985. *Amusing Ourselves to Death: Public Discourse in the Age of Show Business*. New York: Viking.

Prahalad, C. K., and V. Ramaswamy. 2000. "Co-opting Customer Competence." *Harvard Business Review* 78(January-February): 79–87.

———. 2002. "The Co-creation Connection." *Strategy and Business* 27 (2): 51–60.

———. 2004a. "Co-creation Experiences: The Next Practice in Value Creation." *Journal of Interactive Marketing* 18 (3): 5–14.

———. 2004b. *The Future of Competition: Co-creating Unique Value with Customers.* Boston, MA: Harvard Business School.

Pridmore, J. 2013. "Collaborative Surveillance: Configuring Contemporary Marketing Practice." In *The Surveillance Industrial Complex: A Political Economy of Surveillance*, edited by K. Ball and L. Snider. New York: Routledge.

Pridmore, J., and D. Lyon. 2011. "Marketing as Surveillance: Assembling Consumers as Brands." In *Inside Marketing: Practices, Ideologies, Devices*, edited by D. Zwick and J. Cayla, 115–36. Oxford: Oxford University Press.

Pridmore, J., and D. Zwick. 2011. "Editorial: Marketing and the Rise of Commercial Consumer Surveillance." *Surveillance & Society* 8 (3): 269–77.

———. 2012. "The Rise of the Customer Database." In *Routledge Companion to Digital Consumption*, edited by R. W. Belk and R. Llamas, 102–12. New York: Routledge.

Prodnik, J. 2012. "A Note on the Ongoing Processes of Commodification: From the Audience Commodity to the Social Factory." *tripleC* 10 (2): 253–73.

Proulx, M., and S. Shepatin. 2012. *Social TV: How Marketers Can Reach and Engage Audiences by Connecting Television to the Web, Social Media, and Mobile*. Hoboken, NJ: Wiley.

Qualman, E. 2009. *Socialnomics: How Social Media Transforms the Way We Live and Do Business*. Hoboken, NJ: John Wiley & Sons.

Reardon, M. 2013. "Location Information to Make Mobile Ads More Valuable." *CNET*, April 15. http://news.cnet.com/8301-1035_3-57579746-94/location-information-to-make-mobile-ads-more-valuable/.

Reichelt, H. 2007. "Marx's Critique of Economic Categories: Reflections on the Problem of Validity in the Dialectical Method of Presentation in *Capital*." *Historical Materialism* 15: 3–52.

Ray, M., and P. Webb. 1978. *Advertising Effectiveness in a Crowded Television Environment.* Cambridge, MA: Marketing Science Institute.

Rey, R. J. 2012. "Alienation, Exploitation, and Social Media." *American Behavioral Scientist* 56 (4): 399–420.

Reel, F. 1979. *Networks: How they Stole the Show*. New York: Scribner.

Rheingold, H. 1993. *The Virtual Community: Homesteading on the Electronic Frontier.* Reading, MA: Addison-Wesley.

Ricœur, P. 1977. *The Rule of Metaphor: Multi-disciplinary Studies of the Creation of Meaning in Language*. Toronto: University of Toronto Press.

Ritzer, G. 2009. "Correcting an Historical Error." Keynote Address. Conference on Prosumption, Frankfurt, Germany, March, 2009.

Ritzer, G., and N. Jurgenson. 2010. "Production, Consumption, Prosumption." *Journal of Consumer Culture* 10 (1): 13–36.

Ritzer, G., P. Dean, and N. Jurgenson, eds. 2012. "The Coming Age of Prosumption and the Prosumer." *American Behavioral Scientist* 56 (4): 379–640.

Robins, K., and F. Webster. 1999. *Times of the Technoculture: From the Information Society to the Virtual Life*. London and New York: Routledge.

Rose, N. S. 1999. *Powers of Freedom: Reframing Political Thought*. Cambridge, UK: Cambridge University Press.

———. 2001. "The Politics of Life Itself." *Theory, Culture & Society* 18 (6): 1–30.

Rotella, P. 2012. "Is Data the New Oil? *Forbes*, April 2. http://www.forbes.com/sites/perryrotella/2012/04/02/is-data-the-new-oil/.

Rowinski, D. 2011. "What is the Difference Between AdMob and AdSense?" *ReadWrite*, September 6. http://readwrite.com/2011/09/06/what-is-the-difference-between#awesm=~orCSuuALtukjLo

Rowlands, S. 2013. "Juniper: In-app Advertising is Ripe for Growth as Rich Media and RTB Gain Prominence." *Fierce Wireless*, October 29. http://www.fiercewireless.com/story/juniper-app-advertising-ripe-growth-rich-media-and-rtb-gain-prominence/2013-10-29

Rubens, W. S. 1984. "High Tech Audience Measurement for New Tech Audiences." *Critical Studies in Mass Communication* 1 (2): 195–205.

Russell, M.G. 2009. "A Call for Creativity in New Metrics for Liquid Media." *Journal of Interactive Advertising* 9 (2): 44–61.

Sandoval, M. 2012. "A Critical Empirical Case Study of Consumer Surveillance on Web 2.0." In *Internet and Surveillance. The Challenges of Web 2.0 and Social Media*, edited by C. Fuchs, K. Boersma, A. Albrechtslund and M. Sandoval, 147–69. New York: Routledge.

Schiller, D. 1998. "Social Movement in U.S. Telecommunications: Rethinking the Rise of the Public Service Principle, 1894–1919." *Telecommunications Policy* 22(4): 397–408.

———. 1999a. *Digital Capitalism: Networking the Global Market System*. Cambridge, MA: MIT Press.

———. 1999b. "The Legacy of Robert A. Brady: Antifascist Origins of the Political Economy of Communications." *Journal of Media Economics* 12 (2): 89–101.

———. 2007. "The Hidden History of U.S. Public Service Telecommunications, 1919–1956." *Info* 9 (2/3): 17–28.

———. 2013. "Rosa Luxemburg's Internet? State Mobilization and the Movement of Accumulation in Cyberspace." Lecture at the University of Western Ontario, April 5.

Schiller, H. 1973. *The Mind Managers*. Boston: Beacon Press.

———. 1976. *Communication and Cultural Domination*. White Plains, NY: M.E. Sharpe, Inc.

———. 1977. Mass Communications and American Empire. New York: Augustus Kelly.

———. 1981. "Foreword." In D.W. Smythe, *Dependency Road*, xix-xxii. Norwood, NJ: Ablex.

Scholz, T., ed. 2013. *Digital Labor: The Internet as Playground and Factory*. New York: Routledge.

Schor, J. 2004. *Born to Buy: The Commercialized Child and the New Consumer Culture*. New York: Scribner.

Seles, S. 2010. "Turn On, Tune In, Cash Out: Maximizing the Value of Television Audiences." *Convergence Culture Consortium*, July 23. http://convergenceculture.org/research/c3-turnon-full.pdf.

Shah, D.V., L.A. Friedland, C. Wells., Y.M. Kim, and H. Rojas, eds. 2012. "Communication, Consumers, and Citizens: Revisiting the Politics of Consumption. *The Annals of the American Academy of Political and Social Science* 644 (1): 6–293.

Shanks, B. 1977. *The Cool Fire*. New York: Vintage Books.

Shannon, C. E., and W. Weaver. 1949. *The Mathematical Theory of Communication*. Urbana, IL: University of Illinois Press.

Shimpach, S. 2005. "Working Watching: The Creative and Cultural Labor of the Media Audience." *Social Semiotics* 15 (3): 343–60.

Shirky, C. 2008. *Here Comes Everybody: The Power of Organizing Without Organizations*. New York: Penguin.

Shufelt, T. 2013. "Why Canada Needs Europe." *Canadian Business* 85 (16): 54–55.

Singer, N. 2012. "Your Online Attention, Bought in an Instant." *New York Times*, November 17. http://www.nytimes.com/2012/11/18/technology/your-online-attention-bought-in-an-instant-by-advertisers.html?pagewanted=all

Skalen, P., M. Fellesson, and M. Fougere. 2006. "The Governmentality of Marketing Discourse." *Scandinavian Journal of Management* 22 (4): 275–91.

Skornia, H.J. (1965). *Television and Society*. New York: McGraw-Hill.

Slack, J.D. 1983. "The Information Revolution as Ideology." Position paper prepared for the *Seminar on Ideology and Infrastructure in the "Communications/Information Revolution"* at the national conference of the Speech Communication Association, Washington, DC.

Slater, D. 2011. "Marketing as a Monstrosity: the Impossible Place Between Culture and Economy." In *Inside Marketing: Practices, Ideologies, Devices*, edited by D. Zwick and J. Cayla, 23–41. Oxford: Oxford University Press.

Smith, A. 1973. *The Shadow in the Cave*. Urbana: University of Illinois Press.

Smith, A. 2012. "The Best (and Worst) of Mobile Connectivity." Pew Research Center. http://pewinternet.org/~/media/Files/Reports/2012/PIP_Best_Worst_Mobile_113012.pdf

Smythe, D.W. 1950. "A National Policy on Television?" *Public Opinion Quarterly* 14 (3): 461–74.

———. 1951. "An Analysis of Television Programs." *Scientific American* 184 (6): 15–17.

———. 1952. "Facing Facts about the Broadcast Business." *The University of Chicago Law Review* 20 (1): 96–106.

———. 1954. "Reality as Presented by Television." *Public Opinion Quarterly* 18 (2): 143–56.

———. 1957. *The Structure and Policy of Electronic Communication*. Urbana, IL: University of Illinois Press.

———. 1960. "On the Political Economy of Communications." *Journalism & Mass Communication Quarterly* 37 (4): 563–72.

———. 1973/1994. "After Bicycles, What?" In *Counterclockwise: Perspectives on Communication*, edited by T. Guback, 230–44. Boulder, CO: Westview Press.

———. 1977. "Communications: Blindspot of Western Marxism." *Canadian Journal of Political and Social Theory* 1 (3): 1–27.

———. 1978. "Rejoinder to Graham Murdock." *Canadian Journal of Political and Social Theory* 2 (2): 120–29

———. 1981. *Dependency Road: Communications, Capitalism, Consciousness, and Canada*. Norwood, NJ: Ablex.

———. 1984. "New Directions for Critical Communications Research." *Media, Culture & Society* 6 (3): 205–17.

Contributors

Mark Andrejevic is a media scholar at The University of Queensland, Australia. He is interested in the ways in which forms of surveillance and monitoring enabled by the development of new media technologies impact the realms of economics, politics, and culture. His first book, *Reality TV: The Work of Being Watched* (Rowman & Littlefield, 2003), explores the way in which this popular programming genre equates participation with willing submission to comprehensive monitoring. His second book, *iSpy: Surveillance and Power in the Interactive Era* (University of Kansas Press, 2007), considers the role of surveillance in the era of networked digital technology and explores the consequences for politics, policing, popular culture, and commerce. His latest book, *Infoglut* (Routledge, 2013), investigates the social implications of mediated information and "big data."

Alan Bradshaw teaches and learns at Royal Holloway, University of London.

Edward Comor is Professor in the Faculty of Information and Media Studies at the University of Western Ontario. He is editor of *Media, Structures, and Power: The Robert E. Babe Collection* (University of Toronto Press, 2011), and author of *Consumption and the Globalization Project: International Hegemony and the Annihilation of Time* (Palgrave Macmillan, 2008) and *Communication, Commerce and Power: The Political Economy of America and the Direct Broadcast Satellite, 1960–2000* (St. Martins and Macmillan, 1998). His research is published in multidisciplinary journals, including *International Communication Gazette*, *Critical Sociology*, *Canadian Journal of Communication*, and *Topia: Canadian Journal of Cultural Studies*.

Christian Fuchs is Professor of Social Media at the University of Westminster, and the former Chair in Media and Communication Studies at Uppsala University. His fields of work are Critical Social Theory, Critical Media and Communication Studies, and Critical Internet and Information Society Studies. He is co-founder of the ICTs and Society Network, editor of *tripleC: Communication, Capitalism & Critique*, and Chair of the European Sociological Association's Research Network 18—Sociology of Communications and Media Research.

Sut Jhally is Professor of Communication at the University of Massachusetts-Amherst and founder and Executive Director of the Media Education Foundation. His research addresses popular culture and media from the interdependent perspectives of critical cultural studies and political economy. While his focus is advertising and consumer culture, he is broadly concerned with ideology, consciousness, and politics. He has published many important articles on these topics and authored or co-authored several widely taught and cited monographs, including *The Codes of Advertising* and three editions of *Social Communication in Advertising* (with William Leiss, Stephen Kline and Jacqueline Botterill).

Micky Lee (PhD Oregon) is an Associate Professor of Media Studies at Suffolk University, Boston. She is the author of *Free Information? The Case Against Google* and has published more than ten journal articles on feminist political economy, telecommunications, and new information and communication technologies.

Vincent Manzerolle (PhD Media Studies) is a lecturer in the Faculty of Information and Media Studies at the University of Western Ontario, Canada where he teaches courses on the political economy of information, mobile media, search engines and data mining, and media convergence. His research interests center on the history, theory, and political economy of media. He has published articles on a variety of topics including credit and payment technologies, consumer databases, virtual work, mobile devices, apps, and media theory. He is also a member of the editorial board for *triple C: Communication, Capitalism & Critique.*

Lee McGuigan is a PhD student in the Annenberg School for Communication at the University of Pennsylvania. He studies advertising/marketing, commercial television, the sociology of consumption, and the political economy of technology. His research, supported by the Social Sciences and Humanities Research Council of Canada, has been published in the *Journal of Communication Inquiry* and *Television & New Media.*

Eileen R. Meehan is a political economist specializing in media industries and commercial culture. She is the author of *Why TV Is Not Our Fault* (Rowman & Littlefield, 2005). Meehan has co-edited two books: *Sex and Money: Feminism and Political Economy in Media Studies* with Ellen Riordan and, with Janet Wasko and Mark Philips, *Dazzled by Disney?: The Global Disney Audiences Project.* Meehan is a professor in the College of

———. 1989. "Television Deregulation and the Public." *Journal of Communication* 39 (4): 133–37.

———. 1994a. "Exerpts from Autobiography, 'Chapter 4: Mature Immaturity: The Urbana Years, 1948–1963.'" In *Counterclockwise: Perspectives on Communication*, edited by T. Guback, 37–58. Boulder, CO: Westview Press.

———. 1994b. "Outline of a Proposal for Competitive U.S. Broadcast Systems." In *Counterclockwise: Perspectives on Communication*, edited by T. Guback, 86–90. Boulder, CO: Westview Press.

———. 1994c. "The Role of Mass Media and Popular Culture in Defining Development." In *Counterclockwise: Perspectives on Communication*, edited by T. Guback, 247–62. Boulder, CO: Westview Press.

Smythe, D. W., and T. Guback. 1994. *Counterclockwise*. Boulder, CO: Westview Press.

Smythe, D.W., and T. Van Dihn. 1983. "On Critical and Administrative Research: A New Critical Analysis." *Journal of Communication* 33 (3): 117–27.

Solis, B. 2011. *Engage!* Hoboken, NJ: John Wiley & Sons.

Speech Communication Association. 1984. "Policy." *Critical Studies in Mass Communication* 1 (2): not paginated.

Spurgeon, C. 2008. *Advertising and New Media*. New York: Routledge.

Sraffa, Piero. 1960. *The Production of Commodities by Means of Other Commodities*. London: Cambridge University Press.

Steedman, I. 1977. *Marx After Sraffa*. London: New Left Books.

Steedman, I. et al. 1981. *The Value Controversy*. London: Verso.

Steinberg, B. 2009. Comcast Play for NBC Universal a Bet on Future of Advertising. *Advertising Age*, November 9. http://adage.com/article/mediaworks/comcast-bid-nbc-universal-a-bet-future-advertising/140383/.

Stelter, B. 2011. "Twitter and TV Get Close to Help Each Other Grow." *New York Times*, October 25. http://www.nytimes.com/2011/10/26/business/media/twitter-and-tv-get-close-to-help-each-other-grow.html?src=tp&_r=0

———. 2012. "New Nielsen Ratings to Measure TV and Online Together." *New York Times*, March 18. http://mediadecoder.blogs.nytimes.com/2012/03/18/new-nielsen-ratings-to-measure-tv-and-online-ads-together/.

———. 2013. "Twitter Buys Company that Mines Chatter About TV." *New York Times*, February 5. http://mediadecoder.blogs.nytimes.com/2013/02/05/twitter-buys-company-that-mines-chatter-about-tv/.

Stereck, W. 2012. "Citizens as Consumers: Considerations on the New Politics of Consumption." *New Left Review* [Second Series] 76 (July/August): 27–47.

Sterne, J. 1999. "Television Under Construction: American Television and the Problem of Distribution, 1926–1962." *Media, Culture & Society* 21 (3): 503–30.

Stiegler, B. 2010. *For a New Critique of Political Economy*. Malden, MA: Polity.

Stilson, J. 2011. "Rentrak's Influence Growing in Ratings Wars." TVNewsCheck, December 13. http://www.tvnewscheck.com/article/56019/rentraks-influence-growing-in-ratings-wars.

Stone, Catherine. 1974. "The Origins of Job Structures in the Steel Industry." *Review of Radical Political Economy* 6 (2): 113–73.

Stratten, S. 2010. *Unmarketing: Stop Marketing, Start Engaging*. Hoboken, NJ: John Wiley & Sons.

Streeter, T. 1996. Selling the Air: A Critique of the Policy of Commercial Broadcasting. Chicago: University of Chicago Press.

Sunstein, C. R. 2009. *Republic.com 2.0*. Princeton, NJ: Princeton University Press.

Sweezy, P. 1942. *The Theory of Capitalist Development*. New York: Monthly Review Press.

Tapscott, D. 2009. *Grown Up Digital: How the Net Generation is Changing your World*. New York: McGraw Hill.

Tapscott, D., and A. D. Williams. 2008. *Wikinomics: How Mass Collaboration Changes Everything*. New York: Penguin.

Tate, R. 2013. "Creepy Side of Search Emerges on Facebook." *Wired.com*, February 15. http://www.wired.com/business/2013/02/creepy-graph-searchers/.

Tejada, C. 2012. "Google to Shut Music Service Inside China." *Wall Street Journal*, September 22.

Terkel, S. 1974. *Working: People Talk About What They Do All Day and How They Do What They Do*. New York: Pantheon Books.

Terranova, T. 2000. "Free Labour: Producing Culture for the Digital Economy." *Social Text* 18 (2): 33–58.

———. 2004. *Network Culture: Politics for the Information Age*. London and Ann Arbor, MI: Pluto Press.

———. 2012. "Attention, Economy and the Brain." *Culture Machine* 13.

Thomas, S. 1995. "Myths in and About Television." In *Questioning the Media: A Critical Introduction*, edited by J. Downing and A. Mohammadi, 330–344. Thousand Oaks, CA: SAGE.

Thomke, S., and E. von Hippel. 2002. "Customers as Innovators: A New Way to Create Value." *Harvard Business Review* 80 (4): 74–81.

Thompson, E. P. 1978. *The Poverty of Theory*. London: Merlin Press.

Tinic, S. 2006. "(En)Visioning the Television Audience: Revisiting Questions of Power in the Age of Interactive Television." In *The New Politics of Surveillance and Visibility*, edited by R. Ericson, and K. Haggerty, 308–326. Toronto: University of Toronto Press.

Toffler, A. 1980. *The Third Wave*. New York: Morrow.

Tomlinson, J. 2007. *The Culture of Speed*. Thousand Oaks, CA: SAGE.

Tönnies, F. 1973. *Community and Society*. New York: Harper & Row.

Toscano, A. 2008a. "The Culture of Abstraction." *Theory, Culture & Society* 25 (4): 57–75.

———. 2008b. "The Open Secret of Real Abstraction." *Rethinking Marxism* 20 (2): 273–87.

Trendrr.tv 2012. "February 2012 White Paper." http://www.scribd.com/doc/84722043/Trendrr-TV-February-White-Paper.

Troianovski, A. 2012. "Cellphones Are Eating the Family Budget." *Wall Street Journal*, September 28. http://online.wsj.com/article/SB10000872396390444083304578018731890309450.html.

Trottier, D. 2011. "Mutual Transparency or Mundane Transgressions? Institutional Creeping on Facebook." *Surveillance & Society* 9 (1/2): 17–30.

———. 2012. *Social Media as Surveillance: Rethinking Visibility in a Converging World*. Farnham, UK: Ashgate.

Tse-Tung, M. 1967. *Selected Works of Mao Tse-Tung, Vol. 1*. Peking, Foreign Languages Press,

Tucker, P. 2013. "Has Big Data Made Anonymity Impossible?" *MIT Technology Review*, May 7. http://www.technologyreview.com/news/514351/has-big-data-made-anonymity-impossible/.

Turow, J. 1997a. *Breaking Up America*. Chicago, IL: University of Chicago Press.

———. 1997b. *Media Systems in Society: Understanding Industries, Strategies, and Power*, 2nd edition. White Plains, NY: Longman.

———. 2005. "Audience Construction and Culture Production: Marketing Surveillance in the Digital Age." *The Annals of the American Academy of Political and Social Science* 597: 103–21.

———. 2006. *Niche Envy: Marketing Discrimination in the Digital Age*. Cambridge, MA: MIT Press.

———. 2011. *The Daily You: How the New Advertising Industry is Defining Your Identity and Your Worth.* New Haven, CT: Yale University Press.

Turow, J., and M. McAllister. 2002. "New Media and the Commercial Sphere." *Journal of Broadcasting and Electronic Media* 46 (4): 505–14.

Turow, J., M.X. Delli Carpini, N. Draper, and R. Howard-Williams. 2012. *Americans Roundly Reject Tailored Political Advertising*. Philadelphia, PA: Annenberg School for Communication. http://www.asc.upenn.edu/news/Turow_Tailored_Political_Advertising.pdf.

"Twitter on TV: A Producer's Guide." 2012. https://dev.twitter.com/media/twitter-tv.

Tye, L. 1998. *The Father of Spin: Edward L. Bernays and The Birth of Public Relations*. New York: Henry Holt.

Ungerleider, N. 2013. "Mobile Phones Have Fingerprints, Too." *FastCompany*, March 29. http://www.fastcompany.com/3007645/location-location-location/mobile-phones-have-fingerprints-too

Vaidhyanathan, S. 2011. *The Googlization of Everything (And Why We Should Worry).* Berkeley, CA: University of California Press.

Van Dijck, J. 2009. "Users Like You?: Theorizing Agency in User-Generated Content." *Media, Culture & Society* 31 (1): 41–58.

Vasquez, D. 2011. "The Bigger TV Story, Beyond Ratings." *Media Life Magazine*, November 10. http://www.medialifemagazine.com/artman2/publish/Research_25/The-bigger-TV-story-beyond-ratings.asp.

Veblen, T. 1964. *Theory of the Business Enterprise*. New York: Viking Press.

Vollmer, C., and G. Precourt. 2008. *Always On: Advertising, Marketing, and Media in an Era of Consumer Control*. New York, NY: McGraw-Hill.

von Hippel, E. 2005. *Democratizing Innovation*. Cambridge, MA: MIT Press.

Wasko, J. 1982. *Movies and Money: Financing the American Film Industry*. Norwood, NJ: Ablex.

———. 1993. "Introduction." In *Illuminating the Blindspots. Essays Honoring Dallas W. Smythe*, ed. J. Wasko, V. Mosco and M. Pendakur, 1–11. Norwood, NJ: Ablex.

———. 2004. "The Political Economy of Communications." In *The SAGE Handbook of Media Studies*, edited by J.D.H. Downing, D. McQuail, P. Schlesinger, and E. Wartella, 309–30. Thousand Oaks, CA: SAGE.

———, ed. 2005a. *A Companion to Television*. Malden, MA: Wiley-Blackwell.

———. 2005b. "Studying the Political Economy of Media and Information." *Comunicação e Sociedade* 7: 25–48.

———. 2008. "The Commodification of Youth Culture." In *The International Handbook of Children, Media and Culture*, edited by K. Drotner and S. Livingstone, 460–74. Thousand Oaks, CA: SAGE.

Wasko, J., G. Murdock, and H. Sousa, eds. 2011. *The Handbook of Political Economy of Communications*. Malden, MA: Wiley-Blackwell.

Watson, R.T., L.F. Pitt, P. Berthon, and G.M. Zinkhan. 2002. "U-commerce: Expanding the Universe of Marketing." *Journal of the Academy of Marketing Science* 30 (4): 329–43.

Weber, L. 2009. *Marketing to the Social Web: How Digital Customer Communities Build Your Business*. Hoboken, NJ: John Wiley & Sons.

"What is an API? Your Guide to the Internet (R)evolution." 2013. http://www.3scale.net/wp-content/uploads/2012/06/What-is-an-API-1.0.pdf.

Williams, M. 1998. "Money and Labour-Power." *Cambridge Jouranl of Economics* 22: 187–98.

Williams, R. 1974. *Television: Technology and Cultural Form*. New York and London: Routledge.

———. 1977a. *Marxism and Literature*. Oxford: Oxford University Press.

———. 1977b. "Notes on Marxism in Britain since 1945." *New Left Review* 100 (January/February).

———. 1980. "Advertising: The Magic System." In *Problems in Materialism and Culture*, 170–95. London: Verso.

Williamson, J. 1978. *Decoding Advertisements: Ideology and Meaning in Advertising*. London: Boyars.

Winn, M. 1977. *The Plug-in Drug*. New York: Viking.

Winseck, D. 2003. "Convergence, Network Design, Walled Gardens, and Other Strategies of Control in the Information Age." In *Surveillance as Social Sorting: Privacy, Risk, and Digital Discrimination*, edited by D. Lyon, 176–98. New York: Routledge.

———. 2008. "The State of Media Ownership and Media Markets: Competition or Concentration and Why Should we Care?" *Sociology Compass* 2 (1): 34–47.

———. 2011. "The Political Economies of Media and the Transformation of the Global Media Industries." In *The Political Economies of Media: The Transformation of the Global Media Industries*, edited by D. Winseck and D.Y. Jin, 3–48. London: Bloomsbury.

World Economic Forum. 2011. "Personal Data: The Emergence of a New Asset Class—Opportunities for the Telecommunications Industry." World Economic Forum. http://www3.weforum.org/docs/WEF_ITTC_PersonalDataNewAsset_Report_2011.pdf.

———. 2012. "Rethinking Personal Data: Strengthening Trust." World Economic Forum.

Wortham, J. 2012. "How New Yorkers Adjusted to Sudden Smartphone Withdrawal." *New York Times*, November 3. http://bits.blogs.nytimes.com/2012/11/03/how-new-yorkers-adjusted-to-sudden-smartphone-withdrawal/

Wray, R. 2009. "Digital Dividend 'Should Be Used to Bring Broadband to Remote Areas.'" *The Guardian*, February 16.

Wu, T. 2011. *The Master Switch: The Rise and Fall of Information Empires*. New York: Vintage.

Wyatt, E. 2013. "U.S. May Sell Airwaves That Help Broadway Sing." *New York Times*, March 29. http://www.nytimes.com/2013/03/30/business/fcc-has-yen-for-broadways-wireless-spectrum.html

Yao, L. 2010. "Revisiting Critical Scholars' Alternative: A Case Study of Dallas Smythe's Praxis." Paper presented at the 2010 Annual Meeting of the International Communication Association.

Yorke, D. 1931. "The Radio Octopus." *American Mercury* 23: 385–400.

Young, W., and N. Young. 2010. *World War II and the Post War Years in America, A Historical and Cultural Encyclopedia, Vol. I*. Santa Barbara, CA: ABC-CLIO.

Zang, L., and A. Fung. 2013. "Working as Playing? Consumer Labor, Guild and the Secondary Industry of Online Gaming in China." *New Media & Society* (February 28) DOI: 10.1177/1461444813477077.

Zarella, D., and Zarella, A. 2011. *The Facebook Marketing Book.* Sebastopol, CA: O'Reilly Media Inc.

Zhao, Y. 2007. "After Mobile Phones, What? Re-embedding the Social in China's 'Digital Revolution.'" *International Journal of Communication* 1: 92–120.

———. 2011. "The Challenge of China: Contribution to a Transcultural Political Economy of Communication for the Twenty-first Century." In *The Handbook of Political Economy of Communication*, edited by J. Wasko, G. Murdock and H. Sousa, 558–82. Malden, MA: Polity.

Zittrain, J. 2012. "Meme Patrol: 'When Something Online Is Free, You're Not the Customer, You're the Product.'" *The Future of the Internet and How to Stop It* (blog), March 21. http://futureoftheinternet.org/meme-patrol-when-something-online-is-free-youre-not-the-customer-youre-the-product.

Žižek, S. 1996. *The Indivisible Remainder: An Essay on Schelling and Related Matters.* London and New York: Verso.

———. 1997. *The Plague of Fantasies*. London and New York: Verso.

———. 1998. "The Interpassive Subject." *Traverses.* Paris: Centre Georges Pompidou.

———. 2008. *Violence*. London: Profile Books.

Zwick, D., Bonsu, S.K., and Darmody, A. 2008. "Putting Consumers to Work: 'Co-creation' and New Marketing Governmentality." *Journal of Consumer Culture* 8 (2): 163–96.

Zwick, D., and J. Cayla, eds. 2011. *Inside Marketing: Practices, Ideologies, Devices*. Oxford and New York: Oxford University Press.

Zwick, D., and Y. Ozalp. 2011. "Flipping the Neighborhood: Biopolitical Marketing as Value Creation for Condos and Lofts." In *Inside Marketing: Practices, Ideologies, Devices*, edited by D. Zwick and J. Cayla, 234–53. Oxford and New York: Oxford University Press.

Zwick, D., and J.D. Knott. 2009. "Manufacturing Consumers: The Database as New Means of Production." *Journal of Consumer Culture* 9 (2): 221–47.

General Editor: *Steve Jones*

Digital Formations is the best source for critical, well-written books about digital technologies and modern life. Books in the series break new ground by emphasizing multiple methodological and theoretical approaches to deeply probe the formation and reformation of lived experience as it is refracted through digital interaction. Each volume in **Digital Formations** pushes forward our understanding of the intersections, and corresponding implications, between digital technologies and everyday life. The series examines broad issues in realms such as digital culture, electronic commerce, law, politics and governance, gender, the Internet, race, art, health and medicine, and education. The series emphasizes critical studies in the context of emergent and existing digital technologies.

Other recent titles include:

Felicia Wu Song
Virtual Communities: Bowling Alone, Online Together

Edited by Sharon Kleinman
The Culture of Efficiency: Technology in Everyday Life

Edward Lee Lamoureux, Steven L. Baron, & Claire Stewart
Intellectual Property Law and Interactive Media: Free for a Fee

Edited by Adrienne Russell & Nabil Echchaibi
International Blogging: Identity, Politics and Networked Publics

Edited by Don Heider
Living Virtually: Researching New Worlds

Edited by Judith Burnett, Peter Senker & Kathy Walker
The Myths of Technology: Innovation and Inequality

Edited by Knut Lundby
Digital Storytelling, Mediatized Stories: Self-representations in New Media

Theresa M. Senft
Camgirls: Celebrity and Community in the Age of Social Networks

Edited by Chris Paterson & David Domingo
Making Online News: The Ethnography of New Media Production

To order other books in this series please contact our Customer Service Department:

(800) 770-LANG (within the US)
(212) 647-7706 (outside the US)
(212) 647-7707 FAX

To find out more about the series or browse a full list of titles, please visit our website:

WWW.PETERLANG.COM

Mass Communication and Media Arts at Southern Illinois University Carbondale.

William H. Melody is Guest Professor, Center for Communication, Media and IT (CMI), Aalborg University Copenhagen, Denmark, and Emeritus Professor, Delft University of Technology (NL). He is the founder of LIRNE.NET and the World Dialogue on Regulation for Network Economies (WDR) and continues to advise both initiatives. He is former Chief Economist, U.S. FCC, and has held numerous academic appointments, leading new multidisciplinary program development in seven countries. For more than 50 years Melody has contributed to research and policy literature, with more than 150 publications. He is a periodic consultant and advisor to universities and research centres around the world, for the UN and other international organizations, governments and corporate organizations.

Vincent Mosco is Professor Emeritus in the Department of Sociology at Queen's University. He is formerly Canada Research Chair in Communication and Society. Dr. Mosco is the author of numerous books on communication, technology, and society, including *The Political Economy of Communication* (second edition, Sage, 2009), *The Laboring of Communication: Will Knowledge Workers of the World Unite* (co-authored with Catherine McKercher, Lexington Books, 2008), and *The Digital Sublime: Myth, Power, and Cyberspace* (MIT Press, 2004). Dr. Mosco is a founding member of the Union for Democratic Communication and in 2004 he received the Dallas W. Smythe Award for outstanding achievement in communication research.

Graham Murdock is Reader in the Sociology of Culture at Loughborough University. His main interests are in the Sociology and Political Economy of Culture. He has written extensively on the organization of mass media industries, and he was a primary contributor to the "blindspot debate."

Philip M. Napoli teaches in the School of Communication and Information at Rutgers University, and he is Director of the Donald McGannon Communication Research Center at Fordham University in New York. His research focuses on media institutions, audiences, and policy. He is the author of the books *Audience Economics: Media Institutions and the Audience Marketplace* (Columbia University Press, 2003) and *Audience Evolution: New Technologies and the Transformation of Media Audiences* (Columbia University Press, 2011).

Jason Pridmore is Assistant Professor in the Department of Media and Communication at Erasmus University Rotterdam. Prior to joining the department, he was the Senior Researcher on the DigIDeas project, based in Maastricht, the Netherlands, where he examined the social and ethical implications of digital identification. His research focuses on consumer surveillance practices and the use of new media in marketing practice, specifically marketing techniques such as the use of loyalty cards and social media integration as forms of collaborative surveillance. He is the author of numerous texts on consumer surveillance, including the entry on consumer surveillance in the *Routledge Handbook of Surveillance Studies* (2012) and the expert report on the surveillance of consumers and consumption, part of the report on the surveillance society commissioned by the British Information Commissioner in 2006. Prior to joining the DigIDeas team, Jason was a Postdoctoral Fellow with The New Transparency Project, under the auspices of The Surveillance Project. He received his PhD from the Department of Sociology at Queen's University, Canada, in 2008.

Daniel Trottier is a Postdoctoral Fellow in Social and Digital Media at the Communication and Media Research Institute (CAMRI), University of Westminster. Prior to this post, he was a Postdoctoral Fellow in the Department of Informatics and Media at Uppsala University. His research focuses on surveillance practices on social media in general, with a specific focus on applications for law enforcement and investigations. He is currently participating in two European Union research projects on this topic (RESPECT and PACT). He is the author of the forthcoming book, *Social Media as Surveillance* (2012), as well as numerous articles on surveillance practices and digital media. He previously held a Postdoctoral fellowship in the Department of Sociology at the University of Alberta, Canada, and obtained his PhD in the Department of Sociology at Queen's University, Canada, in 2010.

Detlev Zwick teaches and learns at York University, Toronto, Canada. His research critiques the cultural politics of contemporary forms of consumption, marketing and management practice. He is the editor (with Julien Cayla) of *Inside Marketing: Practices, Ideologies, Devices* (Oxford, 2011).

Index

•L•

•M•

•N•

•P•

•V•

•W•

•Y•

•Z•